Decision Methods for Forest Resource Management

Decision Methods for Forest Resource Management

Joseph Buongiorno
University of Wisconsin, Madison

J. Keith Gilless
University of California, Berkeley

ACADEMIC PRESS
An imprint of Elsevier Science

Amsterdam   Boston   London   New York   Oxford   Paris
San Diego   San Francisco   Singapore   Sydney   Tokyo

This book is printed on acid-free paper. ∞

Copyright 2003, Elsevier Science (USA)

All rights reserved.

No part of this publication may be reproduced or transmitted in any form or by any means, electronic or mechanical, including photocopy, recording, or any information storage and retrieval system, without permission in writing from the publisher.

Permissions may be sought directly from Elsevier's Science & Technology Rights Department in Oxford, UK: phone: (+44) 1865 843830, fax: (+44) 1865 853333, e-mail: permissions@elsevier.com.uk. You may also complete your request on-line via the Elsevier Science homepage (http://elsevier.com), by selecting "Customer Support" and then "Obtaining Permissions."

Academic Press
An imprint of Elsevier Science
525 B Street, Suite 1900, San Diego, California 92101-4495, USA
http://www.academicpress.com

Academic Press
An imprint of Elsevier Science
84 Theobald's Road, London WC1X 8RR, UK
http://www.academicpress.com

Academic Press
An imprint of Elsevier Science
200 Wheeler Road, Burlington, Massachusetts 01803, USA
www.academicpressbooks.com

Library of Congress Catalog Card Number: 2002117166

International Standard Book Number: 0-12-141360-8

PRINTED IN THE UNITED STATES OF AMERICA
03 04 05 06   9 8 7 6 5 4 3 2 1

For our families

CONTENTS

Preface   xv

1 Introduction

1.1 Scope of Forest Resource Management   1
1.2 The Nature of Models   2
1.3 Systems Models   3
1.4 The Role of Computers   4
1.5 Good Models   5
     Annotated References   6

2 Principles of Linear Programming: Formulations

2.1 Introduction   9
2.2 First Example: A Poet and His Woods   10
2.3 Second Example: Keeping the River Clean   14
2.4 Standard Formulation of a Linear Programming Problem   17
2.5 Spreadsheet Formulation of Linear Programs   20
2.6 Assumptions of Linear Programming   22
2.7 Conclusion   24
     Problems   24
     Annotated References   27

3 Principles of Linear Programming: Solutions

3.1 Graphic Solution of the Poet's Problem   29
3.2 Graphic Solution of the River Pollution Problem   33
3.3 The Simplex Method   36
3.4 Duality in Linear Programming   42

3.5	Spreadsheet Solution of Linear Programs	47
3.6	Summary and Conclusion	50
	Problems	51
	Annotated References	52

4 Even-Aged Forest Management: A First Model

4.1	Introduction	53
4.2	Definitions	54
4.3	Example: Converting Southern Hardwoods to Pine	55
4.4	Model Formulation	56
4.5	Solution	58
4.6	Maximizing Present Value	59
4.7	A Note on Redundancies	60
4.8	Spreadsheet Formulation and Solution	61
4.9	General Formulation	64
4.10	Conclusion	65
	Problems	66
	Annotated References	67

5 Area- and Volume-Control Management with Linear Programming

5.1	Introduction	69
5.2	Preliminary Definitions	70
5.3	Example: Optimizing the Yield of a Loblolly Pine Even-Aged Forest	71
5.4	Solutions	76
5.5	Adding Constraints and Objectives	77
5.6	Spreadsheet Formulation and Solution	78
5.7	General Formulation	82
5.8	Summary and Conclusion	83
	Problems	84
	Annotated References	86

6 A Dynamic Model of the Even-Aged Forest

6.1	Introduction	89
6.2	Example	90
6.3	A Model of Forest Growth	90
6.4	Sustainability Constraints	93
6.5	Objective Function	95

6.6 Spreadsheet Formulation and Solution	97
6.7 General Formulation	103
6.8 Summary and Conclusion	104
Problems	105
Annotated References	106

7 Economic Objectives and Environmental Policies for Even-Aged Forests

7.1 Introduction	109
7.2 Land Expectation Value and Economic Rotation	110
7.3 Estimating Forest Value by Linear Programming	113
7.4 Spreadsheet Optimization of Forest Value	116
7.5 Even-Flow Policy: Cost and Benefits	119
7.6 Mixed Economic and Environmental Objectives	121
7.7 General Formulation	123
7.8 Summary and Conclusion	125
Problems	126
Annotated References	129

8 Managing the Uneven-Aged Forest with Linear Programming

8.1 Introduction	131
8.2 A Growth Model of the Uneven-Aged Forest Stand	133
8.3 Predicting the Growth of an Unmanaged Stand	136
8.4 Growth Model for a Managed Stand	140
8.5 Optimizing Uneven-Aged Stands	143
8.6 General Formulation	147
8.7 Summary and Conclusion	150
Problems	150
Annotated References	153

9 Economic and Environmental Management of Uneven-Aged Forests

9.1 Introduction	155
9.2 Economic Steady State for Uneven-Aged Stands	156
9.3 Environmental Objectives	164
9.4 Converting a Stand to the Desired Steady State	168
9.5 General Formulation	171

	9.6	Summary and Conclusion	173
		Problems	174
		Annotated References	176

10 Multiple Objectives Management with Goal Programming

	10.1	Introduction	179
	10.2	Example: River Pollution Control Revisited	180
	10.3	Goal-Programming Constraints	181
	10.4	Goal-Programming Objective Function	182
	10.5	Spreadsheet Formulation and Solution	184
	10.6	Objective Functions with Ordinal Weights	187
	10.7	Goal Programming in Even-Aged Forest Management	189
	10.8	Goal Programming in Uneven-Aged Stand Management	192
	10.9	General Formulation	195
	10.10	Conclusion: Goal Versus Linear Programming	198
		Problems	199
		Annotated References	201

11 Forest Resource Programming Models with Integer Variables

	11.1	Introduction	203
	11.2	Shortcomings of the Simplex Method with Integer Variables	204
	11.3	Connecting Locations at Minimum Cost	207
	11.4	Assigning Foresters to Jobs	210
	11.5	Designing an Efficient Road Network	214
	11.6	Models with Integer and Continuous Variables	221
	11.7	Conclusion	226
		Problems	226
		Annotated References	232

12 Project Management with the Critical Path Method (CPM) and the Project Evaluation and Review Technique (PERT)

	12.1	Introduction	235
	12.2	A Slash-Burn Project	236
	12.3	Building a CPM/PERT Network	237
	12.4	Earliest Start and Finish Times	239

12.5 Latest Start and Finish Times	243
12.6 Activity Slack and Critical Path	246
12.7 Gantt Chart	248
12.8 Dealing with Uncertainty	249
12.9 Summary and Conclusion	252
Problems	253
Annotated References	256

13 Multistage Decision Making with Dynamic Programming

13.1 Introduction	259
13.2 Best Thinning of an Even-Aged Forest Stand	260
13.3 General Formulation of Dynamic Programming	269
13.4 Trimming Paper Sheets to Maximize Value	270
13.5 Minimizing the Risk of Losing an Endangered Species	277
13.6 Conclusion	282
Problems	283
Annotated References	287

14 Simulation of Uneven-Aged Stand Management

14.1 Introduction	289
14.2 Types of Simulation	290
14.3 Deterministic Simulation of Uneven-Aged Forest Management	291
14.4 Applications of Deterministic Simulation Model	295
14.5 Stochastic Simulation	300
14.6 Conclusion	308
Problems	308
Annotated References	311

15 Simulation of Even-Aged Forest Management

15.1 Introduction	313
15.2 Deterministic Simulation of Even-Aged Management	314
15.3 Applications of Deterministic Simulation	319
15.4 Simulating Catastrophic Events	324
15.5 Spreadsheet Stochastic Simulation	328
15.6 Conclusion	332
Problems	332
Annotated References	335

16 Projecting Forest Landscape and Income Under Risk with Markov Chains

16.1 Introduction	337
16.2 Natural Forest Growth as a Markov Chain	338
16.3 Predicting the Effects of Management	344
16.4 Summary and Conclusion	351
Problems	352
Annotated References	354

17 Optimizing Forest Income and Biodiversity with Markov Decision Processes

17.1 Introduction	357
17.2 Markov Decision Process	358
17.3 Maximizing Discounted Expected Returns	359
17.4 Maximizing Long-Term Expected Biodiversity	362
17.5 Maximum Expected Biodiversity with Income Constraint	367
17.6 Summary and Conclusion	369
Problems	370
Annotated References	371

18 Analysis of Forest Resource Investments

18.1 Introduction	373
18.2 Investment and Interest Rate	374
18.3 Investment Criteria	378
18.4 Choice of Projects Under Constraints	385
18.5 Inflation and Investment Analysis	388
18.6 Choosing an Interest Rate	392
18.7 Summary and Conclusion	396
Problems	396
Annotated References	399

19 Econometric Analysis and Forecasting of Forest Product Markets

19.1 Introduction	401
19.2 Econometrics	402
19.3 A Model of the United States Pulpwood Market	402
19.4 Data	406
19.5 Model Estimation	406

19.6	Inferences, Forecasting, and Structural Analysis	414
19.7	Conclusion	416
	Problems	416
	Annotated References	420

Appendix A
Compounding and Discounting 421

Appendix B
Elements of Matrix Algebra 425

Index 429

PREFACE

This book on decision methods for forest resource management is a complete revision and extension of our 1987 book, *Forest Management and Economics: A Primer in Quantitative Methods*. There are three main differences between the two books. First, since 1987, much more emphasis has been placed on ecological objectives in forest management. The new book reflects this change by considering many examples of managing forest systems with ecological objectives or constraints, including the landscape diversity of forests, the tree diversity of forest stands, and the use of forests for carbon sequestration. Second, every college student now uses a computer regularly and has been using one at least since high school or earlier. Almost universally, the basic software on a personal computer includes a spreadsheet program, often with a built-in optimizer. Thus, every chapter in this new book shows how to implement the method under study with a spreadsheet. Thirdly, during the past 10 years, considerable research has been done on uncertainty in forest decision making. This has led to the addition of four chapters on modeling uncertainty, two with simulation methods and two on Markov processes and related optimization models.

This book started from a set of notes and laboratory exercises for a course in decision methods for resource management, taught for many years at the University of Wisconsin—Madison. The course is designed for senior undergraduates in forest science and environmental studies, although a few graduate students in forestry and other disciplines often take it. Like the course, this book is meant mostly for undergraduates with little patience for theory. Our students chose forestry to walk the woods, not to dissect arcane equations or to waste their young years staring at computer screens. Smart undergraduates are willing to study abstract methods and principles only if they can clearly see their application in the woods.

Having learned this, not without a few setbacks, we developed a pedagogy in which no more than one lecture would deal with methods or principles without being followed by a solid application. Wherever possible the method and the application are developed together, the application almost calling for a method to solve the problem being posed.

The book follows this general principle. It is meant to be mostly a book in forest management, but with the belief that modeling is a good way to train managers to think in terms of alternatives, opportunity cost, compromises, and best solutions. The mathematics used are kept to a minimum. Proofs of theorems and detailed descriptions of algorithms are avoided. Heuristics are used instead, wherever possible. Nevertheless, mathematics are used freely if they describe a particular concept more succinctly and precisely than lengthy verbosities.

The book does not assume any specific preparation in mathematics or other sciences. Nevertheless, the material presented does require, by its very nature, a certain level of mathematical maturity. This should not be a problem for most forestry seniors, of whom a fair knowledge of mathematics and statistics is expected.

Some parts, such as the end of the Chapters 8 and 9 on uneven-aged management, use elementary linear algebra. However, all the matrix definitions and operations needed are presented in Appendix B. This material can be taught in less than one lecture. It would be a pity if that little extra effort stopped anyone from studying thoroughly what is perhaps the oldest and certainly one of the best-looking forms of forest management.

Modern quantitative methods are inseparable from computers. Again, many forestry students now take formal courses in computer science. Nevertheless, programming knowledge is not required to understand and use the methods discussed in this book. All the examples in the chapters, and the related problems, can be solved with a spreadsheet with a built-in optimizer, such as the Solver within Excel. Even the stochastic simulations are done with a spreadsheet. Simulation being such a powerful modeling tool, we hope that this will be only a prelude for more extensive studies using more specialized software.

In fact, our objectives in planning the book were to awaken the curiosity of students, to expose them to many different methods, and to show with a few examples how these methods can be used in forest resource management. The emphasis throughout is not on how particular methods work, but rather on how they can be used. For example, we do not want to teach how linear programs are solved, but rather how forestry problems can be expressed as linear programs. The task of a forest manager is not to solve models, but rather to express managerial problems in model forms for which solutions exist.

Consequently, this book is more about the art of model formulation than about the science of model solution. Like any art, mastering it takes a long apprenticeship, in which exposure to a variety of methods prepares students to resist the urge to cast every problem in the mold of the technique that they know best. Being aware of different approaches helps ensure that the best method is selected for a particular problem.

The decisions to discuss many different methods and to formulate and solve the related models with spreadsheets meant that each application had to be a simplification of models used in actual decision making. We feel that this is the most appropriate approach for classroom instruction. The students should spend the short time they have learning the principles of a method rather than the intricate details of its application. Nevertheless, the longest exercises at the end of each chapter are realistic enough that they give a good idea of the power and limitations of each method and of the time needed for formulating and solving a practical problem.

The undergraduate course on decision methods for resource management at the University of Wisconsin—Madison covers Chapters 1–12, Chapter 15, Chapter 18, and sometimes Chapter 19. Chapter 14, on uneven-aged simulation, is no more difficult and could also be substituted, or added, time permitting. We feel confident that Chapter 16, on Markov chains, could also be part of an undergraduate course, if so desired. Only Chapters 13 and 17 may be regarded as graduate-level material, although we are confident that bright undergraduates with good guidance can master recursive optimization.

Despite the extensive forest modeling literature, we have deliberately limited the number of references at the end of each chapter, listing only a few that we deemed accessible to undergraduates wanting further study of the concepts addressed in that chapter.

We thank our students for having served as involuntary, though sometimes vocal, guinea pigs for the successive versions of the material presented in this book. We also thank warmly our colleagues who have reviewed the initial drafts, corrected errors, and provided suggestions for improvement, in particular James Turner, Joseph Chang, Lauri Valsta, Larry Leefers, Jing-Jing Liang, Anders Nyrud, Oka Hiroyasu, Bowang Chen, Brian Schwingle, Chris Edgar, Alan Thomson, Xiaolei Li, Mo Zhou, Ronald Raunikar, and Sijia Zhang.

We are also grateful to the people at Academic Press, in particular to Charles Crumly and Angela Dooley for their care in managing this project.

SUPPLEMENTARY MATERIALS: Excel workbooks containing all the spreadsheets shown in this book are available at the web page address: www.forest.wisc.edu/facstaff/buongiorno/book/index.htm.

An instructor manual entitled *Problems and Answers for Decision Methods for Forest Resource Management* is available from Academic Press (ISBN 0-12-141363-2).

We welcome comments and suggestions. Please, send them by e-mail to: jbuongio@facstaff.wisc.edu or gilless@nature.berkeley.edu. Thank you.

Joseph Buongiorno
J. Keith Gilless

CHAPTER 1

Introduction

1.1 SCOPE OF FOREST RESOURCE MANAGEMENT

Forest resource management is the art and science of making decisions with regard to the organization, use, and conservation of forests and related resources. Forests may be actively managed for timber, water, wildlife, recreation, or a combination thereof. Management also includes the "hands-off" alternative: letting nature take its course, which may be the best thing to do in some cases. Forest resource managers must make decisions affecting both the very long-term future of the forest and day-to-day activities. The decisions may deal with very complex forest systems or with simple parts. The geographic area of concern may be an entire country, a region, a single stand of trees, or an industrial facility. Some of the forest resource management problems we shall consider in this book include:

    Scheduling harvesting and reforestation in even-aged forests to best meet production and ecological objectives (Chapters 4–7)
    Determining what trees to harvest in uneven-aged forests and when to harvest them to optimize timber production, revenues, or ecological diversity (Chapters 8 and 9)

Planning the production activities in forest stands and in forest industries to meet goals concerning revenues, employment, and pollution control (Chapters 10 and 13)

Designing efficient road networks to provide access to recreation and timber production projects (Chapter 11)

Managing complex projects in efficient and timely ways, given fixed budgets and other constraints (Chapters 12 and 18)

Recognizing the uncertainty of biological and economic outcomes and dealing with this uncertainty in the best possible way (Chapters 14–17)

Ranking alternative investment projects in such a way that those selected maximize the contribution to private or public welfare (Chapter 18)

Forecasting the demand, supply, and price of forest products (Chapter 19)

1.2 THE NATURE OF MODELS

In tackling problems of this sort and making related decisions, forest managers use models. *Models* are abstract representations of the real world that are useful for purposes of thinking, forecasting, and decision making.

Models may be very informal, mostly intuitive, and supported by experience and information that is not put together in any systematic manner. Nevertheless, in the process of thinking about a problem, pondering alternatives, and reaching a decision, one undoubtedly uses a model, that is, a very abstract representation of what the real-life problem is. Most decisions are made with this kind of informal model. The results may be very good, especially for a smart, experienced manager, but the process is unique to each individual and it is difficult to learn.

Forest managers have long used more concrete models. Some are physically very similar to what they represent. For example, a forest hydrologist may use a sand-and-water model of a watershed that differs from the real watershed only with respect to scale and details. Water or a liquid of higher density is made to flow through the model at varying rates to simulate seasonal variation in precipitation and flooding. The resulting erosion is observed, and various systems of dams and levees can be tested using this model.

A forest map is an example of a more abstract model. There is very little physical correspondence between the map and the forest it represents. Nevertheless, maps are essential in many forestry activities. Few management decisions are made without referring to them to define the location and the extent of activities such as harvesting, reforestation, campground development, and road building.

The models dealt with in this book are even more abstract than maps. They are *mathematical models*. Here, little visual analogy is left between the real world and the model. Reality is captured by symbolic variables and by formal algebraic relations between them. Yet, despite or because of their abstraction, mathematical models are very powerful.

Mathematical models are not new in forestry. For example, tabular and mathematical functions have long been used to express biometric relationships between a stand volume per unit area, its age, and site quality. Forest economists long ago developed formulas to calculate the value of land as a function of its expected production, forest product prices (both timber and nontimber), management costs, and interest rates. These investment models are fundamental to forest resource decision making. We shall study them in detail in Chapter 18.

However, most of the mathematical methods used in this book can deal with much broader questions than the narrow issues just mentioned. They can tackle problems with a very large number of variables and relationships. This makes them well suited to complex, real-life managerial situations.

1.3 SYSTEMS MODELS

Forest resource management problems involve many different variables. Some are biological, like the growth potential of a particular species of trees on a particular soil. Others are economic, like the price of timber and the cost of labor. Still others are social, like the environmental laws that may regulate for whom and for what a particular forest must be managed. Often, these variables are interrelated. Changes in one of them may influence the others.

All these variables and the relationships that tie them together constitute a *system*. Because of the complexity of the real forest resource systems, foreseeing the consequences of a particular decision is not an easy task. For example, to increase the diversity of the trees in a forest, we may think of changing the method, timing, or intensity of harvesting. But what exactly is the relation between harvest, or lack thereof, and diversity? How much does the frequency and the intensity of the harvest matter? What is the effect on the long-term health of the forest of taking some trees and leaving others? What is the effect of changing the harvesting pattern on the timber income from the forest? How much will it cost, if anything, to increase forest diversity?

System models are meant to help answer such questions. They are tools that managers can use to predict the consequences of their actions. In a sense, a model is a device to bring the real world to the laboratory or to the office. Managers can and do carry out experiments with models that would be impossible in reality. For example, they can try several management alternatives on a model of their forest and observe the consequences of each alternative for many future decades, a thing that is impossible to do with a real forest. It is this ability to experiment and predict, to ponder different choices, that makes forest systems modeling such an exciting endeavor.

Some of the first systems models and the methods to solve them were developed during the Second World War, to assist in military operations. This led to a body of knowledge known as *operations research* or *management science*. After the war, operations research methods began to be applied successfully in industry, agriculture, and government. In the United States, the society that coordinates and promotes the activities of professionals in this field is the Institute for Operations Research and Management Sciences.

The first applications of operations research to forest management problems date from the early 1960s. Their number has been growing rapidly since then. The Society of American Foresters has had for many years an active Operations Research Working Group. A similar group exists within the International Union of Forest Research Organizations.

Most of the quantitative methods presented in this book are part of the field of operations research, but investment and econometric models have an even longer history within the field of economics. Several modern systems models in forest resource management combine the methods of operations research and those of economics. Economics remains an essential part of forest resource management. Even when the objectives of management are purely ecological, such as in designing a conservation program, economics are needed to compare the costs, if not the benefits, of alternative approaches.

1.4 THE ROLE OF COMPUTERS

Although systems models are formulated via mathematics, mathematics alone cannot make them work. The reason is that only very simple mathematical models have exact analytical solutions. For example, a simplistic model of the growth of a deer population in a forest would state that the growth proceeds at a rate proportional to the number of animals. That relation can be expressed as a simple equation. A solution of that equation would give the population size as a function of time. In fact, the growth of the population is also a function of the amount of food available in the forest, which itself changes at a rate that depends on the way the forest is managed, and so on. To model these relationships properly one needs a system of equations for which there is no exact solution, only approximate ones.

This example is typical of systems models. By their very nature, they do not have exact analytical solutions. They must be solved by numerical methods, that is to say, essentially by trial and error. But *algorithms* can decrease the number of trials considerably. Algorithms are methods of calculation that ensure that, starting from a rough approximation, a good solution is approached within a reasonable number of steps.

Algorithms have long been used in approximating solutions of equations. But the power of algorithms has been increased immensely by computers.

Introduction

The advent of computers has caused a scientific revolution akin to the discovery of differential and integral calculus. Problems that a mere 50 years ago could not even be considered are now routinely solved in a few seconds on a personal computer. Computers can now easily determine the best solution to problems with several thousand variables and as many constraints on the values of these variables. The search for *optimality*, that is, seeking not just a solution but the *best* solution among a possibly infinite number of solutions, is a recurring theme in operations research and a feature of several of the models studied in this book.

1.5 GOOD MODELS

The availability of powerful and cheap computers is not without dangers. In forestry, as in other fields, it has often led to the development of many awkward, expensive, and cumbersome models. A good roadmap does not need confusing topographical detail. Similarly, the best forest system models are usually the simplest ones that reflect the key elements of the question to be answered. Too many times, models have been sought that could "do everything." It is usually better to precisely define the problem to be solved and to limit a model strictly to that problem.

In this respect, one can recognize three elements in model development: problem definition, model building, and model implementation. There is a tight dependency between them. A well-defined problem is half solved, and the solution of a well-defined problem is likely to be readily understood and implemented. To be any good, models must ultimately help managers make decisions. Thus, it is unfortunate that managers do not usually build models themselves.

A recent development that is helping to bridge the gap between forest resource managers and model builders is the popularity of computer spreadsheets. Most managers are now using spreadsheets routinely for a variety of purposes. Modern spreadsheets have sophisticated built-in functions, including optimizers that avoid the need for specialized computer programming. A spreadsheet is an ideal medium for managers to develop simple, small, purpose-oriented models on their own.

With the spreadsheet as a medium and a basic knowledge of decision methods of the kind presented in this book, forest resource managers should be in a position to develop at least prototype models of their problems and to explore the consequences of various courses of action. A full implementation of their models may require specialized software or programming, but the approach itself ceases to be a "black box" with little managerial input or understanding.

Good modeling is not a way of computing, but rather a way of thinking. More than finding a particular solution, good models should help forest resource managers reason through a problem in a logical manner. Thus, although the quality of data underlying the model is important, it is not critical. Much useful understanding of a problem can be acquired by building a model with very rough data. All-important decisions must often be made quickly. Good models do not need the perfect data set to materialize. Instead, they help make the best decision possible in a timely fashion with whatever data are available.

ANNOTATED REFERENCES

Barlow, R.J., and S.C. Grado. 2001. Forest and wildlife management planning: An annotated bibliography. Forest and Wildlife Research Center Bulletin #FO 165. Mississippi State University, State College. 51 pp. (With keyword index.)

Brack, C.L., and P.L. Marshall. 1996. A test of knowledge-based forest operations scheduling procedures. *Canadian Journal of Forest Research* 26:1193–1202. (Compares the results of timber harvest schedules generated using a variety of the techniques discussed in this book on the basis of scenic beauty, stand health, water quality, and economics.)

Costanza, R., L. Wainger, C. Folke, and K. Maler. 1996. Modeling complex ecological economic systems: Toward an evolutionary, dynamic understanding of people and nature. Pages 148–163 in *Ecosystem Management: Selected Readings*. Springer, New York. 462 pp. (Discusses modeling the interactions between anthropogenic and natural systems.)

Gautier, A., B.F. Lamond, D. Pare, and F. Rouleau. 2000. The Quebec ministry of natural resources uses linear programming to understand the wood-fiber market. *Interfaces* 30(6):32–48. (Description of quantitative methods used to improve understanding of forest product markets.)

Hoekstra, T.W. (Team Leader) *et al*. 1990. Critique of land management planning. Vol. 4: *Analytical Tools and Information*. FS-455. USDA Forest Service, Washington, D.C. 47 pp. (Evaluates the Forest Service's use of analytic tools such as mathematical programming and PERT/CPM models in the preparation of plans for national forests.)

Iverson, D.C., and R.M. Alston. 1986. *The Genesis of FORPLAN: A Historical Review of the Forest Service Planning Models*. USDA Forest Service General Technical Report INT-214. 30 pp. (Evolution of principal forest planning model used for many years by the USDA Forest Service.)

Johnsen, K., L. Samuelson, R. Teskey, S. McNulty, and T. Fox. 2000. Process models as tools in forestry research and management. *Forest Science* 47(1):2–8. (Discusses the trend toward increased complexity in models of ecosystem processes and the resulting reduction in their applicability.)

Kent, B., B.B. Bare, R.C. Field, and G.A. Bradley. 1991. Natural resource land management planning using large-scale linear programs: The USDA Forest Service experience with FORPLAN. *Operations Research* 39(1):13–27. (Describes the U.S. Forest Service's experiences using a linear programming–based decision support system to prepare management plans for the national forests.)

Loomis, J.B. 2002. *Integrated Public Lands Management: Principles and Applications to National Forests, Parks, Wildlife Refuges, and BLM Lands*. Columbia University Press, New York. 594 pp. (Chapter 5 discusses the role of models in natural resource management, and Chapter 9 discusses their use specifically by the U.S. Forest Service.)

Manley, B.R., and J.A. Threadgill. 1991. LP used for valuation and planning of New Zealand plantation forests. *Interfaces* 21(6):66–79. (Models used for forest valuation and negotiations between buyers and sellers.)

Qi, Y., and J.K. Gilless. 1999. Modeling ecosystem processes and patterns for multiple-use management. Pages 14-22 in *Multiple Use of Forests and Other Natural Resources: Aspects of Theory and Application*. Kluwer Academic, Dordrecht, The Netherlands. 244 pp. (Discusses the implications for forest resource decision models of the increasing concern about ecosystem functions and services.)

Reed, W.J. 1986. Optimal harvesting models in forest management—A survey. *Natural Resource Modeling* 1(1):55–79. (Compares the model structures, objective functions, and system constraints for many of the techniques discussed in this book.)

Starfield, A.M. 1997. Pragmatic approach to modeling wildlife management. *Journal of Wildlife Management* 61(2):261–270. (Argues that managers should learn modeling.)

Weintraub, A., and B. Bare. 1996. New issues in forest land management from an OR perspective. *Interfaces* 26:9–25. (Perspective on forest industry problems suitable for an operations research approach.)

White, G.C. 2001. Perspectives. Why take calculus? Rigor in wildlife management. *Wildlife Society Bulletin* 29(1):380–386. (Discusses the importance of quantitative reasoning for modern wildlife managers.)

Williams, B.K. 1989. Review of dynamic optimization methods in renewable natural resource management. *Natural Resource Modeling* 3(2):137–216. (Compares the model structures, objective functionals, and system constraints for many of the techniques discussed in this book.)

Wood, D.B., and S.M. Dewhurst. 1998. A decision support system for the Menominee legacy forest. *Journal of Forestry* 96(11):28–32. (Describes the use of quantitative techniques for very long-term forest planning.)

CHAPTER 2

Principles of Linear Programming: Formulations

2.1 INTRODUCTION

This chapter is an introduction to the method of linear programming. Here we shall deal mostly with simple examples showing how a management problem can be formulated as a special mathematical model called a *linear program*. We shall concentrate on formulation, leaving the question of how to solve linear programs to the next chapter.

Linear programming is a very general optimization technique. It can be applied to many different problems, some of which have nothing to do with forestry or even management science. Nevertheless, linear programming was designed and is used primarily to solve managerial problems. In fact, it was one of the first practical tools to tackle complex decision-making problems common to industry, agriculture, and government.

For our immediate purpose, *linear programming* can be defined as a method to allocate limited resources to competing activities in an optimal manner. This definition describes well the situation faced by forest managers. The resources with which they work, be they land, people, trees, time, or money, are always limited. Furthermore, many of the activities that managers administer compete for these resources. For example, one manager might want to increase the land area that is growing red pine, but then less land would be available for aspen. Another manager might want to assign more of her staff to prepare timber sales, but then fewer people would be available to do stand improvement work. She could hire more people, but then she would have too little money.

No matter what course of action they choose, managers always face constraints that limit the range of their options. Linear programming is designed to help them choose. Not only can the method show which alternatives are possible ("feasible" in linear programming jargon), it can also help determine the best one. But this requires that both the management objective as well as the constraints be defined in a precise mathematical manner. Finding the best alternative is a recurring theme in management science, and most of the methods presented in this book involve optimization models.

The first practical way of solving linear programs, the simplex method, was invented by George Dantzig in the late 1940s. At first, by hand and with mechanical desk calculators, only small problems could be solved. Using computers and linear programming, one can now routinely solve problems with several thousands of variables and constraints.

Linear programming is by far the most widely used operations research method. Although simulation (which we shall examine in Chapters 14 and 15) is also a very effective method, linear programming has been and continues to be used extensively in forest management. Some of the most widely used forest planning models to date, in the United States and abroad, in industry and on National Forests, use linear programming or its close cousin, goal programming, which we shall study in Chapter 10.

2.2 FIRST EXAMPLE: A POET AND HIS WOODS

This first example of the application of linear programming is certainly artificial, too simple to correspond to a real forestry operation. Nevertheless, it will suffice to introduce the main concepts and definitions. Later on we will use this same example to discuss the graphical and simplex methods for solving linear programs. Anyway, the story is romantic.

PROBLEM DEFINITION

The protagonist is a congenial poet-forester who lives in the woods of Northern Wisconsin. Some success in his writing allowed him to buy, about 10 years ago, a cabin and 90 hectares (ha) of woods in good productive condition. The poet needs to walk the beautiful woods to keep his inspiration alive. But the muses do

Principles of Linear Programming: Formulations

not always respond, and he finds that sales from the woods come in very handy to replenish a sometimes-empty wallet. In fact, times have been somewhat harder than usual lately. He has firmly decided to get the most he can out of his woods. But the arts must go on. The poet does not want to spend more than half of his time in the woods; the rest is for prose and sonnets. Our poet has a curious mind. He has even read about linear programming: a method to allocate scarce resources to optimize certain objectives. He thinks that this is exactly what he needs to get the most out of his woods while pursuing his poetic vocation.

DATA

In order to develop his model, the poet has put together the following information:

> About 40 ha of the land he owns are covered with red pine plantations. The other 50 ha contain mixed northern hardwoods.
> Having kept a very good record of his time, he figures that since he bought these woods he has spent approximately 800 days managing the red pine and 1500 days on the hardwoods.
> The total revenue from his forest during the same period was $36,000 from the red pine land and $60,000 from the northern hardwoods.

PROBLEM FORMULATION

Decision Variables

To formulate his model, the poet-forester needs to choose the variables to symbolize his decisions. The choice of proper decision variables is critical in building a model. Some choices will make the problem far simpler to formulate and solve than others. Unfortunately, there is no set method for choosing decision variables. It is part of the art of model building, which can only be learned by practice.

Nevertheless, the nature of the objective will often give some clue as to what the decision variables should be. We noted earlier that the poet's objective is to maximize his revenues from the property. But this has meaning only if the revenues are finite; thus he must mean revenues per unit of time, say, per year (meaning an average year, like any one of the past 10 enjoyable years that the poet has spent on his property). Formally, we begin to write the objective as:

$$\text{Maximize } Z = \$ \text{ of revenues per year}$$

The revenues symbolized by the letter Z arise from managing red pine or northern hardwoods or both. Therefore, a natural set of decision variables is:

X_1 = the number of hectares of red pine to manage

X_2 = the number of hectares of northern hardwoods to manage

These are the unknowns. We seek the values of X_1 and X_2 that make Z as large as possible.

Objective Function

The objective function expresses the relationship between Z, the revenues generated by the woods, and the decision variables X_1 and X_2. To write this function, we need an estimate of the yearly revenues generated by each type of forest. Since the poet has earned $36,000 on 40 ha of red pine and $60,000 on 50 ha of northern hardwoods during the past 10 years, the average earnings have been $90 per ha per year (90 $/ha/y) for red pine and 120 $/ha/y for northern hardwoods. Using these figures as measures of the poet's expected revenues during the coming years, we can now write his objective function as:

$$\max Z = 90\ X_1 + 120\ X_2$$
$$(\$/y)\quad (\$/ha/y)(ha)\quad (\$/ha/y)(ha)$$

where the units of measurement of each variable and constant are shown in parentheses. A good modeling practice is always to check the homogeneity of all algebraic expressions with respect to the units of measurement. Here, Z is expressed in dollars per year; therefore, the operations on the right of the equality sign must also yield dollars per year, which they do.

To complete the model, we must determine what constraints limit the actions of our poet forester and then help him express these constraints in terms of the decision variables, X_1 and X_2.

Land Constraints

Two constraints are very simple. The area managed in each timber type cannot exceed the area available; that is:

$X_1 \leq 40$ ha of red pine

$X_2 \leq 50$ ha of northern hardwoods

Time Constraint

Another constraint is set by the fact that the poet does not want to spend more than half his time, let us say 180 days a year, managing his woods. In order to write this constraint in terms of the decision variables, we note that the time he has spent managing red pine during the past 10 years (800 days for 40 ha of land) averages to 2 days per hectare per year (2 d/ha/y). Similarly, he has spent 3 d/ha/y on northern hardwoods (1500 days on 50 ha).

In terms of the decision variables X_1 and X_2, the total time spent by the poet-forester to manage his woods is:

$$\underset{\text{(d/ha/y)}}{2} \underset{\text{(ha)}}{X_1} + \underset{\text{(d/ha/y)}}{3} \underset{\text{(ha)}}{X_2}$$

and the expression of the constraint limiting this time to no more than 180 days is:

$$\underset{\text{(d/ha/y)}}{2} \underset{\text{(ha)}}{X_1} + \underset{\text{(d/ha/y)}}{3} \underset{\text{(ha)}}{X_2} \underset{\text{(d/y)}}{\leq 180}$$

Nonnegativity Constraints

The last constraints needed to complete the formulation of the problem state that none of the decision variables may be negative, since they refer to areas. Thus:

$$X_1 \geq 0 \quad \text{and} \quad X_2 \geq 0$$

Final Model

In summary, combining the objective function and the constraints, we obtain the complete formulation of the poet-forester problem as: Find the variables X_1 and X_2, which measure the number of hectares of red pine and of northern hardwoods to manage, such that:

$$\max Z = 90X_1 + 120X_2$$

subject to:

$$X_1 \leq 40$$

$$X_2 \leq 50$$

$$2X_1 + 3X_2 \leq 180$$

$$X_1, X_2 \geq 0$$

Note that northern hardwoods are cultivated under a selection system. This requires more time per land area, especially to mark the trees to be cut, than the even-aged red pine. But in exchange, the hardwoods tend to return more per unit of land, as reflected in the objective function. Therefore, the choice of the best management strategy is not obvious.

In the next chapter we will learn how to solve this problem. But before that, let us consider another example.

2.3 SECOND EXAMPLE: KEEPING THE RIVER CLEAN

The purpose of this second example is to illustrate the formulation of a linear programming model that, in contrast to the poet's problem, involves (1) minimizing an objective function and (2) constraints of the greater-than-or-equal-to form.

Also, in this problem we move away from the strict interpretation of constraints as limits on available resources. Here, some of the constraints express management objectives. Furthermore, this example shows that the objective function being optimized does not have to express monetary returns or costs. Indeed, because it is a general optimization method, linear programming has much broader applications than strictly financial ones.

PROBLEM DEFINITION

This story deals with a pulp mill operating in a small town in Maine. The pulp mill makes mechanical and chemical pulp. Unfortunately, it also pollutes the river in which it spills its spent waters. This has created enough turmoil to change the management of the mill completely.

The previous owners felt that it would be too costly to reduce the pollution problem. They decided to sell. The mill has been bought back by the employees and local businesses, who now own the mill as a cooperative. The new owners have several objectives. One is to keep at least 300 people employed at the mill. Another is to generate at least $40,000 of revenue per day. They estimate that this will be enough to pay operating expenses and yield a return that will keep the mill competitive in the long run. Within these limits, everything possible should be done to minimize pollution.

A bright forester who has already provided shrewd solutions to complex wood procurement problems is asked to suggest an operating strategy for the mill that will meet all these objectives simultaneously and in the best possible way. She feels that it could be done by linear programming. Towards this end, she has put together the following data:

> Both chemical and mechanical pulp require the labor of one worker for about 1 day, or 1 workday (wd), per ton produced.

Principles of Linear Programming: Formulations

The chemical pulp sells at some $200 per ton, the mechanical pulp at $100.

Pollution is measured by the biological oxygen demand (BOD). One ton of mechanical pulp produces 1 unit of BOD, 1 ton of chemical pulp produces 1.5 units.

The maximum capacity of the mill to make mechanical pulp is 300 tons per day; for chemical pulp it is 200 tons per day. The two manufacturing processes are independent; that is, the mechanical pulp line cannot be used to make chemical pulp, and vice versa.

Given this, our forester has found that the management objectives and the technical and financial data could be put together into a linear program. Next we show how she did it.

LINEAR PROGRAMMING FORMULATION

Decision Variables

Pollution, employment, and revenues result from the production of both types of pulp. A natural choice for the decision variables then is:

X_1 = amount of mechanical pulp produced (in tons per day, or t/d) and

X_2 = amount of chemical pulp produced (t/d)

Objective Function

The objective function to minimize is the amount of pollution, Z, measured here by units of BOD per day. In terms of the decision variables, this is:

$$\min \underset{\text{(BOD/d)}}{Z} = \underset{\text{(BOD/t)(t/d)}}{1 \; X_1} + \underset{\text{(BOD/t)(t/d)}}{1.5 \; X_2}$$

where the units of measurement are shown in parentheses. Verify that the objective function is homogeneous in units, that is, that the operations on the right-hand side of the equality sign give a result in BOD/d.

Employment Constraint

One constraint expresses the objective to keep at least 300 workers employed. In terms of the decision variables, this is:

$$\underset{\text{(wd/t)(t/d)}}{1 \; X_1} + \underset{\text{(wd/t)(t/d)}}{1 \; X_2} \geq \underset{\text{(workers)}}{300}$$

Revenue Constraint

A second constraint states that at least $40,000 of revenue must be generated every day:

$$100 \underset{(\$/t)\,(t/d)}{X_1} + 200 \underset{(\$/t)\,(t/d)}{X_2} \geq \underset{(\$/d)}{40{,}000}$$

Capacity Constraints

Two other constraints refer to the fact that the daily production capacity of the mill cannot be exceeded:

$$\underset{(t/d)}{X_1} \leq \underset{(t/d)}{300} \quad \text{(mechanical pulp)}$$

$$\underset{(t/d)}{X_2} \leq \underset{(t/d)}{200} \quad \text{(chemical pulp)}$$

Nonnegativity Constraints

The quantity of mechanical and chemical pulp produced must be positive or zero; that is:

$$X_1 \geq 0 \quad \text{and} \quad X_2 \geq 0$$

In summary, the final form of the linear program that models the dilemma of the pulp-making cooperative is to find the values of X_1 and X_2, which measure the amount of mechanical and chemical pulp produced daily, such that:

$$\min Z = X_1 + 1.5 X_2$$

subject to:

$$X_1 + X_2 \geq 300$$

$$100 X_1 + 200 X_2 \geq 40{,}000$$

$$X_1 \leq 300$$

$$X_2 \leq 200$$

$$X_1, X_2 \geq 0$$

Principles of Linear Programming: Formulations

A Note on Multiple Objectives

In this example, although there were several management objectives (pollution, employment, and revenue), only one of them was expressed by the objective function. The other objectives were expressed as constraints. The fact that there is only one objective function is a general rule and not peculiar to linear programming. In any optimization problem, only one function can be optimized.

For example, strictly speaking it makes no sense to say that we want to maximize the amount of timber that a forest produces *and* maximize the recreation opportunities offered in the same forest. As long as timber and recreation conflict, that is, as long as they use common resources, we must choose between two options: Either we maximize timber, subject to a specified amount of recreation opportunities, or we maximize recreation, subject to a certain volume of timber production.

One of the teachings of linear programming is that we must choose which objective to optimize. Later, we will study methods designed to handle several objectives with more flexibility. Goal programming is one such method, but even in goal programming (as we shall see in Chapter 10), the optimized objective function is unique.

2.4 STANDARD FORMULATION OF A LINEAR PROGRAMMING PROBLEM

Any linear program may be written in several equivalent ways. For example, like the poet's problem, the river pollution problem can be rewritten as a maximization subject to less-than-or-equal-to constraints, the so-called *standard form*, as follows:

$$\max(-Z) = -X_1 - 1.5X_2$$

subject to:

$$-X_1 - X_2 \leq -300$$

$$-100X_1 - 200X_2 \leq -40{,}000$$

$$X_1 \leq 300$$

$$X_2 \leq 200$$

$$X_1, X_2 \geq 0$$

The minimization of the objective function has been changed to a maximization of its opposite, and the direction of the first two inequalities has been reversed by multiplying both sides by −1.

Strict equality constraints can also be expressed as less-than-or-equal-to constraints. For example if the cooperative wanted to employ exactly 300 workers, the first constraint would be:

$$X_1 + X_2 = 300$$

which is equivalent to these two inequalities:

$$X_1 + X_2 \leq 300 \quad \text{and} \quad X_1 + X_2 \geq 300$$

the second of which can be rewritten as a less-than-or-equal-to constraint:

$$-X_1 - X_2 \leq -300$$

Furthermore, if any variable in a problem, say, X_3, might take a negative value (for example, if X_3 designated the deviation with respect to a goal), then it could be replaced in the model by the difference between two nonnegative variables:

$$X_3 = X_4 - X_5 \quad \text{with} \quad X_4 \geq 0 \quad \text{and} \quad X_5 \geq 0$$

Thus, a linear programming problem may have an objective function that is maximized or minimized, constraints may be inequalities in either direction or strict equalities, and variables may take positive or negative values. Still, the problem can always be recast in the equivalent standard form, with an objective function that is maximized, inequalities that are all of the less-than-or-equal-to type, and variables that are all nonnegative. The general expression of this standard form is: Find the values of n variables $X_1, X_2, \ldots, X_n$ (referred to as decision variables or activities) such that the objective function, Z, is maximized. The objective function is a linear function of the n decision variables:

$$\max Z = c_1 X_1 + c_2 X_2 + \cdots + c_n X_n$$

where $c_1, \ldots, c_n$ are all constant parameters. Each parameter, c_j, measures the contribution of the corresponding variable, X_j, to the objective function. For example, if X_1 increases (decreases) by one unit, then, other variables remaining equal, Z increases (decreases) by c_1 units.

The values that the variables can take in trying to maximize the objective function are limited by m constraints. The constraints have the following

Principles of Linear Programming: Formulations

general expression:

$$a_{11}X_1 + a_{12}X_2 + \cdots + a_{1n}X_n \leq b_1$$
$$a_{21}X_1 + a_{22}X_2 + \cdots + a_{2n}X_n \leq b_2$$
$$\vdots$$
$$a_{m1}X_1 + a_{m2}X_2 + \cdots + a_{mn}X_n \leq b_m$$

where $b_1, b_2, \ldots, b_m$ are constants. These constants often reflect the amounts of available resources. For example, b_1 could be the land area that a manager can use, b_2 the amount of money available to spend. In that case, each a_{ij} is a constant that measures how much of resource i is used per unit of activity j. For example, keeping the interpretation of b_2 just given and assuming that X_1 is the number of hectares planted in a given year, a_{21} is the cost of planting one hectare.

More generally, this interpretation means that the product $a_{ij}X_j$ is the amount of resource i used when activity j is at the level X_j. Adding these products up over all activities leads to the following general expression for the total amount of resource i used by all n activities:

$$R_i = a_{i1}X_1 + a_{i2}X_2 + \cdots + a_{in}X_n$$

In linear programming R_i is referred to as the row activity i, in symmetry with the column activity, X_j.

Adding the nonnegativity constraints completes the standard form:

$$X_1, X_2, \ldots, X_n \geq 0$$

The standard linear programming model can be expressed in a more compact form by using the Greek capital letter sigma (Σ) to indicate summations. The general linear programming problem is then to find X_j ($j = 1, \ldots, n$) such that:

$$\max Z = \sum_{j=1}^{n} c_j X_j$$

subject to:

$$\sum_{j=1}^{n} a_{ij} X_j \leq b_i \quad \text{for } i = 1, \ldots, m$$

$$X_j \geq 0 \quad \text{for } j = 1, \ldots, n$$

2.5 SPREADSHEET FORMULATION OF LINEAR PROGRAMS

Much of the power of mathematical models stems from the ability to formulate and solve them quickly with computers. For ease of learning and application, modern spreadsheets have become the ideal software to handle many management models. Throughout this book, we shall give examples of modeling with the Excel software. Like several other spreadsheets, Excel contains a Solver to find the best solution of linear programs and other problems.

SPREADSHEET FORMULATION OF THE POET'S PROBLEM

Figure 2.1 shows how the poet's problem can be formulated in a spreadsheet. All the fixed parameters, that is, the data, are in bold characters, while the variables, or the cells that depend on the variables, are not. The decision variables, or activities, X_1 and X_2 are in cells B3:C3. The amounts of land and time available are in cells F6:F8. The data in cells B6:C8 are the amounts of resources used per unit of each activity. The data in cells B10:C10 are the revenues per unit of each activity.

Cells D6:D8 contain formulas expressing the amount of resource used by the activities (the row activities). For example, the formula in cell D6 is the

	A	B	C	D	E	F	G
1	POET PROBLEM						
2		Red pine	Hardwoods				
3	Managed area	10	10				
4		(ha)	(ha)			Resources	
5		Resources required		Total		available	
6	Red pine land	1		10	<=	**40**	(ha)
7	Hardwoods land		1	10	<=	**50**	(ha)
8	Poet's time	2	3	50	<=	**180**	(d/y)
9			Objective function	Total			
10	Returns	**90**	**120**	2,100	Max		
11		($/ha/y)	($/ha/y)	($/y)			
12							
13			Key Formulas				
14	Cell		Formula			Copied to	
15	D6		=SUMPRODUCT(B6:C6,B$3:C$3)			D6:D8	
16	D10		=SUMPRODUCT(B10:C10,B$3:C$3)				

FIGURE 2.1 Spreadsheet formulation of the poet's problem. Excel workbooks containing all the spreadsheets shown in this book are available at the web page address: www.forest.wisc.edu/facstaff/buongiorno/book/index.htm

Principles of Linear Programming: Formulations

equivalent of $1X_1 + 0X_2$, expressing the amount of red pine managed by the poet. The "<=" symbols in cells E6:E8 remind us that the amounts of resources used should not exceed the amounts available.

Cell D10 contains the formula of the objective function, the equivalent of $Z = 90X_1 + 120X_2$.

The spreadsheet in Figure 2.1 shows that by managing only 10 ha of red pine and 10 ha of hardwoods, the poet would obtain yearly revenues of $2,100. He would be using only 50 days of his time to do this.

Set up this simple model on your own spreadsheet, and explore the effect of different values of the decision variables. In each case, check to see whether the decision is feasible or whether it uses too much of some resource. You might find the best solution by trial and error. In the next chapter we will learn a way to find the best solution easily and surely with Excel's optimization program, Solver.

SPREADSHEET FORMULATION OF THE RIVER POLLUTION PROBLEM

Figure 2.2 shows the spreadsheet for the river pollution problem. The two variables X_1 and X_2, defining the production levels for mechanical and chemical pulp, are in cells B3 and C3.

	A	B	C	D	E	F	G
1	RIVER POLLUTION PROBLEM						
2		Mech pulp	Chem pulp				
3	Production	100	100				
4		(t/d)	(t/d)				
5		Constraints		Total			
6	Employment	1	1	200	>=	300	(workers)
7	Revenues	100	200	30000	>=	40000	($/d)
8	Mech capacity	1		100	<=	300	(t/d)
9	Chem capacity		1	100	<=	200	(t/d)
10		Objective function		Total			
11	Pollution	1	1.5	250	Min		
12		(BOD/t)	(BOD/t)	(BOD/d)			
13							
14		Key Formulas					
15	Cell	Formula			Copied to		
16	D6	=SUMPRODUCT(B6:C6,B$3:C$3)		D6:D9			
17	D11	=SUMPRODUCT(B11:C11,B$3:C$3)					

FIGURE 2.2 Spreadsheet formulation of the river pollution problem.

The formula in cell D6 corresponds to $1X_1 + 1X_2$, the total number of workers employed. Cell F6 contains the lower bound for the number of workers employed. The revenues constraint is set up in the same way in cells D7 and F7.

The last two constraints refer to the limits on production capacity. Cells F8:F9 contain data on the mill capacity for producing each type of pulp, while cells D8:D9 contain formulas expressing the amount of each type of capacity being used.

Cell D11 contains the formula for the daily amount of BOD produced, corresponding to the objective function $Z = 1X_1 + 1.5X_2$. The "Min" label in cell E11 is a reminder that we are trying to minimize the amount of daily BOD. The ">=" and "<=" symbols in cells E6:E9 remind us of the direction of each constraint.

The spreadsheet in Figure 2.2 is set up with a production of mechanical and chemical pulp of 100 tons per day. However, this is not a feasible solution. Although the production of each product is less than capacity, employment and revenue generated are too low. Set up this problem in a spreadsheet, and find a solution by trial and error that would meet all of the constraints while possibly keeping pollution very low. You may not get the best solution; however, you will learn how to do that in the next chapter.

2.6 ASSUMPTIONS OF LINEAR PROGRAMMING

Before proceeding to study the solutions and applications of linear programming, it is worth stressing the assumptions that it makes. A linear programming model is a satisfactory representation of a particular management problem when all these assumptions are warranted. They will never hold exactly, but they should be reasonable. The determination of what is reasonable or not is part of the art of management and model building. Keep in mind that bold assumptions are more useful in understanding the world than are complicated details.

PROPORTIONALITY

A linear programming model assumes that the contribution of any activity to the objective function is directly proportional to the level of that activity. As the level of the activity increases or decreases, the change in the objective function due to a unit change of the activity remains the same. For example, in the poet-forester problem, the contribution of red pine management to revenues is directly proportional to the area of red pine being managed.

In a similar manner, the amount of resource used by each activity is assumed to be directly proportional to the level of that activity. For example, the time the poet must put in managing his land is directly proportional to the area being managed. If, as the managed area increased, each additional hectare required an increasing amount of time, then the linear programming model would not be valid, at least not without some modification.

ADDITIVITY

A linear programming model assumes that the contribution of all activities to the objective function is just the sum of the contributions of each activity considered independently. Similarly, the total amount of a resource used by all activities is assumed to be the sum of the amounts used by each individual activity considered independently. This means that the contribution of each variable does not depend on the presence or absence of the others.

In our example, regardless of what the poet-forester does with his northern hardwoods, he will always get $90 per hectare from each hectare of managed red pine, and it will still take him 2 days per hectare per year to manage.

DIVISIBILITY

A linear programming model assumes that all activities are continuous and can take any positive value. This means that linear programming models are not generally suitable in situations where the decision variables can take only integer values. For example, management decisions may require yes or no answers: Should we build this bridge or not?

For some problems that involve integer variables, it may be enough to compute a continuous solution by ordinary linear programming and then round the variables to the nearest integer. But this is not always appropriate. We will study programming models that use integer variables in Chapter 11.

DETERMINISM

A linear programming model is deterministic. In computing a solution, it does not take into account that all of the coefficients in the model are only approximations. For this reason, it is wise when using linear programming to compute not only one solution, but several. Each solution corresponds to different, but reasonable, assumptions regarding the values of the parameters. Such *sensitivity analysis* shows how sensitive a solution is to changes in the values of parameters. In order to arrive at good decisions, one should examine carefully those parameters that have the most impact.

Most of the models examined in this book are deterministic. Stochastic models, in which the random nature of some parameters is considered explicitly, will be examined in Chapters 12 and 13 (on network analysis and dynamic programming), in Chapters 14 and 15 (on simulation), and in Chapters 16 and 17 (on Markov chains). Interestingly, the linear programming method will turn out to be useful even to solve some stochastic problems.

2.7 CONCLUSION

The two examples considered in this chapter have shown the flexibility of linear programming. Problems involving the optimization of a specific objective, subject to constraints, can be cast as linear programs. The objective may be to minimize or maximize something. The constraints may represent the limited resources that the manager can work with, but they may also refer to objectives. Only one objective can be optimized.

Formulating a forest management problem so that it could be solved by linear programming is not always easy. It takes ingenuity and much practice, plus some courage. The world, to be understood, must be simplified; this is what models are all about. Linear programming is not different. It makes some drastic assumptions, but the assumptions are not so critical as to render the method useless. On the contrary, we shall discover in the forthcoming chapters that linear programming is so flexible that it can be usefully applied to a wide array of forest management problems, from harvest scheduling and multiple-use planning to investment analysis. It can even help deal with uncertainty. There is almost no limit, except our imagination.

PROBLEMS

2.1 Several management problems are listed here. For each, what kind of objective function would be appropriate in a linear programming model? What kinds of decision variables? What kinds of constraints?

(a) A farmer wants to maximize the income he will receive over the next 20 years from his woodlot. The woodlot is covered with mature sugar maple trees that could be sold as stumpage or managed to produce maple syrup.

(b) The manager of a hardwood sawmill wants to maximize the mill's net revenues. The mill can produce pallet stock, dimension lumber, or some combination of the two. Pallet stock commands a lower price than dimension lumber, but it can be produced from less expensive logs, and the daily capacity of the mill to produce pallet stock exceeds its capacity to produce dimension lumber.

(c) A logging contractor wants to minimize the cost of harvesting a stand of timber. She can use mechanical fellers, workers with chainsaws, or some of both. Leasing and operating a mechanical feller is more expensive than hiring a worker with a chainsaw, but the feller can do more work per hour. On the other hand, a mechanical feller cannot be used to harvest some of the largest and most valuable trees in the stand.

2.2 Consider the linear programming model of the poet and his woods in Section 2.2.

(a) If the poet had received $50,000 over the last 10 years from managing his red pine plantations and $30,000 from managing his hardwoods, how would the coefficients of X_1 and X_2 change in the objective function?

(b) Suppose that the poet found that time spent pruning branches had a particularly inhibiting effect on his literary endeavors and that two-thirds of the time devoted to managing hardwoods had to be spent pruning. If he wanted to limit the time he spent pruning to no more than 70 days per year, what constraint would have to be added to the model?

(c) If one-half of the time devoted to managing red pine plantations had to be spent pruning, how would this constraint have to be further modified?

2.3 Consider the linear programming model of pulp mill management in Section 2.3. The mill management might prefer to maximize gross revenues while limiting pollution to no more than 300 BOD per day. Reformulate the model to reflect this new management orientation, leaving the employment and capacity constraints unchanged.

2.4 Consider the linear programming model of the poet and his woods in Section 2.2.

(a) If the poet decided to manage 25 acres of red pine and 35 acres of hardwoods, how much income would he receive from his lands each year?

(b) How much of his time would he need to manage his lands?

2.5 Consider the linear programming model of pulp mill management in Section 2.3.

(a) If the mill's management decided to produce 150 tons of chemical pulp and 190 tons of mechanical pulp per day, how much pollution would result?

(b) How much revenue would this decision generate?

(c) How many people would be employed?

2.6 A logging contractor wants to maximize net revenues per day from the operation of her four tractor-skidders and six wheeled-skidders. From her records, she estimates her net revenue per day of operation for a tractor-skidder at $300 and for a wheeled-skidder at $600. Only 18 people trained to operate this kind of logging equipment are available in the local labor market, and it takes two people to run a wheeled-skidder and three to run a tractor-skidder.

(a) Formulate this problem as a linear program, defining the units of all decision variables, coefficients, and parameters in the model.

(b) What logical constraints must be placed on the values each decision variable can take?

(c) Can this problem be solved as an ordinary linear program?

2.7 A logging contractor wants to allocate her logging equipment between two logging sites to maximize daily net revenues. She has determined that the net revenue from a cubic meter (m^3) of wood is $19 from Site 1 and $21 from Site 2. At her disposal are two feller-skidders, one brancher-slasher, and one truck. Each type of equipment can be used for nine hours per day, and this time can be divided in any proportion between the two sites. The equipment hours needed to produce a m^3 of wood from each site varies as shown in the following table. Formulate this problem as a linear program, defining the units of all decision variables, coefficients, and parameters in the model. (*Hint*: You will need two variables, plus one constraint for each kind of equipment.)

Equipment Hours Needed to Produce a Cubic Meter of Wood

Feller-skidder	Brancher-slasher	Truck
0.30	0.30	0.17
0.40	0.15	0.17

2.8 You and your partner own a ranch on which you raise sheep and cows. You love cows, but your partner is all for sheep. The ranch produces 1200 animal-unit months (AUM) of forage per year. An AUM is the amount of forage necessary for the sustenance of one cow for a month. Sheep require only 0.20 AUM per month. The average profit from a cow is $300/y, compared to $100/y from a sheep. After intense discussion, you and your partner have agreed to a compromise in which you will keep at least 100 sheep and 50 cows on the ranch, but no more than 200 animals in total.

(a) Formulate this problem as a linear program to find the number of cows and sheep that would maximize total profit from the ranch, defining the units of all decision variables, coefficients, and parameters in the model.

(b) Set up this linear program in a spreadsheet, and explore some feasible solutions.

2.9 Consider a ponderosa pine forest that could be managed either as a multiple-use area for recreation and timber or as a wilderness that would allow only for recreational activities. The forest consists of 1,600 ha of high-site (i.e., high-productivity) land and 2,400 ha of low-site land. The expected outputs from the forest, by site and management option, are given in the following table. (Note that the sediment going into streams in the forest is higher with multiple-use management than with wilderness management.)

Outputs per Hectare, by Site and Management Option

Output	High-site land		Low-site land	
	Wilderness	Multiple use	Wilderness	Multiple use
Timber (m^3/ha/y)		3.5		1.2
Sediment (m^3/ha/y)	0.06	0.12	0.03	0.06
Recreation (vd/ha/y)	1	0.25	0.6	0.15

(a) Formulate this problem as a linear program to find the management plan that would maximize the amount of recreation (in visitor-days per year, vd/y) while producing at least 1,400 m³/y of timber and keeping sediments less than 200 m³/y, defining the units of all decision variables, coefficients, and parameters in the model. (*Hint:* use decision variables X_{hw} = hectares of high-site land assigned to wilderness, X_{hm} = hectares of high-site land assigned to multiple use, X_{lw} = hectares of low-site land assigned to wilderness — you should be able to guess the last decision variable you need.)

(b) Set up this linear program in a spreadsheet, and explore some feasible solutions.

2.10 You are the manager of two paper mills that manufacture three grades of paper. You have contracts to supply at least 1600 tons of low-grade paper, 500 tons of medium-grade paper, and 200 tons of high-grade paper. It costs $1,000/day to operate the first mill, and $2,000/day to operate the second mill. Mill 1 produces 8 tons of low-grade paper, 1 ton of medium-grade paper, and 2 tons of high-grade paper per day. Mill 2 produces 2 tons of low-grade paper, 1 ton of medium-grade paper, and 10 tons of high-grade paper per day.

(a) Formulate this problem as a linear program to determine how many days each mill should operate to satisfy the order at least cost, defining the units of all decision variables, coefficients, and parameters in the model.

(b) Set up this linear program in a spreadsheet, and explore some feasible solutions.

(c) If you had signed a labor contract that specifies that both mills must operate the same number of days, how would this change the problem?

2.11 Gifford Pinchot, one of the founding fathers of American forestry, once said: "Where conflict of interest must be reconciled, the question will always be decided from the standpoint of *the greatest good for the greatest number* in the long run." How does this statement fit with what you learned about optimization at the end of Section 2.3?

ANNOTATED REFERENCES

Davis, L.S., K.N. Johnson, P.S. Bettinger, and T.E. Howard. 2001. *Forest Management: To Sustain Ecological, Economic, and Social Values.* McGraw-Hill, New York. 804 pp. (Chapter 6 discusses formulating linear programming models.)

Dykstra, D.P. 1984. *Mathematical Programming for Natural Resource Management.* McGraw-Hill, New York. 318 pp. (Chapter 2 presents a simple two-variable land management linear programming model.)

Ells, A., E. Bulte, and G.C. van Kooten. 1996. Uncertainty and forest land use allocation in British Columbia: Vague priorities and imprecise coefficients. *Forest Science* 43(4):509–520. (Discusses situations in which the determinism assumption of linear programming models is problematic.)

Foster, B.B. 1969. Linear programming: A method for determining least-cost blends or mixes in papermaking. *Tappi* 52(9):1658-1660. (Determines the least-cost blend of pulpwood species that will produce a paper with certain characteristics.)

Hanover, S.J., W.L. Hafley, A.G. Mullin, and R.K. Perrin. 1973. Linear programming analysis for hardwood dimension production. *Forest Products Journal* 23(11):47–50. (Determines the size and grade of lumber that will maximize the profits of a sawmill, subject to technological and contractual constraints.)

Kent, B.M. 1989. *Forest Service Land Management Planners' Introduction to Linear Programming.* USDA Forest Service, General Technical Report RM-173. Fort Collins, Co. 36 pp. (Discusses simple linear programming models for forest resource management.)

Kotak, D.B. 1976. Application of linear programming to plywood manufacture. *Interfaces* 7(1):56–68. (Optimizes the wood mix used by a plywood mill.)

Little, R.L., and T.E. Wooten. 1972. *Product Optimization of a Log Concentration Yard by Linear Programming.* Department of Forestry, Clemson University, Clemson, SC. Forest Research Series No. 24. 14 pp. (Discusses using a linear programming model for sorting logs for resale.)

Ragsdale, C.T. 1998. *Spreadsheet Modeling and Decision Analysis: A Practical Introduction to Management Science.* South-Western College Publishing, Cincinnati, OH. 742 pp. (Chapter 2 discusses formulating linear programming models.)

Winston, W.L. 1995. *Introduction to Mathematical Programming.* Duxbury Press, Belmont, CA. 818 pp. (Chapter 3 discusses formulating linear programming models.)

CHAPTER 3

Principles of Linear Programming: Solutions

After a forest management problem has been formulated as a linear program, the program must be solved to determine the most desirable management strategy. This chapter deals with two different methods of solution. The simplest procedure is graphic, but it can be used only with very small problems. Computers use a more general technique, the simplex method. After an optimum solution has been obtained, one can explore how sensitive it is to the values of the parameters in the model. To this end we shall study duality, a powerful method of sensitivity analysis in linear programming.

3.1 GRAPHIC SOLUTION OF THE POET'S PROBLEM

Large linear programming models that represent real managerial problems must be solved with a computer. However, the small problem that we developed in Section 2.2 for the poet-forester can be solved with a simple graphic procedure. The technique illustrates well the nature of the general linear programming solution. Recall the expression of that problem:

$$\max Z = 90X_1 + 120X_2 \quad (\$/y)$$

subject to:

$$X_1 \le 40 \quad \text{(ha of red pine)}$$
$$X_2 \le 50 \quad \text{(ha of hardwoods)}$$
$$2X_1 + 3X_2 \le 180 \quad \text{(days of work)}$$
$$X_1, X_2 \ge 0$$

where the variable X_1 is the number of hectares of red pine that the poet should manage and X_2 is the number of hectares of northern hardwoods. The object is to find the values of these two variables that maximize Z, which measures the poet's annual revenue from the property. There are 40 hectares of red pine on the property and 50 hectares of hardwoods, and the poet is willing to use up to 180 days per year to manage his forest.

Because the problem has only two decision variables, it can be represented graphically as in Figure 3.1(a). The number of hectares of red pine is measured on the horizontal axis, that of hardwoods on the vertical axis. Each point on this graph represents a management decision. For example, the point P corresponds to the decision to manage 15 hectares of red pine and 20 hectares of hardwoods.

However, given the resource constraints, not all points on the diagram correspond to a possible (feasible) decision. The first task in solving a linear program is to find all the points that are feasible; among those points we then seek the point(s) that maximize the objective function.

FEASIBLE REGION

Since both X_1 and X_2 cannot be negative, only the shaded portion of Figure 3.1(a) can contain a feasible solution. In addition, the constraint $X_1 \le 40$ means that a feasible point (X_1, X_2) cannot lie to the right of the vertical line $X_1 = 40$. This is reflected in Figure 3.3(b), where the shaded area contains only the values of X_1 and X_2 that are permissible thus far.

Next, the constraint $X_2 \le 50$ eliminates all the points above the horizontal line $X_2 = 50$; the feasible region now consists of the points within the shaded rectangle in Figure 3.1(c).

The last constraint is set by the poet's time: $2X_1 + 3X_2 \le 180$. Only the points that lie on one side of the line $2X_1 + 3X_2 = 180$ satisfy this restriction. To plot that line on our figure, we need two of its points. For example, if $X_1 = 0$, then $X_2 = 60$. Similarly, if $X_2 = 30$, then $X_1 = (180 - 90)/2 = 45$. To find on which side of the line $2X_1 + 3X_2 = 180$ the feasible region lies, we need to check for one

Principles of Linear Programming: Solutions

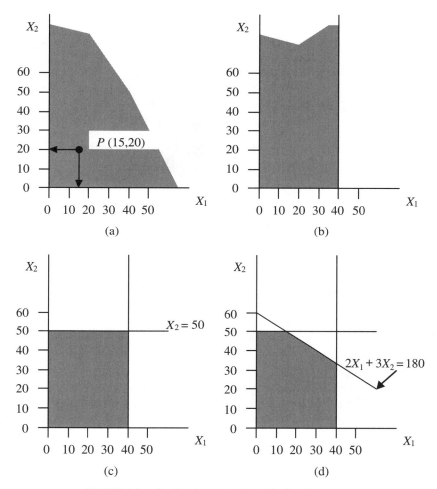

FIGURE 3.1 Graphic determination of the feasible region.

point only. For example, at the origin, both X_1 and X_2 are zero and the time constraint holds; therefore, all the points on the same side of $2X_1 + 3X_2 = 180$ as the origin satisfy the poet's time constraint.

In summary, the feasible region is represented by the shaded polygon in Figure 3.1(d). The coordinates of any point within that region simultaneously satisfy the land constraints, the poet's time constraint, and the nonnegativity constraints. In the next step we shall determine which point(s) in the feasible region maximize the objective function.

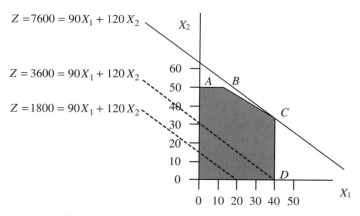

FIGURE 3.2 Graphic determination of the best solution.

BEST SOLUTION

To find the optimum solution graphically, we first determine the position of the line that represents the objective function for some arbitrary value of the objective. For example, let $Z = \$1{,}800$ per year. All the combinations of X_1 and X_2 that lead to these returns lie on the line

$$1800 = 90X_1 + 120X_2$$

This line has been plotted in Figure 3.2. Many of its points lie in the feasible region. Therefore, it is indeed possible for the poet to get this amount of revenue from his property, and there are many ways in which he can do it. But could he get more? For example, could he double his income? This question is readily answered by plotting the line

$$3600 = 90X_1 + 120X_2$$

Again, there are many points on this line that are feasible. Note that this line is parallel to the previous one but farther from the origin.

It is clear that the best solution will be obtained by drawing a straight line that is parallel to those we have just plotted that has at least one point within the feasible region and that is as far from the origin as possible. Thus, the optimum solution must correspond to point C in Figure 3.2.

Reading the coordinates of C on the graph ($X_1 = 40$, $X_2 = 33$) gives an approximation of the best solution. A more precise solution can be obtained by solving the system of equations of the two lines that intersect at C:

$$X_1 = 40 \quad \text{and} \quad 2X_1 + 3X_2 = 180$$

Principles of Linear Programming: Solutions

Therefore, the best value of X_1 is:

$$X_1^* = 40 \text{ ha}$$

Substituting X_1^* in the second equation leads to:

$$X_2^* = \frac{180-80}{3} = 33.33 \text{ ha}$$

Therefore, the best strategy for the poet is to cultivate all the red pine he has but only 33 ha of the hardwoods, leaving the rest idle. That it may be best in some circumstances not to use all of the available resources is an important lesson of linear programming.

The best value of the objective function, that is, the maximum revenue that the poet can obtain from his land, is, then:

$$90X_1^* + 120X_2^* = 7600 \quad (\$/y)$$

Sensitivity Analysis

As mentioned in Section 2.6, linear programming assumes that the parameters of the model are known exactly. The solution just obtained is best only if the parameters are correct. This may not be true. It is therefore useful to do a sensitivity analysis; that is to explore how the best solution changes with different values of the parameters. The simplest form of sensitivity analysis is to observe how the best solution responds to a change in one single parameter, keeping all other things equal.

For example, assume that the returns to hardwood management were $150/ha/y, instead of $120/ha/y, while everything else stays the same. Show how this would change the objective function and produce the best solution at point B instead of C in Figure 3.2. The new best solution would then be $X_1^* = 15$ hectares of red pine, $X_2^* = 50$ hectares of hardwoods, and $Z^* = \$8,850/\text{year}$ of revenue.

Note that one of the resources would still not be fully used in this new best solution; the poet would now be better off by not managing 25 hectares of his red pine.

3.2 GRAPHIC SOLUTION OF THE RIVER POLLUTION PROBLEM

The problem of the cooperative owning the pulp mill (as described in Section 2.3) consisted of finding X_1 and X_2, the daily production of mechanical and chemical pulp, such that the river pollution from mill effluents would be as small as possible:

$$\min Z = X_1 + 1.5X_2 \quad \text{(BOD units/day)}$$

subject to:

$X_1 + X_2 \geq 300$ (employment target, workers)

$100X_1 + 200X_2 \geq 40{,}000$ (revenue target, \$/day)

$X_1 \leq 300$ (mechanical-pulping capacity, tons/day)

$X_2 \leq 200$ (chemical-pulping capacity, tons/day)

$X_1, X_2 \geq 0$

The graphic solution of this linear program proceeds as follows. There are only two decision variables in the problem; these are measured along the axes of Figure 3.3. We first determine the possible values of X_1 and X_2 (feasible region) and then find the point in this region that maximizes the objective function (best solution).

Feasible Region

The nonnegativity constraints ($X_1 \geq 0$, $X_2 \geq 0$) limit the possible solution to the positive part of the plane defined by the axes in Figure 3.3. In addition, the employment constraint ($X_1 + X_2 \geq 300$ workers) limits the solution to the half

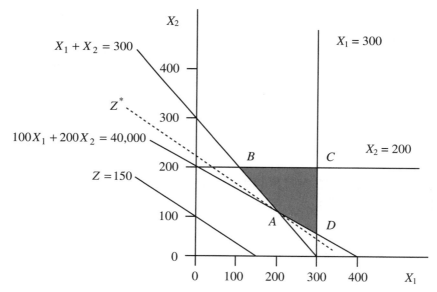

FIGURE 3.3 Graphic solution of the river pollution problem.

Principles of Linear Programming: Solutions

plane to the right of the boundary line $X_1 + X_2 = 300$, which goes through the points $(X_1 = 0, X_2 = 300)$ and $(X_1 = 300, X_2 = 0)$. This can be verified by observing that for any point to the left of that line, say, the origin, the employment constraint is not satisfied.

The feasible region is limited further by the revenue constraint ($100X_1 + 200 X_2 \geq 40{,}000$ \$/day). The boundary line of this constraint goes through the points $(X_1 = 0, X_2 = 200)$ and $(X_1 = 400, X_2 = 0)$. For the origin the constraint is false; therefore the feasible region lies to the right of the boundary line.

Last, the possible solutions must satisfy the capacity constraints ($X_1 \leq 300$ tons per day of mechanical pulp and $X_2 \leq 200$ tons per day of chemical pulp). Thus, the feasible area lies below the line $X_2 = 200$ and to the left of the line $X_1 = 300$.

In summary, the feasible region is inside the polygon $ABCD$ in Figure 3.3. The figure shows that any solution to the problem requires the production of some of both kinds of pulp. More precisely, all objectives can be achieved simultaneously only if at least 100 tons per day of mechanical pulp are produced (point B in Figure 3.3) along with at least 50 tons per day of chemical pulp (point D).

BEST SOLUTION

We find the best solution graphically by first finding the slope of the family of straight lines that correspond to the objective function. This is done by drawing the objective function for an arbitrary level of pollution, say, $Z = 150$ units of BOD per day. The corresponding line, $150 = X_1 + 1.5X_2$, goes through the points $(X_1 = 0, X_2 = 100)$ and $(X_1 = 150, X_2 = 0)$.

At the origin, $Z = 0$; thus, the value of the objective function decreases the closer the line $Z = X_1 + 1.5X_2$ is to the origin. Consequently, the point in Figure 3.3 that leads to the smallest possible value of Z while satisfying all the constraints is A.

The coordinates of A can be read directly from the graph. Alternatively, one can solve the system of equations that define the coordinates of A:

$$X_1 + X_2 = 300 \quad \text{and} \quad 100X_1 + 200X_2 = 40{,}000$$

We eliminate X_1 by first multiplying the first equation by 100 and then subtracting it from the second. Solving this leads to:

$$X_2^* = 100 \text{ tons/day of chemical pulp}$$

Substituting this result in the first equation then gives:

$$X_1^* = 200 \text{ tons/day of mechanical pulp}$$

The value of the objective function that correspond to this optimum operating strategy is:

$$Z^* = X_1^* + 1.5X_2^* = 350 \text{ units of BOD/day}$$

This is the minimum amount of pollution that the pulp mill can produce while satisfying all other objectives.

3.3 THE SIMPLEX METHOD

The graphic method that we have used to solve the two previous examples is limited to cases where there are at most two or three decision variables in the model. For larger problems, a more general technique is needed. The *simplex* method is an algebraic procedure that, when programmed on a computer, can solve problems with thousands of variables and constraints quickly and cheaply.

This section will give only an overview of the method. The objective is to show the principles involved rather than the laborious arithmetic manipulations. The principles of the simplex method are straightforward and elegant. The arithmetic is best left to a computer.

SLACK VARIABLES

The first step of the simplex method is to transform all inequalities in a linear programming model into equalities. This is done because equalities are much easier to handle mathematically. In particular, a lot is known about the properties and solutions of systems of linear equations.

As an example, let's recall the formulation of the poet's problem (Section 2.2): Find the areas of red pine, X_1, and of hardwoods, X_2, to manage such that:

$$\max Z = 90X_1 + 120X_2 \quad (\$/y)$$

subject to:

$$X_1 \leq 40 \quad \text{(ha of red pine)}$$
$$X_2 \leq 50 \quad \text{(ha of hardwoods)}$$
$$2X_1 + 3X_2 \leq 180 \quad \text{(days of work)}$$
$$X_1, X_2 \geq 0$$

The first constraint can be changed into an equality by introducing one additional variable, S_1, called a *slack variable*, as follows:

$$X_1 + S_1 = 40 \quad \text{and} \quad S_1 \geq 0$$

Note that S_1 simply measures the area of red pine that is not managed. We proceed in similar fashion with each constraint and obtain the following transformed model: Find X_1, X_2, S_1, S_2, S_3 such that:

$$\max Z = 90X_1 + 120X_2$$

subject to:

$$X_1 + S_1 = 40$$
$$X_2 + S_2 = 50$$
$$2X_1 + 3X_2 + S_3 = 180$$
$$X_1, X_2, S_1, S_2, S_3 \geq 0$$

where S_2 is the slack variable measuring unused hardwoods land and S_3 is the slack variable measuring unused poet time.

BASIC FEASIBLE SOLUTIONS

Let us return to the geometric representation of the feasible solutions for this linear program. For convenience, it is reproduced in Figure 3.4. The feasible region is the entire area inside the polygon OABCD. The equations of the boundary lines are shown on the figure.

A *basic feasible solution* of this linear program corresponds to the corners of the polygon OABCD; we shall call these corners the *extreme points* of the feasible region. For example, extreme point O corresponds to the basic feasible

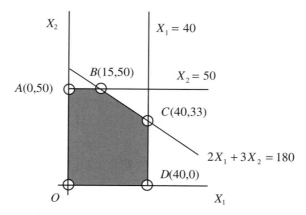

FIGURE 3.4 Extreme points and basic feasible solutions.

solution:

$$(X_1, X_2, S_1, S_2, S_3) = (0, 0, 40, 50, 180)$$

since at O, $X_1 = X_2 = 0$ and thus, from the constraints $S_1 = 40$, $S_2 = 50$, and $S_3 = 180$.

Similarly, the extreme point A corresponds to the basic feasible solution:

$$(X_1, X_2, S_1, S_2, S_3) = (0, 50, 40, 0, 30)$$

Note that in each basic feasible solution, there are as many positive variables as there are constraints. Positive variables are called *basic* variables, while those equal to zero are called *nonbasic* variables. In this example, there are always three basic variables and two nonbasic variables. Verify that this is true for the basic feasible solutions corresponding to extreme points B, C, and D.

This property of basic feasible solutions is general. In a linear program with n variables and m independent constraints, a basic feasible solution has m basic variables and $n - m$ nonbasic ones. Constraints are independent if none can be expressed as a linear combination of the others; that is, no constraint is a direct consequence of the others and thus unnecessary.

THEOREM OF LINEAR PROGRAMMING

The fundamental theorem of linear programming, which we give without proof, states that if a best solution exists, then one of them is a basic feasible solution.

This theorem implies that in a linear program, there may be one, many, or no solution. The theorem is fundamental because it means that to solve a linear program one needs to consider only a *finite* number of solutions—the basic feasible solutions corresponding to the extreme points of the feasible region.

Since the best solution of a linear program is a basic feasible solution, it has exactly as many positive variables as there are independent constraints. If a problem has 10 independent constraints and 10,000 variables, only 10 variables in the best solution have positive values; all the rest are zero.

There may be even fewer positive variables in the best solution if all constraints are not independent. Assume there are 10 constraints in a linear program and we get only 8 positive variables in the best solution. Then two of the constraints must be redundant; they result necessarily from the others, and thus they can be omitted from the model without altering the results.

SOLUTION ALGORITHM

Given the theorem of linear programming, a possible solution procedure (an *algorithm*) would be to calculate all the basic feasible solutions and to find the one that maximizes or minimizes the objective function. But this is impractical

Principles of Linear Programming: Solutions

for large problems, because the number of basic feasible solutions may still be too large to examine all of them, even with a fast computer.

The simplex method uses, instead, a *steepest-ascent* algorithm. It consists of moving from one extreme point to the next adjacent extreme point of the feasible region in the direction that improves the objective function most.

The process can be visualized in this way: Think of the feasible region as a mountain, the peak of which corresponds to the optimum solution. A climber is lost in the fog and can barely see her feet. To reach the summit, she proceeds cautiously but surely. Keeping one foot fixed at one point, she moves the other foot around her to find the direction of the next step that will raise her most. When she has found it she moves in that direction. If no step in any direction lifts the climber, she has reached the summit.

The flowchart in Figure 3.5 summarizes the various steps of the simplex method. Step (a) consists in finding an initial feasible solution. In step (b) we move from one extreme point to an adjacent extreme point in the direction that most increases the objective function Z. If step (b) has improved the objective function, step (b) is repeated. The iterations continue until no improvement in

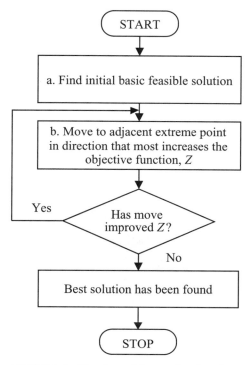

FIGURE 3.5 Flowchart of the simplex algorithm.

Z occurs, indicating that the optimum solution was obtained in the penultimate iteration.

EXAMPLE

To illustrate the principles of the simplex method we will solve the poet's problem by following the steps just described (see Figure 3.6).

Step a: Find an initial basic feasible solution. The simplest one corresponds to point O in Figure 3.6; that is:

Nonbasic variables: $X_1 = 0, X_2 = 0$
Basic variables: $S_1 = 40, S_2 = 50, S_3 = 180$
Objective function: $Z_0 = 0$

The three slack variables are basic in this initial solution.

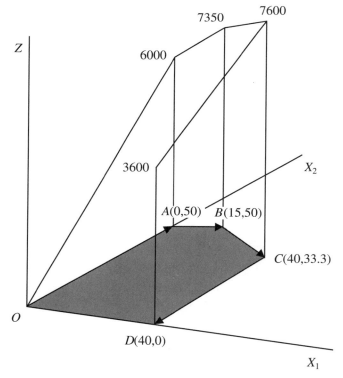

FIGURE 3.6 Iterations of the simplex algorithm.

Principles of Linear Programming: Solutions

Step b1: Since the coefficient of X_1 in the objective function is $90 per hectare while the coefficient of X_2 is $120 per hectare, the objective function increases most by moving from O, in the direction of OX_2, to the adjacent extreme point A, which corresponds to the new basic feasible solution:

Nonbasic variables: $S_2 = 0, X_1 = 0$
Basic variables: $X_2 = 50, S_1 = 40, S_3 = 30$
Objective function: $Z_A = \$6,000/\text{year}$

In the movement from extreme point O to A, the variable X_2 that was nonbasic has become basic, and the variable S_2 that was basic has become nonbasic. This is general; the algebraic equivalent of an adjacent extreme point is a basic feasible solution with a single different basic variable. The steepest ascent chooses as the new basic variable the one that increases the objective function the most.

Since the value of the objective function for this new basic feasible solution is higher than for the last one, we try another iteration.

Step b2: From extreme point A we now move in the direction OX_1, since this is the only way the objective function may be increased. The adjacent extreme point is B, corresponding to the following basic feasible solution:

Nonbasic variables: $S_2 = 0, S_3 = 0$
Basic variables: $X_1 = 15, X_2 = 50, S_1 = 25$
Objective function: $Z_B = \$7,350/\text{year}$

Since the last iteration has increased the objective function, we try another one.

Step b3: The only way the objective function may be increased is by moving to the adjacent extreme point C, which corresponds to the basic feasible solution:

Nonbasic variables: $S_1 = 0, S_3 = 0$
Basic variables: $X_1 = 40, X_2 = 33.3, S_2 = 17.7$
Objective function: $Z_C = \$7,600/\text{year}$

The last iteration having increased the objective function, we try another one.

Step b4: The next adjacent extreme point is D, corresponding to the basic feasible solution:

Nonbasic variables: $S_1 = 0, X_2 = 0$
Basic variables: $X_1 = 40, S_2 = 50, S_3 = 100$
Objective function: $Z_D = \$3,600/\text{year}$

This iteration has decreased the value of the objective function; therefore, the optimum solution is the basic feasible solution corresponding to extreme point C, reached in the previous iteration.

3.4 DUALITY IN LINEAR PROGRAMMING

Every linear programming problem has a symmetric formulation that is very useful in interpreting the solution, especially to determine how the objective function changes if one of the constraints changes slightly, everything else remaining equal. This symmetric formulation is called the *dual* problem. It contains exactly the same data as the original (primal) problem, but rearranged in a symmetric fashion. This different way of looking at the same data yields very useful information.

GENERAL DEFINITION

Recall the standard formulation of the linear programming problem given in Chapter 2: Find $X_1, X_2,\ldots, X_n$, all nonnegative, such that:

$$\max Z = c_1 X_1 + c_2 X_2 + \cdots + c_n X_n$$

subject to:

$$a_{11}X_1 + a_{12}X_2 + \cdots + a_{1n}X_n \leq b_1$$
$$a_{21}X_1 + a_{22}X_2 + \cdots + a_{2n}X_n \leq b_2$$
$$\vdots$$
$$a_{m1}X_1 + a_{m2}X_2 + \cdots + a_{mn}X_n \leq b_m$$

The dual of this problem is a linear program with the following characteristics:

The objective function of the dual is minimized (it would be maximized if the primal problem were a minimization).
It has as many variables (dual variables) as there are constraints in the primal, and all dual variables are positive or zero.
It has as many constraints as there are variables in the primal.
The coefficients a_{ij} in each column of the primal problem become coefficients in corresponding rows of the dual (first column becomes first row, second column second row, etc.)
The coefficients of the objective function in the primal become the coefficients on the right-hand side of the constraints in the dual, and vice versa.
The direction of the inequalities is reversed.

Principles of Linear Programming: Solutions

Consequently, the dual of the standard linear program given earlier is to find Y_1 to Y_m, all nonnegative, such that:

$$\min Z' = b_1 Y_1 + b_2 Y_2 + \cdots + b_m Y_m$$

subject to:

$$a_{11} Y_1 + a_{21} Y_2 + \cdots + a_{m1} Y_m \geq c_1$$
$$a_{12} Y_1 + a_{22} Y_2 + \cdots + a_{m2} Y_m \geq c_2$$
$$\vdots$$
$$a_{1n} Y_1 + a_{2n} Y_2 + \cdots + a_{mn} Y_m \geq c_n$$

Duality is symmetric in that the dual of the dual is the primal. You can verify this by applying the definition of duality to the dual, thereby recovering the primal formulation.

APPLICATIONS OF DUALITY

The duality theorem, one of the most important of linear programming, states that a solution of the dual exists if and only if the primal has a solution. Furthermore, the optimum values of the objective functions of the primal and of the dual are *equal*. In our notations, $Z^* = Z'^*$. We shall see the usefulness of this theorem in the following two examples.

Dual of the Poet's Problem

Recall the linear programming model we formulated for the poet who wanted to find the areas of red pine and hardwoods he should manage (X_1, X_2) in order to maximize his annual revenues (Z) while spending no more than half of his time in the woods.

$$\max Z = 90 X_1 + 120 X_2 \quad (\$/y)$$

subject to:

$X_1 \leq 40$ (ha of red pine)

$X_2 \leq 50$ (ha of hardwoods)

$2X_1 + 3X_2 \leq 180$ (days of work per year)

$X_1, X_2 \geq 0$

Applying the duality definition leads to the following dual problem:

$$\min Z' = 40Y_1 + 50Y_2 + 180Y_3$$

subject to:

$$Y_1 + 0Y_2 + 2Y_3 \geq 90$$
$$0Y_1 + Y_2 + 3Y_3 \geq 120$$
$$Y_1, Y_2, Y_3 \geq 0$$

Shadow Prices

We know from Section 3.3 that the best value of the objective function for the primal problem is $Z = \$7{,}600$ per year. The duality theorem states that the best value of the objective function of the dual must be equal to the best value of the objective function of the dual:

$$Z'^* = Z^* = \$7{,}600/y$$

Thus, Z', the objective function of the dual, must be measured in dollars per year. In addition, we know the units of measurement of the coefficients of the objective function of the dual because they are the coefficients of the right-hand side of the primal. Consequently, one can infer the units of measurement of the dual variables by making the objective function of the dual homogeneous in its units. This leads to:

$$\underset{(\$/y)}{Z'} = \underset{(ha)}{40} \underset{(\$/ha/y)}{Y_1} + \underset{(ha)}{50} \underset{(\$/ha/y)}{Y_2} + \underset{(d/y)}{180} \underset{(\$/d)}{Y_3}$$

where y and d refer to year and day, respectively. Verify that, with these units for Y_1, Y_2, and Y_3, the two constraints are also homogeneous in their units.

It is now apparent that Y_1 expresses the value of using red pine land, in dollars per hectare per year. Similarly, Y_2 is the value of using hardwoods land, and Y_3 is the value of the poet's time, in dollars per day. In linear programming terminology, Y_1, Y_2, and Y_3 are *shadow prices*.

The qualifier *shadow* is a reminder that these prices are not necessarily equal to the market prices of the resources. For example, Y_3 is not the value of the poet's time for hire; it is only an implicit value that reflects the activities in which the poet can engage (in this problem), managing red pine and hardwoods.

The duality theorem indicates that when all resources are used in an optimal manner, the total implicit value of the resources is equal to the annual returns.

Principles of Linear Programming: Solutions

The shadow prices are very useful in getting the most out of a linear programming model. To see this, assume that the dual of the poet's problem has been solved. Designate the value of the shadow prices at the optimum by Y_1^*, Y_2^*, and Y_3^*. Then the expression of the objective function of the dual problem at the optimum is:

$$Z'^* = \underset{(\$/y)}{40} \underset{(ha)(\$/ha/y)}{Y_1^*} + \underset{(ha)(\$/ha/y)}{50\ Y_2^*} + \underset{(d/y)(\$/d)}{180\ Y_3^*}$$

Thus, if the red pine land available increased or decreased by 1 hectare (from 40 to 41 or 39 ha) while the amounts of hardwoods land and poet time remained fixed, the objective function would increase or decrease by Y_1^* ($/y). Similarly, if the amount of hardwoods land available changed from 50 to 51 or 49 ha, the objective function would increase or decrease by Y_2^* ($/y). And if the amount of time available to the poet increased or decreased by 1 day per year, the objective function would increase or decrease by Y_3^* ($/y).

In summary, the shadow prices measure by how much the best value of the objective function would change if the right-hand side of a constraint changed by one unit, other things being equal.

To obtain the shadow prices it is not necessary to formulate and solve the dual separately. Modern versions of the simplex method give simultaneously the optimal primal and dual solution. The next section shows how to get the dual solution with the Excel Solver. It turns out that the shadow prices for the poet's problem are:

$Y_1^* = 10$ ($/y/ha of red pine)

$Y_2^* = 0$ ($/y/ha of hardwoods)

$Y_3^* = 40$ ($/day of poet's time)

These shadow prices show that one additional hectare of land would increase the poet's annual revenues by $10. On the other hand, extra hardwoods would be worth nothing. This is consistent with the fact that in the best primal solution we found that about 16.7 ha of hardwoods were not used. The third shadow price shows that one additional day working in the woods is worth $40 to the poet. This is the most revenue that he could get by managing his woods optimally with that additional time. This information should be most useful for the poet to decide if the financial and esthetic benefits of versification are worth that much.

In interpreting dual solutions, keep in mind that shadow prices are strictly marginal values. They measure changes in the objective function that result from small changes in each of the constraints. For example, in the poet's problem, the shadow price Y_2^* is zero as long as the hardwoods constraint is not binding. The best solution found in Section 3.3 showed that 16.7 ha of hardwoods should be left idle. Thus, were the poet to sell more than 16.7 ha of his land, the hardwoods constraint would become binding and the shadow price Y_2^* would become positive.

Dual of the River Pollution Problem

In using the shadow prices of a linear program one must keep in mind the direction of the inequalities and whether the objective function is minimized or maximized. As an example of a slightly more involved interpretation of shadow prices, let us recall the river pollution problem formulated in Section 2.3. The primal problem was: Find X_1 and X_2, the tonnages of mechanical and chemical pulp produced daily, such that:

$$\min Z = X_1 + 1.5X_2 \quad \text{(units of BOD per day)}$$

subject to:

$X_1 + X_2 \geq 300$	(workers employed)
$100X_1 + 200X_2 \geq 40{,}000$	(daily revenue, \$)
$X_1 \leq 300$	(mechanical pulping capacity, t/d)
$X_2 \leq 200$	(chemical pulping capacity, t/d)
$X_1, X_2 \geq 0$	

This primal problem is not in the standard format, so the interpretation of the shadow prices requires some care.

Solving the river pollution problem with a computer program (see next section and problem 3.4) gives the following shadow prices:

$Y_1^* = 0.5$	(BOD units/day/worker)
$Y_2^* = 0.005$	(BOD units/\$)
$Y_3^* = 0$	(BOD units/t)
$Y_4^* = 0$	(BOD units/t)

We have inferred the units of each shadow price by dividing the unit of the objective function by the units of the constraint to which the shadow price applies.

The two easiest shadow prices to interpret are Y_3^* and Y_4^*. They are both zero because at the optimum solution there is excess capacity for both pulp-making processes. This can be checked in Figure 3.3. Additional capacity would have no effect on pollution.

The workers' constraint is binding. Its shadow price shows that pollution would increase by 0.5 units of BOD per day for each additional worker that the cooperative might employ. Similarly, pollution would increase by 0.005 units of BOD for each additional dollar of daily revenues that the cooperative earned.

In many linear programming problems, some careful thinking will bring useful information out of the dual solution. Nevertheless, there are situations in which the shadow prices are either difficult to interpret or do not have any economic meaning because of the structure of the problem.

3.5 SPREADSHEET SOLUTION OF LINEAR PROGRAMS

In Section 2.5 we learned how to formulate linear programming problems with a computer spreadsheet and to use the spreadsheet to explore the effects of different choices of variables. We can then use the Solver optimization program of Excel to find the best solution.

Spreadsheet Solution of the Poet's Problem

To invoke the Solver in Excel, choose the Solver command from the Tools menu. This displays a Solver Parameters dialog box. The Solver Parameters dialog box with the parameters for the poet's problem is shown in Figure 3.7. The target cell, D10, is the cell that contains the objective function in the spreadsheet formulation of the poet's problem.

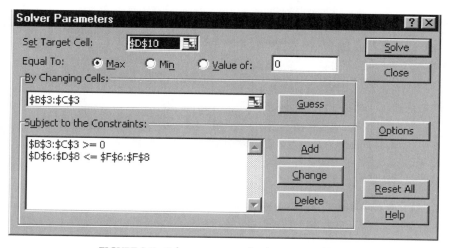

FIGURE 3.7   Solver parameters for the poet's problem.

Selecting the Max button directs the Solver to maximize the objective function. The Solver seeks the best solution by changing cells B3:C3, which contain the two decision variables.

The first line in the Subject to the Constraints window shows that the decision variables must be nonnegative. The second line indicates that the cells D6:D8 must be less than or equal to the corresponding cells F6:F8. This means that the red pine land managed must be at most 40 ha, the hardwoods land managed must be at most 50 ha, and the poet's time used must be at most 180 days per year. The Add, Change, and Delete buttons allow you to add, change, or delete constraints.

Before launching the Solver, click on the Options button, and check Assume Linear Model in the Solver Options dialog box (Figure 3.8). This directs the Solver to use the simplex method to solve the problem.

Launch the Solver by clicking on the Solve button in the Solver Parameters dialog box. In an instant, the program lets you know that it has found a solution. This solution (Figure 3.9) prescribes that the poet should cultivate 40 ha of red pine and about 33.3 ha of hardwoods. The maximum annual return would then be $7,600 per year.

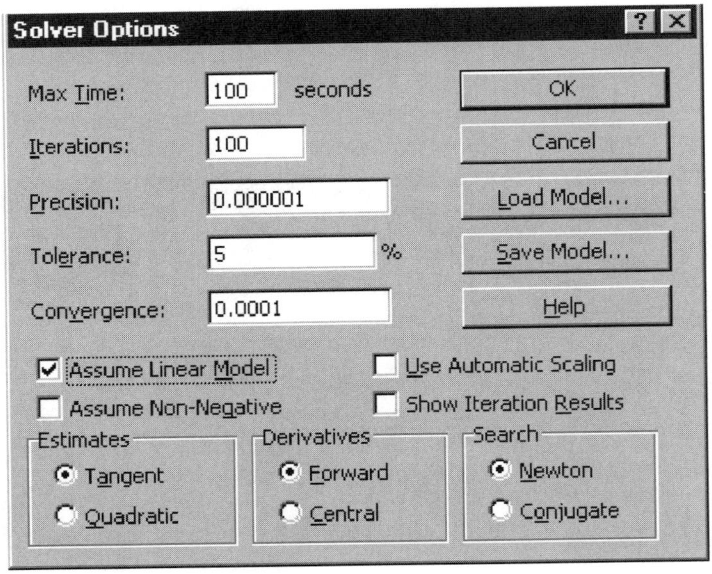

FIGURE 3.8 Setting Solver options for a linear model.

Principles of Linear Programming: Solutions

	A	B	C	D	E	F	G
1	POET PROBLEM						
2		Red pine	Hardwoods				
3	Managed area	40	33.333333				
4		(ha)	(ha)			Resources	
5		Resources required		Total		available	
6	Red pine land	1		40	<=	40	(ha)
7	Hardwoods land		1	33	<=	50	(ha)
8	Poet's time	2	3	180	<=	180	(d/y)
9		Objective function		Total			
10	Returns	90	120	7,600	Max		
11		($/ha/y)	($/ha/y)	($/y)			
12							
13		Key Formulas					
14	Cell	Formula				Copied to	
15	D6	=SUMPRODUCT(B6:C6,B$3:C$3) D6:D8					
16	D10	=SUMPRODUCT(B10:C10,B$3:C$3)					

FIGURE 3.9 Solver best solution for the poet's problem.

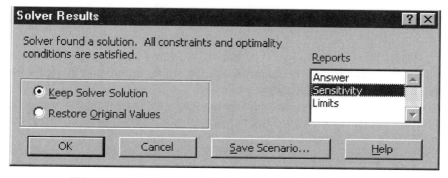

FIGURE 3.10 Solver results dialog box set to get the shadow prices.

GETTING SHADOW PRICES WITH THE SOLVER

After the Solver has found a solution, you can get the shadow prices by choosing Sensitivity in the Solver Results dialog box (Figure 3.10). This commands the Solver to do a series of sensitivity analyses, to show how the solution responds to changes in the problem parameters. The shadow prices are the most useful part of this sensitivity analysis. They show how the objective function changes with slight changes of the constraints.

	A	B	C	D	E	F
12		Constraints				
13				Final	Shadow	Constraint
14		Cell	Name	Value	Price	R.H. Side
15		$D$6	Red pine land Total	40	10	40
16		$D$7	Hardwoods land Total	33	0	50
17		$D$8	Poet's time Total	180	40	180

FIGURE 3.11 Shadow prices in the Solver sensitivity report.

Figure 3.11 shows the sensitivity report for the poet's problem. The shadow price for the red pine land constraint, which is in cell D6, is $10/ha/y. The final value of the constraint, the land managed, is 40 ha, while the constraint right-hand side, the land available, is also 40 ha. For the hardwoods land constraint in cell D7 instead, the land managed is 33 ha, while the land available is 50 ha. As a result, the shadow price is zero. Can you interpret the sensitivity report data for cell D8?

3.6 SUMMARY AND CONCLUSION

Linear programs with many constraints but no more than two decision variables can be solved graphically. This graphic solution illustrates nicely the key steps in finding a solution: First we determine the feasible region, that is, the set of all possible values of the decision variables. Then we find the point within the feasible region where the objective function is highest or lowest.

The simplex algorithm can be applied to a linear program of any size. It uses the fact that if optimum solutions exist, one of them is at a corner point of the feasible region. The simplex method consists, then, in moving from one extreme point of the feasible region to the next in the direction that most increases the objective function. When the objective function ceases to increase, the objective function has been found.

Every linear program has a dual formulation. The dual has one variable for each constraint of the primal and one constraint for each variable of the primal. At the optimum, the objective functions of primal and dual are equal. The dual variables, or shadow prices, have a very useful interpretation: they indicate by how much the optimum value of the objective function would change if the right-hand side of the constraint of the primal changed by one unit.

Linear programs can be formulated and solved efficiently with spreadsheets. The Solvers use variants of the simplex method and calculate simultaneously the primal problem and dual solution. To interpret the dual solution correctly requires a full understanding of the meaning of the primal problem.

Principles of Linear Programming: Solutions

PROBLEMS

3.1 Consider the problem of the poet and his woods that was solved graphically in Section 3.1. Suppose that the price of hardwood lumber goes up, increasing the return from managing northern hardwoods from $120 to $180/ha/y.
(a) How would this change the objective function of this problem?
(b) Use the graphic solution method to determine the best way for the poet to allocate his time between managing red pine and northern hardwoods land given this change in his economic environment.
(c) Perform the same analysis assuming that the return from managing northern hardwoods increases, but this time to only $135/ha/y. Is there still a unique best way for the poet to allocate his time?

3.2 Consider the river pollution problem that was solved graphically in Section 3.2. Suppose that the pulp mill installs chemical recycling equipment that reduces the pollution resulting from producing chemical pulp from 1.5 to 0.9 units of BOD/ton.
(a) How would this change the objective function?
(b) Use the graphic solution method to determine the best way for the mill owners to allocate productive capacity between mechanical and chemical pulp given this new technology.
(c) Perform the same analysis assuming that the pulp mill installs, instead of chemical recycling, solid waste treatment equipment that reduces the pollution due to mechanical pulp production from 1.0 to 0.6 units of BOD/ton.

3.3 Consider the linear programming model developed for Problem 2.7.
(a) Use the graphic solution method to determine the best way for the logging contractor to allocate her logging equipment between the two logging sites.
(b) Use the simplex algorithm demonstrated in Section 3.3 to solve the same problem.
(c) Compare the solutions you obtained with these two different solution methods.

3.4 Consider the river pollution problem that was solved graphically in Section 3.2 and for which a spreadsheet model is shown in Figure 2.2.
(a) Use the Excel Solver to solve this problem, and compare your solution to the solution obtained by solving the problem graphically.
(b) Use the Solver Results dialog box to do a sensitivity analysis (see Figure 3.10 showing how this was done for the problem of the poet and his woods). What is the shadow price for the employment constraint?
(c) To analyze the effects of increasing the pulp mill's employment target, change the right-hand-side constant in the employment constraint from 300

to 302 workers, and use the Excel Solver to solve this revised problem. How much does the best value of the objective function change?

(d) Explain this change in terms of the shadow price you obtained in part (c).

3.5 Consider the problem of the poet and his woods, for which a spreadsheet model is shown in Figure 2.1 and Excel Solver parameters are shown in Figure 3.7. The dual of this problem was solved in Section 3.4, and shadow prices for the constraints are shown in Figure 3.11.

(a) Assume that the poet is willing to spend more time managing his forest. Change the right-hand-side constant in the time constraint from 180 to 182 days, and then use the Excel Solver to solve this revised problem. How much does the best value of the objective function change?

(b) Explain this change in terms of the fact that the shadow price for the poet's time constraint in the original problem was $40/day.

(c) Assume that the poet is considering buying more land. Change the right-hand-side constant in the hardwood land constraint from 50 ha to 55 ha, and then use the Excel Solver to solve this revised problem. How much does the best value of the objective function change?

(d) Explain this change in terms of the fact that the shadow price for the northern hardwoods land constraint in the original problem was $0/ha.

ANNOTATED REFERENCES

Blanning, R.W. 1974. The sources and uses of sensitivity information. *Interfaces* 4(4):32–38. (Describes four different methods of evaluating the sensitivity of a model to its parameters.)

Davis, L.S., K.N. Johnson, P.S. Bettinger, and T.E. Howard. 2001. *Forest Management: To Sustain Ecological, Economic, and Social Values.* McGraw-Hill, New York. 804 pp. (Chapter 6 discusses solving linear programming models.)

Hof, J., M. Bevers, and J. Pickens. 1995. Pragmatic approaches to optimization with random yield coefficients. *Forest Science* 41(3):501–512. (Discusses several ways to account for uncertainty in harvest yields in linear programming models.)

Kent, B.M. 1989. *Forest Service Land Management Planners' Introduction to Linear Programming.* USDA Forest Service, Gen. Tech. Report RM-173. Fort Collins, CO. 36 pp. (Discusses simple linear programming models for forest resource management.)

Perry, C., and K.C. Crellin. 1982. The precise management meaning of a shadow price. *Interfaces* 12(2):61–63. (Precise explanation of shadow prices and how they can be used in decision making.)

Ragsdale, C.T. 1998. *Spreadsheet Modeling and Decision Analysis: A Practical Introduction to Management Science.* South-Western College Publishing, Cincinnati, OH. 742 pp. (Chapters 2 and 3 discuss solving linear programming models graphically and with spreadsheets, and Chapter 4 discusses sensitivity analysis of solutions.)

Weintraub, A., and A. Abramovich. 1995. Analysis of uncertainty of future timber yields in forest management. *Forest Science* 41(2):217–234. (Discusses an approach to handling uncertainty in harvest yields in linear programming models by formulating constraints as probability statements.)

Winston, W.L. 1995. *Introduction to Mathematical Programming.* Duxbury Press, Belmont, CA. 818 pp. (Chapter 4 discusses solving linear programming models, and Chapters 5 and 6 discuss sensitivity analysis of solutions.)

CHAPTER 4

Even-Aged Forest Management: A First Model

4.1 INTRODUCTION

In the next six chapters we shall study applications of linear programming to timber harvest scheduling. Chapters 4 through 7 deal with even-aged forests, Chapters 8 and 9 with uneven-aged, or selection, forests.

Planning the future sequence of harvests on a forest is only one of the numerous tasks of a forest resource manager, but it is essential for industrial forests and it is important for multiple-use forests as well. One of the main purposes of forestry is still to produce wood. Indeed, this is often the dominant goal for industrial forests. On public forests, and many private forests as well, an adequate balance between timber, recreation, water, and wildlife is almost always required. The nontimber goals may constrain timber production, or they may define the objective function. Thus, timber harvesting may sometimes be primarily a tool to achieve nontimber objectives, such as landscape diversity. Regardless, it is useful to be able to plan harvest operations to reach the desired objectives in the best possible way. The harvest-scheduling models in this and the following chapters will help us do that, first in even-aged forests and then in uneven-aged forests. Then, in Chapter 10, we will study goal-programming techniques that may be used to weigh timber production against other goals.

4.2 DEFINITIONS

Even-aged management deals with forests composed of even-aged stands. In such stands, individual trees originate at about the same time, either naturally or artificially. In addition, stands have a specific termination date at which time all remaining trees are cut. This complete harvest is called a *clear-cut*.

Regeneration of even-aged stands may be done by planting or seeding. The latter may be natural. For example, in a *shelterwood* system, a few old trees are left during the period of regeneration to provide seed and protect the young seedlings. Natural regeneration may continue for a few years after initial planting or seeding. Nevertheless, the basic management remains the same, it leads to a total harvest and a main crop when the stand has reached rotation age. Light cuts called *thinnings* are sometimes done in even-aged stands before the final harvest.

An even-aged forest consists of a mosaic of even-aged stands of different age and size called *management units* or *compartments*. Each unit must be big enough for practical management, but size may vary greatly, depending on the management objectives.

Even-aged management is used widely. Many valuable commercial species grow best in these full-light conditions. Furthermore, even-aged management has many economic advantages. Site preparation and planting can be done economically over large areas, using machinery and fire. Artificial regeneration allows the foresters to control the quality of the trees they use and to select the best trees. As trees grow they all have approximately the same size within each compartment. This standardization of products helps in mechanizing harvest and simplifies processing later on at the sawmill or pulp mill. Logging costs per unit of timber removed are lower in a clear-cutting operation than in a selection harvest, because mechanical harvesting is easier and the area that must be covered to extract a specific amount of timber is smaller. Finally, the fresh vegetation that succeeds a clear-cut is a favorite food for some wildlife, such as deer.

Nevertheless, even-aged management has some disadvantages. Clear-cut land is ugly. This has caused considerable opposition to clear-cutting on public forests. This esthetic problem is sometimes reduced by clear-cutting only small tracts of land and by leaving a screen of trees around the clear-cut areas, at least until the young trees cover the ground. In this and the following chapters we shall learn how to alleviate some of the negative aspects of

4.3 EXAMPLE: CONVERTING SOUTHERN HARDWOODS TO PINE

In this example we shall study a linear programming model, originally proposed by Curtis (1962), to manage the very simple forest represented in Figure 4.1(a). There are only two compartments on this forest, labeled 1 and 2. Compartment 1 has an area of 120 ha, compartment 2 has 180 ha. Southern hardwoods of low quality currently cover the two compartments. However, they are on distinct soils, and timber grows better in compartment 1 than in compartment 2.

One objective of the owner of this property is to convert the entire area to a pine plantation during a period of 15 years. The forest created at the end of this period should be *regulated*, with a rotation age of 15 years. That is to say, one-third of the forest should be covered with trees 0–5 years old, a third with trees 6–10 years old, and another third with trees 11–15 years old. This would lead to a pattern of age classes like that shown in Figure 4.1(b). Note, however, that the age classes do not have to be contiguous.

Finally, the owner desires to maximize the amount of wood that will be produced from his forest during the period of conversion to pine. However, the owner will not cut any of the pine stands before they are 15 years old.

We shall learn how to represent this problem as a linear program with decision variables, constraints, and an objective function and how to solve it to find the best solution.

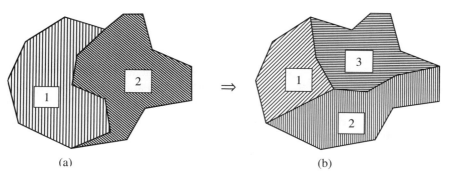

FIGURE 4.1 Hardwood forest with (a) two initial age classes converted to a regulated pine plantation with (b) three age classes.

4.4 MODEL FORMULATION

DECISION VARIABLES

The harvest-scheduling problem consists of deciding *when*, *where*, and *how much* timber to cut in order to reach all management objectives.

Since we use an even-aged silviculture, a natural decision variable should measure the area cut. More precisely, let X_{ij} be the area to be cut from compartment i in period j, where i and j are integer subscripts. Here, i may take the value 1 or 2, since there are only two compartments in the initial forest. To keep the number of variables reasonably small, we must work with time units that are often longer than 1 year. Let us use a time unit of 5 years in this example. Therefore, j can take the values 1, 2, or 3, depending on whether a cut occurs during the first 5 years of the plan, the second 5 years, or the third. As soon as an area is cut over, it is immediately replanted with pine trees, so X_{ij} is also the area replanted in compartment i during period j.

Thus, all the possible harvests and reforestations in compartment 1 are defined by the three decision variables—X_{11}, X_{12}, and X_{13}—while those possible in compartment 2 are X_{21}, X_{22}, and X_{23}. Naturally, all the decision variables must be positive or zero.

CONSTRAINTS

One set of constraints expresses the fact that, no matter what the choice of variables is, the entire forest must have been cut once during the management plan. This is necessary and sufficient to convert the entire forest to southern pine. Therefore, for the first initial compartment we must have:

$$X_{11} + X_{12} + X_{13} = 120 \text{ ha}$$

For the second we must have:

$$X_{21} + X_{22} + X_{23} = 180 \text{ ha}$$

The second set of constraints expresses the requirement that the sequence of harvests be such that it leads to a regulated pine forest at the end of the 15-year plan. To achieve this, one must cut one-third of the forest every 5 years and then plant it immediately with southern pine. In terms of the decision variables we are using, this means that:

$$X_{11} + X_{21} = \frac{180 + 120}{3} = 100 \text{ ha}$$

This is the area cut during the first 5 years, which by the end of the conversion will be covered with trees 10–15 years older than when they were planted.

Similarly, during the second period we must cut and reforest:

$$X_{12} + X_{22} = 100 \text{ ha}$$

and during the third period:

$$X_{13} + X_{23} = 100 \text{ ha}$$

OBJECTIVE FUNCTION

One of the management objectives is to maximize the total *amount* of hardwoods produced during the conversion. Therefore, the objective function must express the amount harvested in terms of the decision variables, X_{ij}. Since the decision variables measure areas cut, we need data on the amount of hardwoods available in each compartment, per unit area, throughout the 15 years of management. These data are shown in Table 4.1. The expected timber yield is given in tons because we assume that the hardwoods are of poor quality, useful only to make pulp. For example, Table 4.1 states that each hectare cut from compartment 1 in period 2 will yield 23 tons of hardwoods (on average). Note that the yield per unit area increases over time, because the trees in compartments 1 and 2 grow.

With these yield data, we can now express the objective function as a linear function of the decision variables. The expected total forest output during the 15-year management plan is:

$$Z = 16X_{11} + 23X_{12} + 33X_{13} + 24X_{21} + 32X_{22} + 45X_{23} \text{ (tons)}$$

where $16X_{11} + 23X_{12} + 33X_{13}$ is the expression of the tonnage of hardwood cut from compartment 1 and $24X_{21} + 32X_{22} + 45X_{23}$ is the tonnage cut from compartment 2.

In similar fashion, one can readily calculate the tonnage produced during any one of the three periods, in terms of the decision variables. For example, the tonnage produced during the last 5 years of the plan is $33X_{13} + 45X_{23}$.

In summary, the model that expresses the problem posed in this example has the following expression: Find $X_{11}, X_{12}, X_{13}, X_{21}, X_{22}, X_{23}$, all nonnegative, such

TABLE 4.1 Projected Yield by Compartment and 5-Year Management Period

Compartment		Tons per hectare		
No.	Area (ha)	Period 1	Period 2	Period 3
1	120	16	23	33
2	180	24	32	45

TABLE 4.2 Linear Programming Tableau for the Hardwoods Conversion Example

	X_{11}	X_{12}	X_{13}	X_{21}	X_{22}	X_{23}	
Z	16	23	33	24	32	45	
COM1	1	1	1				= 120
COM2				1	1	1	= 180
AGC1	1			1			= 100
AGC2		1			1		= 100
AGC3			1			1	= 100

that:

$$\max Z = 16X_{11} + 23X_{12} + 33X_{13} + 24X_{21} + 32X_{22} + 45X_{23}$$

subject to:

$$X_{11} + X_{12} + X_{13} = 120$$
$$X_{21} + X_{22} + X_{23} = 180$$
$$X_{11} + X_{21} = 100$$
$$X_{12} + X_{22} = 100$$
$$X_{13} + X_{23} = 100$$

A convenient way of displaying this model is shown in Table 4.2. This tableau form gives all the information necessary to prepare the input for most linear programming computer programs. Each column has a name (of a variable), and each row has a name (of a constraint). For example COM1 indicates that the first constraint refers to compartment 1, while AGC3 indicates that the last constraint refers to the third period. In this manner, the position of each coefficient in the tableau is clearly defined by the name of the column and the name of the row in which it is located.

An equivalent way of formulating this problem is as a spreadsheet, as presented in upcoming Section 4.8.

4.5 SOLUTION

The solution of the model developed in the previous section and solved with a spreadsheet or some other computer program is given in Table 4.3. It shows the sequence of harvests that would maximize the tonnage of wood produced in converting the initial hardwood forest into a regulated pine forest. For example, 100 ha of the first compartment are cut and replanted during the first 5 years, and the other 20 ha during the next 5 years. We would start cutting and replanting in

Even-Aged Forest Management: A First Model 59

TABLE 4.3 Harvest Plan That Maximizes Tonnage: Area Data

Compartment	Hectares per 5-year period			Total ha
	1	2	3	
1	100	20	0	120
2	0	80	100	180
Total	100	100	100	300

TABLE 4.4 Harvest Plan That Maximizes Tonnage: Tonnage Data

Compartment	Tons per 5-year period			Total tons
	1	2	3	
1	1,600	460	0	2,060
2	0	2,560	4,500	7,060
Total	1,600	3,020	4,500	9,120

compartment 2 only in the second period, on 80 ha, and finishing that compartment in the last period.

The column totals show clearly that three blocks of 100 ha each are created every 5 years, thus leading to the regulated forest that is desired. Note that the three age classes that have been created cover land of different productivity (see the yield data in Table 4.1). Therefore, regulation done in this way does not ensure that the periodic yield of the forest created at the end of the conversion will be constant. This may not be important to some owners. If it is important, the model should be changed to reflect this objective. We shall learn how to do that in the following chapters.

Table 4.4 shows the tonnage of wood harvested in each compartment of the forest, by period. The total amount produced by the forest is 9,120 tons. This is the maximum that can be obtained, given the constraints specified. But it should be kept in mind that this may not be the only way of getting that amount. As the theorem of linear programming told us (see Section 3.3), the solution we get may just be one of several that would lead to the same value of the objective function.

4.6 MAXIMIZING PRESENT VALUE

The column totals in Table 4.4 show an increasing production, from 1,600 tons in the first five years to 3,020 tons in the next five and to 4,500 tons in the last five. This may be just what the owner wants, but someone who is selling wood for profit would prefer, other things being equal, to harvest earlier and to invest

TABLE 4.5 Harvest Plan That Maximizes Present Value: Tonnage Data

Compartment	Tons per 5-year period			Total tons
	1	2	3	
1	0	2,300	660	2,960
2	2,400	0	3,600	6,000
Total	2,400	2,300	4,260	8,960

the returns at an interest rate better than what is offered by the growth of the poor hardwoods.

The linear program we have developed earlier can be used to determine the harvesting plan that will best meet this new objective. To do this, the objective function must be changed. The coefficients of the decision variables should now express the *present value* of the timber cut on each hectare of land and in each time period.

Since the timber is similar in both compartments, the price per unit of timber is the same over the whole forest. We will also assume that the price will remain the same in every period. In this example, the timber is arbitrarily assigned a value of $1 per ton.

The calculations will be done with an interest rate of 5% per year. The tonnage cut is accounted for at the middle of the 5-year period during which it is harvested. For example, one hectare cut from compartment 2 in period 3 yields $45/ha when cut in period 3, i.e., 12.5 years from now on average. But its present value at an interest rate of 0.05 per year is only:

$$\frac{45}{(1+0.05)^{12.5}} = 24.5 \ (\$/ha)$$

This becomes the coefficient of X_{23} in the objective function. Doing this present-value calculation for each compartment and period produces the final expression of the new objective function:

$$\max Z' = 14.2X_{11} + 16.0X_{12} + 17.9X_{13} + 21.2X_{21} + 22.2X_{22} + 24.5X_{23}$$

The new optimal harvesting strategy, expressed in tonnage produced, is shown in Table 4.5. The total volume produced by the forest is smaller than when tonnage was maximized. However, more is produced in the first 5 years. Also, the entire cut in the first 5 years is in the second compartment instead of the first.

4.7 A NOTE ON REDUNDANCIES

A careful examination of the solutions of the two linear programs in Tables 4.3–4.5 shows that there are only four positive variables in the optimum solution, yet there are five constraints in the linear program (see Table 4.2).

In Section 3.3 we saw that the best solution found by the simplex procedure, if it exists, is a basic feasible solution. We also observed that a basic feasible solution has as many basic (nonzero) variables as there are independent constraints. The key word here is *independent*. In fact, in the hardwoods conversion example the five constraints are not independent. Any one of the constraints in Table 4.2 can be obtained by algebraic manipulation of the other four.

For example, consider the constraint named AGC1 in Table 4.2. It can be obtained in the following steps:

1. Add constraints COM1 and COM2, left and right, thus getting:

$$X_{11} + X_{12} + X_{13} + X_{21} + X_{22} + X_{23} = 300 \text{ ha}$$

2. Add constraints AGC2 and AGC3 in the same manner:

$$X_{12} + X_{13} + X_{22} + X_{23} = 200 \text{ ha}$$

3. Then subtract the second equation from the first, left and right. This yields: $X_{11} + X_{21} = 100$ ha, which is the constraint AGC1.

Verify that this can be done for any one of the constraints in this model. Any one results necessarily from the others; in other words, any one constraint is redundant and can be eliminated without changing the problem.

Redundancy is not very important in this case since it leads to the elimination of only one equation. However, in some situations hundreds of redundant equations can be present in a model. This increases unnecessarily the cost of solving a model. More important, it is useful to look for redundancies because this gives additional insight on the structure of the model.

4.8 SPREADSHEET FORMULATION AND SOLUTION

Figure 4.2 shows how the forest conversion problem studied in this chapter can be formulated in a spreadsheet. The example is the same as the one treated earlier. Entries in bold characters are the given data. The others either are variables or are related by formulas to the variables and the parameters.

SPREADSHEET FORMULAS

Cells B7:D8 contain the areas cut and reforested in each compartment in every period. These are the decision variables, corresponding to the X_{ij} in the mathematical formulation. The cells E7:E8 contain the formulas for the total area cut and reforested in each compartment during the three periods. Cells B9:D9 contain the formulas for the total area cut and reforested in each period.

	A	B	C	D	E	F	G
1	CONVERSION PLAN THAT MAXIMIZES PRODUCTION						
2	Price	$1	(per ton)	D		5 (years)	
3	Rate	5.0%	(per year)				
4			Period				Area
5		1	2	3	Total		available
6		Area converted (ha)					
7	Comp1	100	20	0	120	=	120
8	Comp2	0	80	100	180	=	180
9	Total	100	100	100	300		
10	Area wanted	**100**	**100**	**100**	300		
11		Yield (tons/ha)					
12	Comp1	16	23	33			
13	Comp2	24	32	45			
14		Production (tons)					
15	Comp1	1600	460	0	2060		
16	Comp2	0	2560	4500	7060		
17	Total	1600	3020	4500	9120	max	
18		Present value ($/ha)					
19	Comp1	14.2	16.0	17.9			
20	Comp2	21.2	22.2	24.5			
21		Present value ($)					
22	Comp1	1416	319	0	1735		
23	Comp2	0	1775	2445	4221		
24	Total	1416	2095	2445	5956		
25							
26		*Key cell formulas*					
27	Cell	Formula				Copied to	
28	E7	=SUM(B7:D7)				E7:E10, E15:E17,	
29						E22:E24	
30	B9	=SUM(B7:B8)				B9:D9, B17:D17,	
31						B24:D24	
32	B15	=B7*B12				B15:D16	
33	B19	=B12*Price/(1+Rate)^(B$5*D-D/2)				B19:D20	
34	B22	=B19*B7				B22:D23	

FIGURE 4.2 Spreadsheet model to plan conversion of hardwoods to pine plantations.

Cells G7:G8 contain the initial area in each compartment. Cells B10:D10 contain the area we want to convert in each period.

Cells B12:D13 contain the data in Table 4.1, the tonnage per unit area in each compartment and period. Cells B15:D16 contain the formulas for the volume produced in each compartment and period.

Even-Aged Forest Management: A First Model

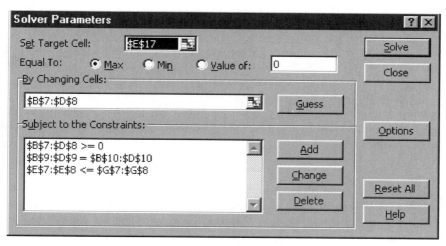

FIGURE 4.3 Solver parameters for conversion that maximizes production.

Cells B19:D20 contain the formulas for the present value generated by harvesting 1 ha in each compartment and period. "Price," "Rate," and "D" are the names of the cells that contain, respectively, the price of hardwoods per ton, the interest rate per year, and the duration of each period, in years.

Cells B22:D23 contain the formulas for the total present value generated from each compartment in each period.

Solver Parameters

The spreadsheet in Figure 4.2 may be used to maximize either production (in tons) or present value (in $). The figure shows the solution that maximizes production. Figure 4.3 shows the corresponding Solver parameters. The objective function is in the target cell, E17. This is the total production, in tons, that must be maximized. The solver tries to maximize production by changing cells B7:D8, which contain the decision variables. This is done while respecting the following constraints:

B7:D8 >= 0, meaning that the decision variables must be nonnegative.
B9:D9 = B10:D10, meaning that the total hardwoods area cut and
  reforested in each period must equal the area that must be converted to
  pine.
E7:E8 = G7:G8, because the area cut from each compartment must equal
  the area available in that compartment.

To get the conversion plan that maximized net present value, instead, one would change the target cell to E24.

4.9 GENERAL FORMULATION

The model we have used to solve the small example of conversion of a southern hardwoods forest to a regulated pine plantation can be generalized to handle as many initial compartments as necessary and as many time periods as desired.

In general, there are m initial compartments in the forest, and the management plan is established for p periods. Each decision variable, X_{ij}, measures the area cut from compartment i in period j. Therefore, the complete set of mp decision variables is:

$$X_{11}, X_{12}, \ldots, X_{1p}, X_{21}, \ldots, X_{m1}, X_{m2}, \ldots, X_{mp} \qquad \text{(all positive or zero)}$$

The area in each initial compartment is indicated by a_i. A first set of constraints states that the area cut and reforested in each compartment throughout the planning horizon must equal the area of that compartment; that is:

$$\sum_{j=1}^{p} X_{ij} = a_i \qquad i = 1, \ldots, m$$

There are as many constraints of this kind as there are compartments.

A second set of constraints specifies the area that must be cut and reforested during each period from the entire forest. Let this area be designated by y_j; then:

$$\sum_{i=1}^{m} X_{ij} = y_j \qquad j = 1, \ldots, p$$

There are as many constraints of this kind as there are periods in the plan. Any one of the constraints is redundant, due to the fact that

$$\sum_{i=1}^{m} a_i = \sum_{j=1}^{p} y_j = A$$

where A is the total area of the forest.

To write the general form of the objective function, let c_{ij} be the increase of the objective function due to cutting 1 hectare of land from compartment i in period j. For example, these coefficients might measure tons of wood per hectare or present value of the timber cut per hectare. Then each harvest X_{ij} contributes the amount $c_{ij}X_{ij}$ to the objective function. The amount contributed by compartment i alone throughout the plan is:

$$\sum_{j=1}^{p} c_{ij} X_{ij}$$

Even-Aged Forest Management: A First Model

and the amount contributed by the entire forest is:

$$Z = \sum_{i=1}^{m} \sum_{j=1}^{p} c_{ij} X_{ij}$$

In summary, the general form of the model is: Find $X_{ij} \geq 0$ for $i = 1,\ldots, m$ and $j = 1,\ldots, p$ such that:

$$\max Z = \sum_{i=1}^{m} \sum_{j=1}^{p} c_{ij} X_{ij}$$

subject to:

$$\sum_{j=1}^{p} X_{ij} = a_i \quad i = 1,\ldots, m$$

$$\sum_{i=1}^{m} X_{ij} = y_j \quad j = 1,\ldots, p$$

In our example, all y_j were equal to A/p, where A is the total area of the forest. This was done to obtain a regulated forest with rotation age p by the end of the plan.

4.10 CONCLUSION

The simple programming model for even-aged management presented in this chapter is useful when one is able to decide in advance what area of the forest will be cut and reforested periodically throughout the duration of the plan being drawn up. The best solution will then show the location and timing of harvests that lead to the highest production or to the highest present value.

For example, the model would be suitable in a situation where the dominant goal is to create a regulated forest in a specified amount of time. However, unless one is very firm about the regulation objective and on how to achieve it, this may not be the best way to manage a forest, especially if the goal is to maximize production or present value.

Another limitation of the model is that, once an area of a compartment is reforested, the plantation that is created is not reconsidered for harvest during the entire duration of the plan. This is adequate as long as the planning horizon is not too long, certainly not much longer than the desired rotation on the target forest. The models we will study in the next two chapters will help circumvent some of these limitations.

PROBLEMS

4.1 Consider the even-aged management model in Section 4.4 that maximizes production.
(a) Would changing the constraints for each compartment from equalities to less-than-or-equal-to inequalities change the best solution? Why?
(b) Would changing the conversion constraints for each period from equalities to greater-than-or-equal-to inequalities change the best solution? Why?
(c) What about changing the conversion constraints for each period from equalities to less-than-or-equal-to inequalities?

4.2 Consider the even-aged management model in Section 4.4 that maximizes production and the model in Section 4.6 that maximizes present value.
(a) Would you expect the shadow prices for the corresponding constraints in the two models to be the same? If not, why not?
(b) Verify your intuition by solving the two problems and then comparing the shadow prices for the corresponding constraints.

4.3 Consider the even-aged management model in Section 4.4 that maximizes production.
(a) Remove the first-period conversion constraint ($X_{11} + X_{21} = 100$), and solve the revised model to verify that this constraint is redundant.
(b) Do the same with the other constraints to test that anyone of the constraints is redundant, given the others.

4.4 Consider the even-aged management model in Section 4.6 that maximizes present value.
(a) Calculate new objective function coefficients assuming that the guiding rate of interest is 10% per year instead of 5%.
(b) Substitute these new coefficients into the objective function, and solve the revised model. How does the best harvesting plan change? Why does it change in this way?

4.5 The forester responsible for managing of 3,038 ha of southern hardwoods wants to convert this land to a regulated pine plantation. The pine plantation will be managed with a 20-year rotation. The conversion will be done in 20 years. The forester intends to maximize pulpwood production while doing this conversion.

The conversion will require that one-quarter of the total forest area be harvested and replanted in each of four (5-year) planning periods. The initial forest consists of five compartments. The area and expected yield of hardwood pulpwood by compartment is shown in the following table.

Write the objective function and constraints of a linear program that the forester could use to determine which areas to harvest and replant in each period.

Projected Hardwood Pulpwood Production (Green Tons/ha)

Compartment	Area (ha)	Period			
		1	2	3	4
1	722	35.6	52.2	74.2	101.8
2	621	53.4	73.3	104.9	148.4
3	469	113.9	162.7	225.3	303.3
4	545	92.8	135.0	185.4	258.2
5	681	39.9	56.5	76.7	102.0

4.6 Formulate and solve the linear programming model developed for Problem 4.5 with a spreadsheet, and describe the solution in terms of the production in each period and of the area and volume harvested from each compartment in each period.

4.7 Assume a discount rate of 5% per year and a pulpwood price of $20 per green ton.

(a) Calculate new objective function coefficients for the spreadsheet model developed for Problem 4.6 so that the revised model could be used to determine which areas to harvest and replant in each period to maximize the present value of pulpwood production over 20 years.

(b) How does the harvest plan differ from that obtained in Problem 4.6? Why?

ANNOTATED REFERENCES

Curtis, F.H. 1962. Linear programming in the management of a forest property. *Journal of Forestry* 60(9):611–616. (The original source of the model described in this chapter and one of the earliest applications of linear programming to forest management.)

Davis, L.S., K.N. Johnson, P.S. Bettinger, and T.E. Howard. 2001. *Forest Management: To Sustain Ecological, Economic, and Social Values*. McGraw-Hill, New York. 804 pp. (Chapter 10 discusses "Model I and Model II" approaches, the former being similar to that presented in this chapter.)

Dykstra, D.P. 1984. *Mathematical Programming for Natural Resource Management*. McGraw-Hill, New York. 318 pp. (Chapter 6 discusses a "Model I" linear programming model with regulatory constraints that is similar to the model used in this chapter).

Hof, J., and M. Bevers. 2000. Optimal timber harvest scheduling with spatially defined sediment objectives. *Canadian Journal of Forest Research* 30:1494–1500. (Describes a simple linear programming with the objective of minimizing sedimentaion into a watershead system while meeting goals for timber harvest.)

Johnson, K.N., and H.L. Scheurman. 1977. Techniques for prescribing optimal timber harvest and investment under different objectives—Discussion and synthesis. *Forest Science* Monograph No. 18. 31 pp. (Discusses the theoretical basis for different linear programming model structures for forest resource management. In their terminology, the model used in this chapter is of the "Model I" type, because its variables preserve the identity of the initial harvesting units throughout the planning horizon.)

Leak, W.B. 1964. *Estimating Maximum Allowable Timber Yields by Linear Programming*. U.S. Forest Service Research Paper NE-17. Northeastern Forest and Range Experiment Station, Upper Darby, PA. 9 pp. (Linear programming model similar to that used by Curtis.)

Leefers, L.A., and J.W. Robinson. 1990. FORSOM: A spreadsheet-based forest planning model. *Northern Journal of Applied Foresty* 7(1):46–47. (The model can be used in simulation mode to predict effects of a management option, or in optimization mode to find the best option.)

Nyland, R.D. 2002. *Silviculture: Concepts and Applications*. McGraw-Hill, New York. 682 pp. (Chapter 2 compares even- and uneven-aged management systems.)

CHAPTER 5

Area- and Volume-Control Management with Linear Programming

5.1 INTRODUCTION

The model we shall study in this chapter was first developed by Loucks (1964). Part of it is similar to Curtis' model studied in Chapter 4. The initial condition of the forest is represented in the same manner. There are several distinct compartments, each treated as a separate management unit. The decision variables are the same; they refer to the area cut from each compartment in different periods of the management plan.

The difference lies in the way the regulation objective is handled. In Curtis' model, this was done in a very rigid manner. In every time period, a prespecified fraction of the entire forest area was cut and reforested immediately. By the end of the plan, the entire forest had been cut and regenerated. In Louck's model, regulation is pursued less rigidly. Only part of the forest may be cut during the management plan, and constraints are used to ensure that the amount of timber removed does not exceed the long-term sustainable output of the forest. Two specifications of these constraints will be considered, based on the traditional management methods of area control and volume control.

Another difference in this chapter's model lies in the attention given to the periodic production of the forest throughout the management plan. In Curtis' model, production could vary from period to period as long as production was maximized and regulation achieved. However, foresters often want harvests that do not vary too much over time. For example, managers of national forests must, by law, make sure that the forest produces a nondeclining *even flow* of timber. That is to say, the amount produced may increase over time, but it should not decline. Departures may be allowed from year to year from the planned even-flow level due to special circumstances; but decade after decade, even flow is still the rule. It is also common, for ecological and esthetic reasons, to limit the area or volume of harvests to a prescribed level.

5.2 PRELIMINARY DEFINITIONS

Before laying out the model to be used in this chapter, let us define the concepts of *area control* and *volume control* as we shall use them. We should stress that these concepts mean slightly different things to different people. The definition given here is a somewhat flexible interpretation. We shall use another interpretation in the simulation model of Chapter 15.

The concepts of area- and volume-control management are based on the model of the *regulated forest*. Let A be the area of a forest. This forest is regulated if it consists of r blocks of equal area, A/r, each one covered by a stand of trees in a single age class. Therefore, if the time unit is one year, the youngest age class consists of trees that are 0–1 year of age, the second youngest age class has trees 1–2 years old, and so on to the oldest age class, which consists of trees $r-1$ to r years old.

Every year the oldest age class is cut, starting with the oldest trees, and then reforested immediately. For that reason, r is referred to as the *rotation age*. Let v_r be the volume per hectare of timber in the oldest age class, in cubic meters (m^3). Then the yearly production of the regulated forest is:

$$Q_r = \frac{A}{r} v_r$$
$$(m^3/y) \quad (ha/y)(m^3/ha)$$

With this formula it is easy to determine the best rotation, in the sense that it maximizes the yearly timber production. It is r^*, the value of r for which v_r/r is highest. That is to say, the best rotation to maximize timber production per unit of time is equal to the age of the stand at which its "mean annual increment" is highest. Of course, quite different rotations are possible, depending on the objective. We shall see in Chapter 7 that the economic rotation is usually shorter than the rotation that maximizes the volume produced per unit of time.

Conversely, much longer rotations may be needed to obtain stands with large trees and snags for purpose of biodiversity.

No matter what rotation age is used, the yearly production of the regulated forest can be maintained perpetually, and it remains constant as long as the productivity of the land, measured by v_r, does not change.

It is this image of a rotation, so similar to the movement of a clock, and the idea of a perpetual and constant output, akin to perpetual celestial motion, that may explain the fascination of pioneer foresters of the sixteenth and seventeenth centuries with the model of a regulated forest. Even now, foresters use the model frequently, almost automatically. Although few forests are regulated or even approach regulation, the concept is useful as a target for the future condition of the forest, because regulation is a sufficient condition of sustained production. We have already used the rotation concept briefly in the previous chapter. It also plays a key role in the methods of area control and volume control. The purpose of these methods is to ensure that no more is cut from a particular forest during the time spanned by the management plan than what the forest could produce in the long run, if it were regulated.

Area-control management is based on the observation that, were a regulated forest of area A managed for y years, the total area cut during that period would be exactly $y(A/r)$ ha. Area-control management proceeds then to suggest that, regardless of the status of the current forest, the area cut from it during a period of y years should be no more than $y(A/r)$ ha.

Volume-control management applies the same logic to standing volume instead of area. It suggests that at most the fraction y/r of the original standing volume be cut from the existing forest during the y years for which the harvest schedule is being drawn up.

Both area-control and volume-control methods are largely rules of thumb. In no way are they optimal guides. It should be clear, for example, that area control makes little sense in a forest that has very little initial volume. In that respect, volume control has the advantage that it takes into consideration the condition of the growing stock, but it still lacks rigor. Nevertheless, both management principles are widely used. We shall see in the next sections how they can be integrated into a linear programming model.

5.3 EXAMPLE: OPTIMIZING THE YIELD OF A LOBLOLLY PINE EVEN-AGED FOREST

Consider a loblolly pine forest of 500 ha, with two age classes, one of trees 20–25 years old, covering an area of 200 ha, and the other one of trees 40–45 years old, covering an area of 300 ha.

The owner of the forest wants to maximize the volume that the forest will produce during the next 15 years. In addition, the amount of timber produced by the forest should increase regularly by 10% every 5 years. The silviculture is even-aged management, with clear-cutting followed by planting. The initial plan is to use area-control management, but the consequences of volume control should also be investigated.

Decision Variables

As in Curtis' model in Chapter 4, the decision variables that define the future harvest schedule are the areas cut from each initial compartment in every period of the plan. To keep the number of variables small we must choose a sufficiently long lapse of time for each period, let us say 5 years. In this example this leads to the following decision variables: X_{11}, X_{12}, X_{13}, X_{21}, X_{22}, X_{23}, where the first subscript of each variable refers to the compartment where the cut occurs and the second refers to the time period.

Objective Function

The objective function expresses the total volume of timber cut during the 15 years of the plan as a linear function of the decision variables. To write it, we need the quantity per hectare that is expected from each compartment at different points in time in the future. These data are presented in Table 5.1. With these data, we can now express the objective function as:

$$Z = 120X_{11} + 230X_{12} + 250X_{13} + 310X_{21} + 320X_{22} + 350X_{23} \ (\text{m}^3)$$

The object of the problem is to find the values of the decision variables that make this function as large as possible, subject to the constraints expressing the other management objectives and the amount of resources available.

There are three kinds of constraints in this model: (1) constraints that refer to the limited land available; (2) constraints expressing the desired pattern of production during the plan; (3) constraints subjecting the management alternatives to either area control or volume control.

TABLE 5.1 Expected Volumes for a Loblolly Pine Forest

Compartment	Volume (m³/ha)		
	Period 1	Period 2	Period 3
1	120	230	250
2	310	320	350

Land Availability Constraints

The area of land that is cut in each compartment cannot exceed the area available. As in the harvest-scheduling model of Chapter 4, land that is planted during the management plan will not be cut again before the end of the plan. However, in contrast with that model, the entire area of each compartment does not need to be cut. Thus, the land availability constraints take the form:

$$X_{11} + X_{12} + X_{13} \leq 200 \text{ ha} \quad \text{for the first compartment}$$

and

$$X_{21} + X_{22} + X_{23} \leq 300 \text{ ha} \quad \text{for the second compartment}$$

Timber Flow Constraints

These constraints express the fact that the amount of timber produced by the forest should increase regularly by at least 10% every 5 years. Let V_1, V_2, and V_3 be the amount of timber cut during the first, second, and third 5-year periods of the plan, respectively. Then, a pattern of production that satisfies this management objective must be such that:

$$V_2 = 1.10 V_1 \quad \text{and} \quad V_3 = 1.10 V_2$$

The expressions of V_1, V_2, and V_3 in terms of the decision variables are:

$$V_1 = 120 X_{11} + 310 X_{21}$$
$$V_2 = 230 X_{12} + 320 X_{22}$$
$$V_3 = 250 X_{13} + 350 X_{23}$$

The final expression of the timber flow constraints is then:

$$230 X_{12} + 320 X_{22} = 1.1(120 X_{11} + 310 X_{21})$$

and

$$250 X_{13} + 350 X_{23} = 1.1(230 X_{12} + 320 X_{22})$$

or, more compactly:

$$230 X_{12} + 320 X_{22} - 132 X_{11} - 341 X_{21} = 0$$

and

$$250 X_{13} + 350 X_{23} - 253 X_{12} - 352 X_{22} = 0$$

AREA-CONTROL CONSTRAINT

If the forest is managed under the area-control system for y years, only the fraction y/r of the entire forest area may be cut during the entire duration of the plan, where r is the rotation age (in our example, $y = 15$ years).

Because the objective of the owner is to maximize the volume of timber produced by this forest, a suitable value of the rotation is the age that maximizes the mean annual increment. For loblolly pine on this site the mean annual increment is highest at an age of 30 years. The productivity is then about 8.3 m³/ha/y.

Given this rotation age, and following the area-control principle, the allowable cut for this loblolly pine forest is $500 \times (15/30) = 250$ ha. The expression of the area control constraint is then:

$$X_{11} + X_{12} + X_{13} + X_{21} + X_{22} + X_{23} \leq 250 \text{ ha}$$

VOLUME-CONTROL CONSTRAINT

If, instead, we choose to manage the forest according to the volume-control principle, the allowable cut would be 1/30 of the initial standing volume during the average year of the plan, which would amount to half of the initial volume in 15 years. However, Loucks suggests a correction for the fact that the standing volume of the forest will grow during the 15 years. For this reason, Louck's allowable-cut formula is based on the average expected standing volume in any 5-year period of the entire plan rather than on just the initial volume. That is:

$$\frac{200(120 + 230 + 250) + 300(310 + 320 + 350)}{3} = 138{,}000 \text{ m}^3$$

The allowable cut during 15 years under volume control is then:

$$\left(\frac{15}{30}\right) 138{,}000 = 69{,}000 \text{ m}^3$$

The final form of the volume control constraint for this example is:

$$120X_{11} + 230X_{12} + 250X_{13} + 310X_{21} + 320X_{22} + 350X_{23} \leq 69{,}000 \text{ m}^3$$

In summary, here is the linear programming model for the management plan of the loblolly pine forest in the example.

Find $X_{11}, X_{12}, \ldots, X_{23}$ such that:

$$\max Z = 120X_{11} + 230X_{12} + 250X_{13} + 310X_{21} + 320X_{22} + 350X_{23} \text{ (m}^3\text{)}$$

subject to:

Land availability:

$$X_{11} + X_{12} + X_{13} \leq 200 \text{ ha}$$
$$X_{21} + X_{22} + X_{23} \leq 300 \text{ ha}$$

Timber flow:

$$230X_{12} + 320X_{22} - 132X_{11} - 341X_{21} = 0$$
$$250X_{13} + 350X_{23} - 253X_{12} - 352X_{22} = 0$$

Area control:

$$X_{11} + X_{12} + X_{13} + X_{21} + X_{22} + X_{23} \leq 250 \text{ ha}$$

or

Volume control:

$$120X_{11} + 230X_{12} + 250X_{13} + 310X_{21} + 320X_{22} + 350X_{23} \leq 69{,}000 \text{ m}^3$$

This model is presented in Table 5.2 in a form appropriate for linear programming packages. Each column is defined by the name of a variable, and each row or constraint is defined by a four-character name. For example, FLO1 is the constraint that imposes the 10% increase in production between period 1 and period 2. Each parameter in the linear program is then defined by the name of the row and of the column to which it belongs.

In Table 5.2, the blank line that separates the last two constraints indicates that usually only one of them is imposed at a time. That is, the forest is managed under either area control or volume control but not both. In the next section we will examine the best solutions corresponding to these two management strategies. As a practical matter, both constraints can be left in the model, and in general only one of them will be binding at the optimum.

TABLE 5.2 Linear Programming Tableau for Area Control or Volume Control

	X_{11}	X_{12}	X_{13}	X_{21}	X_{22}	X_{23}	Control	
Z	120	230	250	310	320	350		Max
COM1	1	1	1				$\leq$	200 ha
COM2				1	1	1	$\leq$	300 ha
FLO1	−132	230		−341	320		=	
FLO2		−253	250		−352	350	=	
AREC	1	1	1	1	1	1	$\leq$	250 ha
VOLC	120	230	250	310	320	350	$\leq$	69,000 m³

The constraint AREC is used for area-control management only, and VOLC for volume control only.

5.4 SOLUTIONS

The solutions of the loblolly pine example are in Table 5.3 for the case of area-control management and in Table 5.4 for volume control. The data in the two tables show that the two management strategies do not lead to very different results. Under both systems the first compartment remains completely untouched. More land is cut under area control (250 ha instead of 212 ha) and more total volume is also produced. In both cases, the quantity of timber cut every 5 years rises regularly by 10%, as required by the timber flow constraints.

Nevertheless, it is unclear how good these control procedures really are. The usefulness of any regulatory system depends in part on the long-term objectives of the forest owners and on the suitability of the forest that is left at the end of the plan to meet these objectives.

For example, Table 5.4 shows that if the owners of this forest opt for volume control, at the end of the 15 years they will have a forest with 72 ha in the youngest age class (trees 0–5 years old), 72 ha in the second age class (trees 6–10 years old), and 68 ha of trees 11–15 years old. In addition, there will remain $300 - 212 = 88$ ha of land from compartment 2 covered with trees 56–60 years old, which would be well beyond the rotation age of maximum mean annual increment (30 years). The 200 ha of compartment 1 will be covered by trees 36–40 years old.

The timber value of such a forest, including the land on which it grows, is a function of its long-term ability to produce timber. However, what it can produce

TABLE 5.3 Best Management Plan Under Area Control

Compartment		Period			Total
		1	2	3	
1	Area cut (ha)	0	0	0	0
	Volume (m^3)	0	0	0	0
2	Area cut (ha)	80	85	85	250
	Volume (m^3)	24,702	27,173	29,890	81,765

TABLE 5.4 Best Management Plan Under Volume Control

Compartment		Period			Total
		1	2	3	
1	Area cut (ha)	0	0	0	0
	Volume (m^3)	0	0	0	0
2	Area cut (ha)	67	72	72	212
	Volume (m^3)	20,846	22,931	25,224	69,000

may not be easy to forecast. For that purpose it would be useful to require that the forest be in some form of a steady-state regime when the end of the plan is reached. We will study a model that permits one to do this in a flexible way in the next chapter.

Meanwhile, the area- and volume-control formulas remain simple and useful rules to lay out timber-harvesting schedules. They do, at least in a rough manner, ensure that the long-term productive potential of the forest is not exceeded.

5.5 ADDING CONSTRAINTS AND OBJECTIVES

The core model just presented may be changed in several ways to adapt to different biophysical conditions of the forest or to represent other management objectives. Here we give two examples: (1) to recognize different land productivity in various compartments; (2) to control the size of the harvests for esthetic reasons.

ALLOWING FOR DIFFERENT ROTATIONS

In the example used earlier, it was assumed that the two compartments differed only by the age of the stands they were carrying but that the site quality was the same in both. This led to the choice of the same rotation to manage the entire forest (30 years, the age of highest mean annual increment). However, if the sites or the species grown on the two compartments are so different that they justify different rotation ages, then the model must be modified.

Assume, for example, that 30 years is the appropriate rotation for the second compartment only but that it should be 40 years on the first compartment because of its lower site quality. Then there should be two area-control constraints in the model:

For compartment 1:

$$X_{11} + X_{12} + X_{13} \le \left(\frac{15}{40}\right) 200 = 75 \text{ ha}$$

For compartment 2:

$$X_{21} + X_{22} + X_{23} \le \left(\frac{15}{30}\right) 300 = 150 \text{ ha}$$

These two constraints would now replace the area-control constraint AREC in the linear programming tableau in Table 5.2.

Similarly, if volume control were applied, we would now have two constraints:

For compartment 1:

$$120X_{11} + 230X_{12} + 250X_{13} \le \left(\frac{15}{40}\right)\left(200\frac{120+230+250}{3}\right) \le 15{,}000 \text{ m}^3$$

For compartment 2:

$$310X_{21} + 320X_{22} + 350X_{23} \le \left(\frac{15}{30}\right)\left(300\frac{310+320+350}{3}\right) \le 49{,}000 \text{ m}^3$$

CONTROLLING THE AREA OF THE CLEAR-CUTS

As pointed out earlier, one of the main drawbacks of even-aged management is clear-cuts are unsightly. One natural way to limit this negative visual impact is to limit the size of the clear-cut. This would be a natural thing to try in our example, because with either area control or volume control, the harvest is concentrated in compartment 2, which contains the oldest trees, and the area cut is large relative to the total area of that compartment. One may control the area cut in each compartment and period with constraints of the form:

$$X_{11} \le X_{11}^u, X_{12} \le X_{12}^u, \ldots, X_{23} \le X_{23}^u$$

where X_{ij}^u is an upper bound for the area that may be clear-cut from compartment i during period j. The next section shows the consequences of applying such a constraint in the example of the loblolly pine forest.

5.6 SPREADSHEET FORMULATION AND SOLUTION

Figure 5.1 shows a variant of the problem studied in this chapter, formulated and solved in a spreadsheet. The setting is the same as that of the loblolly pine forest considered earlier. Area control is applied; but to correct for the concentration of harvests in compartment 2 noted in Table 5.3, we shall limit the area clear-cut in each compartment and period to at most 50 ha.

SPREADSHEET FORMULAS

The entries in bold characters in Figure 5.1 are the given data; the other entries either are variables or are related by formulas to the variables and the parameters.

Area- and Volume-Control Management with Linear Programming

	A	B	C	D	E	F	G
1	HARVEST PLAN WITH AREA-CONTROL						
2	Rotation	30	(years)		Largest cut	50	(ha)
3	Plan length	15	(years)		Allowable cut	250	(ha)
4			Period				Area
5		1	2	3	Total		available
6		Area cut (ha)					
7	Comp1	42	29	29	100	<=	200
8	Comp2	50	50	50	150	<=	300
9	Total	92	79	79	250		
10		Yield (m³/ha)					
11	Comp1	120	230	250			
12	Comp2	310	320	350			
13		Production (m³)					
14	Comp1	5035	6589	7348	18972		
15	Comp2	15500	16000	17500	49000		
16	Total	20535	22589	24848	67972	max	
17	Periodic change		0.10	0.10			
18	Timber flow constraint		0	0			
19							
20			Key cell formulas				
21	Cell	Formula			Copied to		
22	E7	=SUM(B7:D7)			E7:E9,E14:E16		
23	B9	=SUM(B7:B8)			B9:D9, B16:D16		
24	F3	=SUM(G7:G8)*B3/B2					
25	B14	=B7*B11			B14:D15		
26	C18	=C16-(1+C17)*B16			C18:D18		

FIGURE 5.1 Spreadsheet solution of the loblolly pine forest plan, with area control and an upper bound on the size of clear-cuts.

Cells B7:D8 contain the decision variables, the area cut and reforested in each compartment and period, corresponding to the X_{ij} in the mathematical formulation. Cell F2 contains the upper bound on the area that may be cut and reforested in each compartment and period.

Cells E7:E8 contain the formulas for the total area cut and reforested in each compartment. Cells B9:D9 contain the formulas for the total area cut and reforested in each period. The formula in cell E9 gives the total area cut and reforested from all compartments over all periods.

Cells G7:G8 contain the initial area in each compartment. The formula in cell F3 uses these data, together with the rotation age in cell B2 and the length of the plan in cell B3, to compute the allowable cut according to the area-control rule.

Cells B11:D12 contain the same data as Table 5.1, the volume of standing timber per hectare in each compartment and period.

Cells B14:D15 contain the formulas for the volume harvested in each compartment and period. Cells B16:D16 contain the formulas for the total volume harvested in each period. The formulas in cells E14:E15 give the volume harvested from each compartment, over all periods. The formula in cell E16 gives the volume harvested from all compartments, over all periods.

Cells C17:D17 contain the desired relative change in production from period 1 to period 2 and from period 2 to period 3, expressed as fractions. The formulas in cells C18:D18, when forced to equal zero, ensure that the relative change in production from period 1 to period 2 and from period 2 to period 3 is equal to the desired relative change.

SOLVER PARAMETERS

Figure 5.2 shows the Solver parameters to get the solution that maximizes production, subject to area control and an upper bound on the size of the harvest areas. The objective function is in the target cell, E16. This is the total production, in cubic meters. The solver tries to maximize production by changing cells B7:D8, which contain the decision variables. It does this while respecting the following constraints:

B7:D8 <= F2, meaning that the area harvested in each compartment and period may not exceed 50 ha

B7:D8 >= 0, because the areas harvested cannot be negative

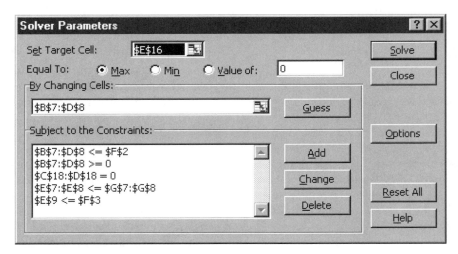

FIGURE 5.2 Solver parameters for the loblolly pine forest plan, with area control and an upper bound on the size of the clear-cuts.

C18:D18 = 0, to force the volume harvested to increase by 10% from period 1 to period 2 and from period 2 to period 3

E7:E8 <= G7:G8, because the area harvested from each compartment may not exceed the area of that compartment

E9 <= F3, to force the total area harvested to less than the allowable cut according to the area-control formula

PRIMAL SOLUTION

The spreadsheet in Figure 5.1 shows the best solution for the primal problem obtained with the Solver. In contrast with the solution shown in Table 5.3, harvests occur in both compartments, and only 50 ha are now harvested in the second compartment in any one period. This would certainly reduce the visual impact of the harvest. However, this improvement comes at a cost, because production is now only 67,972 m^3 instead of the 81,765 m^3 obtained without constraining the size of the harvested areas.

DUAL SOLUTION

Figure 5.3 shows the dual solution produced by the sensitivity report of the solver. The shadow price of the land in compartments 1 and 2 is zero, because the land constraints are not binding. At the optimum, only 100 ha of the 200 ha in compartment 1 are used, as are only 150 ha of the 300 ha available in compartment 2. So, having more land in either compartment would not improve the objective function. The shadow prices of the timber flow constraints are also zero. However, the shadow price of the area-control constraint is 184 m^3/ha. Because the area-control constraint is binding (all of the allowed 250 ha are cut), increasing the allowable cut by 1 ha would increase production by 184 m^3.

Constraints

Cell	Name	Final Value	Shadow Price	Constraint R.H. Side
$E$7	Comp1 Total	100	0	200
$E$8	Comp2 Total	150	0	300
$C$18	Timber flow constraint Period	0	0	0
$D$18	Timber flow constraint	0	0	0
$E$9	Total Total	250	184	250

FIGURE 5.3 Solver dual solution for the loblolly pine forest plan, with area control and an upper bound on the size of the clear-cuts.

5.7 GENERAL FORMULATION

The model we have just used in the small example of the loblolly pine forest can be generalized to deal with a forest that has many compartments and with a management plan as long as desired.

Using the same general notation as in Chapter 4, we may have initially m compartments. The plan is for p periods. Then the set of decision variables is X_{ij}, with $i = 1,\ldots,m$ and $j = 1,\ldots,p$. In all, there are $m \times p$ variables. Each variable refers to the area cut from compartment i in period j.

The object of the problem is to find the values of the X_{ij} such that the total amount of timber produced throughout the plan is maximized. The general form of the objective function is then:

$$\max Z = \sum_{i=1}^{m}\sum_{j=1}^{p} c_{ij}X_{ij}$$

where c_{ij} is the expected volume per hectare in compartment i and period j. The summation over j expresses the volume produced by a specific compartment, i, throughout the plan. The summation over both i and j expresses, then, the volume produced by the whole forest. The objective function may be given another meaning by changing the coefficients c_{ij}. For example, a manager may be interested in maximizing the discounted value of the timber produced. In that case, c_{ij} would be the expected discounted value resulting from harvesting one hectare of land in compartment i during period j.

The land availability constraints state that the area cut in each compartment during the plan cannot exceed the area available:

$$\sum_{j=1}^{p} X_{ij} = a_i \quad i = 1,\ldots,m$$

where a_i is the area of compartment i. There are m constraints of this kind, one for each compartment. In this model, an area can be cut only once during the plan. Thus, the model is suitable only as long as the length of the plan is shorter than the rotation.

The timber flow constraints express the relationship between the volume cut in successive periods. Let f_j be the percentage by which the cut in period j must exceed that in period $j - 1$. Then the general expression of the flow constraint is:

$$\sum_{i=1}^{m} c_{ij}X_{ij} - (1+f_j)\sum_{i=1}^{m} c_{ij-1}X_{ij-1} = 0 \quad j = 2,\ldots,p$$

There are $p - 1$ constraints of this kind.

If area control is used, the total area cut cannot exceed $y(A/r)$, where y is the length of the plan and r is the rotation, expressed in the same unit of time, and

A is the total area of the forest. In terms of the decision variables, this gives:

$$\sum_{i=1}^{m}\sum_{j=1}^{p} X_{ij} \leq y \frac{A}{r}$$

If volume control is used, the volume cut cannot exceed the fraction y/r of the expected growing stock throughout the plan. The expected growing stock in any period j is:

$$S_j = \sum_{i=1}^{m} c_{ij} a_i$$

and the average growing stock over the entire plan is:

$$\bar{S} = \frac{1}{p}\sum_{j=1}^{p} S_j$$

So the final expression of the volume-control constraint is:

$$\sum_{i=1}^{m}\sum_{j=1}^{p} c_{ij} X_{ij} \leq y \frac{\bar{S}}{r}$$

As noted in the example, there should be more than one area- or volume-control constraint if the rotation age is not the same throughout the forest. In that case, there is one constraint for each compartment or group of compartments that is managed under the same rotation age. The expression of the constraints remains the same, except the subscript i in the area and volume constraints refers only to compartments that are managed under the same rotation.

Last, for esthetic reasons the largest area of the clear-cut in any compartment and period may have a prespecified upper bound:

$$X_{ij} \leq X_{ij}^{u} \quad i = 1,\ldots,m; \quad j = 1,\ldots,p$$

where X_{ij}^{u} is the largest acceptable clear-cut size. This constraint may be uniform over the forest, or it may vary by compartment and period.

5.8 SUMMARY AND CONCLUSION

In this chapter we have formulated and applied a model of timber harvest scheduling in an even-aged forest that uses the old concepts of area control and volume control, embedded within a linear program. Area control and volume control are expressed as constraints limiting either the total area or the total volume cut during the duration or the plan. Additional constraints may regulate the volume of timber produced over time or the size of the clear-cuts. The linear program was used to maximize the volume produced, within these constraints.

Several variants of this model are possible. As in Chapter 4, one could try to maximize the present value of the harvest instead of its volume. Or one could minimize the total area harvested, subject to producing the amount of timber allowed by the volume-control principle, of some fraction thereof. We leave the formulation of these variants as an exercise (see Problems 5.4–5.6).

This model does not allow cutting an area twice during the period of the plan, a limitation that applied also to the model of Chapter 4. This may not be a serious problem as long as the management plan is shorter than one rotation. A more serious criticism, perhaps, is that area control and volume control are somewhat arbitrary rules of thumb. It is hard to determine the value of the forest left at the end of the plan under these policies. Furthermore, they may be limiting the cutting alternatives too much, eliminating possibilities that would increase the achievement of the management objectives. Or they may suggest excessive harvests that would not be truly sustainable in the long run. The model we shall study in the next chapter will correct some of these limitations.

PROBLEMS

5.1 A forest is composed of three even-aged compartments of ponderosa pine. The area in each compartment is shown in the table, along with projected per-hectare volumes on each during the next three 5-year periods.

Projected Lumber Volume (m^3/ha) per Compartment and Period for a Ponderosa Pine Forest

Compartment	Area (ha)	Period 1	Period 2	Period 3
1	2,450	88	99	117
2	3,760	117	157	198
3	8,965	82	93	111

(a) With volume control, how much volume could be harvested over the next 15 years, based on the average forest volume over all three periods? Assume a 65-year rotation.

(b) For a harvest-scheduling model based on 5-year periods, how would you express this volume-control constraint? (Use X_{ij} to represent the area harvested from compartment i in period j.)

Area- and Volume-Control Management with Linear Programming 85

(c) With area control, how much area could be harvested over the next 15 years? Assume a 65-year rotation.

(d) For a harvest-scheduling model based on 5-year periods, how would you express this area-control constraint?

5.2 Consider the ponderosa pine forest described in Problem 5.1. Assume that the higher volumes per hectare in compartment 2 are a reflection of a higher site quality than in compartments 1 and 3 so that it might be appropriate to manage compartment 2 on a shorter rotation. Assume that compartment 2 will be managed on a 55-year rotation and compartments 1 and 3 on a 75-year rotation.

(a) With area control, what area could be harvested over the next 15 years in compartment 2 alone?

(b) With area control, what area could be harvested in compartments 1 and 3 together?

(c) For a harvest-scheduling model based on 5-year periods, how would you express these area-control constraints? (Use X_{ij} to represent the area harvested from compartment i in period j.)

(d) With volume control, how much volume could be harvested over the next 15 years in compartment 2 alone?

(e) With volume control, how much volume could be harvested in compartments 1 and 3 together?

(f) For a harvest-scheduling model based on 5-year periods, how would you express these volume-control constraints?

5.3 Consider the ponderosa pine forest described in Problem 5.1. In a harvest-scheduling model with a 15-year planning horizon and 5-year periods, how would you constrain harvests to do the following?

(a) Have an equal periodic volume. (Use X_{ij} to represent the area harvested from compartment i in period j.)

(b) Increase by 25% per period.

(c) Vary by no more than 10%, plus or minus, from the volume harvested in the previous period. (*Hint:* This requires the use of two inequalities for each period.)

5.4 The forester for a paper company in Wisconsin is developing a long-term harvesting plan for 6,000 ha of aspen forestland. This plan covers a 20-year planning horizon and is broken down into five 4-year operating periods. The forest consists of four compartments. The area in each compartment is shown in the table, along with the projected volume per hectare of aspen pulpwood during the next five 4-year periods. Compartment 4 cannot be harvested during the first two periods because it has been cut over recently. Company policy dictates that this land must supply a constant periodic output of pulpwood. Furthermore, production of pulpwood should be maximized; thus the land should be managed on the rotation that maximizes mean annual increment. For all four compartments, this implies a 40-year rotation.

Projected Pulpwood Volume (m³/ha)

Compartment	Area (ha)	Period				
		1	2	3	4	5
1	1,750	217	238	259	273	287
2	1,610	308	315	308	343	294
3	1,340	56	91	119	147	168
4	1,300	—	—	175	203	224

(a) Applying the area-control principle *to each period*, what area could be harvested in any 4-year period?
(b) Applying the volume-control principle *to each period*, what volume could be harvested in any 4-year period?
5.5 For the aspen forest management problem (5.4), do the following:
(a) Write the equations of a linear program to determine the harvest and reforestation schedule that would maximize total harvest over the planning horizon, with area control.
(b) Formulate and solve this model with a spreadsheet to determine the production in each 4-year period and the area and volume harvested from each compartment in each period.
(c) Add a constraint requiring that no more than 300 ha be clear-cut in any compartment and period. How does this change the solution you obtained in part (b)?
5.6 For the aspen forest management problem (5.4), do the following:
(a) Write the equations of a linear program to determine the harvest and reforestation schedule that would harvest the allowable cut set by volume control while minimizing the total area harvested over the planning horizon.
(b) Formulate and solve this model with a spreadsheet to determine the production in each 4-year period and the area and volume harvested from each compartment in each period.

ANNOTATED REFERENCES

Clutter, J.L., J.C. Fortson, L.V. Pienaar, G.H. Brister, and R.L. Baily. 1983. *Timber Management: A Quantitative Approach*. Wiley, New York. 333 pp. (The "XYZ Timber Co. Problem" in Section 10.2 is a harvest-scheduling problem similar to that of this chapter, but with the harvest volume constraints determined by contractual obligations instead of biological productivity.)

Davis, L.S., K.N. Johnson, P.S. Bettinger, and T.E. Howard. 2001. *Forest Management: To Sustain Ecological, Economic, and Social Values*. McGraw-Hill, New York. 804 pp. (Chapter 10 discusses volume control and area control.)

Dykstra, D.P. 1984. *Mathematical Programming for Natural Resource Management*. McGraw-Hill, New York. 318 pp. (Model I in Chapter 6, with inventory and flow constraints, is similar to the model of this chapter.)

Hoganson, H.M., and M.E. McDill. 1993. More on forest regulation: An LP perspective. *Forest Science* 39(2):321–347. (Sensitivity analysis of the economic costs of achieving a regulated forest structure.)

Sharma, M., and A.L. Hammett. 2002. Making forest projects sustainable: Optimal harvesting plan for the Sagarnath plantation in Nepal. *Journal of Sustainable Forestry* 14(2/3):129–143. (Presents a simple linear programming model with harvest flow constraints for a sissoo and eucalyptus plantation.)

Johnson, K.N., and H.L. Scheurman. 1977. Techniques for prescribing optimal timber harvest and investment under different objectives—Discussion and synthesis. *Forest Science* Monograph No. 18. 31 pp. (Discusses the theoretical basis for different linear programming model structures for forest resource management. In their terminology, the models used in Chapter 4 and in this chapter are of the "Model I" type, because they preserve the identity of the initial harvesting units throughout the planning horizon.)

Loucks, D.P. 1964. The development of an optimal program for sustained-yield management. *Journal of Forestry* 62(7):485–490. (This is the original formulation of the harvest-scheduling problem treated in this chapter.)

Macmillan, D.C., and S.E. Fairweather. 1988. An application of linear programming for short-term harvest scheduling. *Northern Journal of Applied Forestry* 5:145–148. (Describes a simple linear programming model similar to that used in this chapter, but with a focus on NPV maximization over a 5-year planning horizon.)

Navon, D.I. 1971. *Timber RAM: A Long-Range Planning Method for Commercial Timber Lands Under Multiple-Use Management*. U.S. Forest Service Research Paper PSW-70. Pacific Southwest Forest and Range Experiment Station, Berkeley, CA. 22 pp. (One of the first computer systems for building harvest-scheduling models of the kind discussed in this chapter. Designed with public forests, but applicable to industrial forests as well.)

Thompson, E.F., B.G. Halterman, T.J. Lyon, and R.L. Miller. 1973. Integrating timber and wildlife management planning. *Forestry Chronicle* 49(6):247–250. (Harvest-scheduling problem similar to the one in this chapter, but incorporating both area- and volume-control constraints to achieve multiple-use goals.)

Ware, G.O., and J.L. Clutter. 1971. A mathematical programming system for the management of industrial forests. *Forest Science* 17(4):428–445. (Another early computer system for building harvest-scheduling models of the kind discussed in this chapter. Designed for industrial forests.)

CHAPTER 6

A Dynamic Model of the Even-aged Forest

6.1 INTRODUCTION

The models we studied in the two previous chapters did not allow a tract of land to be harvested twice during the planning period. Consequently, the objective function was influenced only by the cut from stands that existed at the beginning of the plan. New stands, arising from the reforestation done after cutting, did not affect the solution. Their potential productivity was ignored.

Nevertheless, the young stands that are created by a particular harvesting and reforestation schedule determine the long-term production of a forest. Therefore, the longer the planning horizon, the more important it is to take into account how the forest left at the end of the plan influences the forest performance. The models we studied so far had only fairly short planning horizons (15–50 years), but regulations and owner objectives may require planning horizons of a century or more. In that case, a truly dynamic model is needed, one in which young stands can be harvested to produce timber and generate more young stands as many times as necessary to achieve all management objectives during the planning period.

The purpose of this chapter is to study such a dynamic model. It is based on a model originally developed by Nautiyal and Pearse (1967). This model also describes the condition of the forest at the end of the plan in a way that is more flexible than that used in Chapter 3 and more rigorous than the area- or volume-control approach of Chapter 4. It turns out that it is not necessary to regulate a forest, in the sense used in Chapter 4, to ensure sustainability. We shall present some applications of this model here; more applications will follow in Chapter 7.

6.2 EXAMPLE

Consider a small even-aged forest of short-leaf pine that consists of four distinct age classes. The land is of the same quality throughout the forest (site index of 20 m at age 50 years). Age class 1 occupies 100 ha, age class 2 covers 200 ha, and age classes 3 and 4 cover 50 ha and 150 ha, respectively. The trees of age class 1 are 1–10 years old, those of age class 2 are 11–20 years, and those of age class 3 and 4 are 21–30 years old and 31–40 years old, respectively.

The silvicultural program for the forest consists of clear-cutting from any age class and reforesting immediately with trees of the same species. The owners want to keep this property for an indefinite length of time. Therefore, the very long-term consequences of management decisions must be predicted. Initially, we shall assume that the owners' objective is to maximize the landscape diversity of the forest. We shall later study the effect of economic objectives and alternative timber flow policies.

6.3 A MODEL OF FOREST GROWTH

To monitor the evolution of the forest over time, the forest is inventoried at the beginning of each decade. The state of the forest at each inventory is represented with state variables, and the interventions between inventories (the areas harvested and reforested) are designated by decision variables, as described next.

STATE VARIABLES

Let A_{ij} be the area in age class i at the beginning of decade j. Stand age is measured in decades. This set of variables defines the state of the forest at that time. Thus, in our example, the initial forest state is: $A_{11} = 100$ ha, $A_{21} = 200$ ha, $A_{31} = 50$ ha, $A_{41} = 150$ ha. Age classes may be contiguous or dispersed throughout the forest.

A Dynamic Model of the Even-aged Forest

Decision Variables

Let X_{ij} be the area cut and reforested from age class i during decade j. In contrast with the models studied in the last two chapters, the subscript i does not refer to a geographic location. For example, X_{34} refers to harvest in age class 3 (trees 21–30 years old) during the 4th decade, without specifying where that age class is located. One could keep track of location with another subscript, but we shall not do so, to keep the notation simple.

Growth Equations

With these definitions, it is possible to predict the state of the forest at each future inventory. Table 6.1 shows all the state and decision variables needed to describe the evolution of our sample forest over 30 years and the equations that link them. The table assumes that trees 31 years old or older are lumped into age class 4.

1st Inventory The area in each age class at the beginning of the first decade is:

$$A_{11} = 100 \text{ ha}, \quad A_{21} = 200 \text{ ha}, \quad A_{31} = 50 \text{ ha}, \quad A_{41} = 150 \text{ ha}$$

2nd Inventory The forest state at the beginning of the second decade depends on the state at the beginning of the first decade and on the harvest and reforestation during the first decade. In particular, the area in age class 1 is the total area cut and reforested during the first decade:

$$A_{12} = X_{11} + X_{21} + X_{31} + X_{41}$$

The area in age class 2 is what remains of the area that was in age class 1 at the beginning of the first decade:

$$A_{22} = A_{11} - X_{11}$$

Similarly, the area in age class 3 is:

$$A_{32} = A_{21} - X_{21}$$

The oldest age class, 4, consists of the unharvested areas that were in age classes 3 and 4 at the beginning of the first decade:

$$A_{42} = (A_{31} - X_{31}) + (A_{41} - X_{41})$$

3rd and 4th Inventory The equations to predict the forest state at the beginning of the third and the fourth decades are similar to those for the

TABLE 6.1 Growth Equations for an Even-Aged Forest

Age class	Decade 1		2		3		4
	Stock	Cut	Stock	Cut	Stock	Cut	Stock
1	$A_{11} = 100$	X_{11}	$A_{12} = X_{11} + X_{21} + X_{31} + X_{41}$	X_{12}	$A_{13} = X_{12} + X_{22} + X_{32} + X_{42}$	X_{13}	$A_{14} = X_{13} + X_{23} + X_{33} + X_{43}$
2	$A_{21} = 200$	X_{21}	$A_{22} = A_{11} - X_{11}$	X_{22}	$A_{23} = A_{12} - X_{12}$	X_{23}	$A_{24} = A_{13} - X_{13}$
3	$A_{31} = 50$	X_{31}	$A_{32} = A_{21} - X_{21}$	X_{32}	$A_{33} = A_{22} - X_{22}$	X_{33}	$A_{34} = A_{23} - X_{23}$
4	$A_{41} = 150$	X_{41}	$A_{42} = (A_{31} - X_{31}) + (A_{41} - X_{41})$	X_{42}	$A_{43} = (A_{32} - X_{32}) + (A_{42} - X_{42})$	X_{43}	$A_{44} = (A_{33} - X_{33}) + (A_{43} - X_{43})$

2nd inventory. Only the subscript referring to the current decade changes from decade to decade (see Table 6.1).

The calculations have been continued in Table 6.1 to show the status of the forest at the beginning of the fourth decade. This description of the cut and stock can be pursued into the future as long as desired. The results clearly show that the area in each age class at any point in time is a linear function of the initial condition of the forest and of all subsequent harvesting and reforestation. Thus, the evolution of the forest is defined entirely by the initial state and by the decision variables. The state variables showing the stock are needed only as accounting devices.

The process of writing all the equations may seem laborious, but because they are fully recursive, it is easy to implement them in a spreadsheet (see upcoming Section 6.6).

Before continuing, we note that any feasible solution of the equations in Table 6.1 must be such that the area cut in each age class is less than the corresponding stock:

$$X_{11} \leq A_{11}$$
$$X_{12} \leq A_{12}$$
$$\vdots$$
$$X_{43} \leq A_{43}$$

and the areas cut cannot be negative:

$$X_{11}, X_{12}, \ldots, X_{43} \geq 0$$

Note that these inequalities are enough to ensure that the stock in every age class and period is nonnegative. For example, $X_{11} \leq A_{11}$ and $X_{11} \geq 0$ necessarily imply that $A_{11} \geq 0$.

6.4 SUSTAINABILITY CONSTRAINTS

The quality of a forest plan depends in part on the values of the goods and services that the forest will continue to produce beyond the end of the plan. Some goods such as timber, are tied to the trees that are cut; others, such as esthetics, protection, and biodiversity, depend on the growing stock. Regardless of the objectives, a primary requirement of a forest management plan is that the forest left at the end be in a sustainable condition. One way to ensure sustainability is with steady-state constraints.

STEADY-STATE CONSTRAINTS

A forest of the type described in this example is in a steady state if the growing stock in each age class remains constant under the prevailing interventions (harvest followed by reforestation). If the stock in each age class is the same in two successive inventories, then the decisions between the two inventories must necessarily constitute a sustainable management regime. The time it takes to convert the initial forest into a steady-state structure is called the *conversion period*. Beyond this period, production could continue at a constant level, and the growing stock would remain unchanged.

In the example of Table 6.1, to say that the forest has reached a steady state by the end of the third decade means that:

$$A_{13} = A_{14}$$
$$A_{23} = A_{24}$$
$$A_{33} = A_{34}$$
$$A_{43} = A_{44}$$

That is, the stock in each age class is the same at the beginning of the fourth decade as it was at the beginning of the third decade. If this is true, then the growth of the forest during the third decade is just enough to replace the harvest during the third decade. Thus, the harvest symbolized by X_{13}, X_{23}, X_{33}, and X_{43} is sustainable, and applying it again in the fourth decade would produce a mosaic of age classes identical to the one at the beginning. This process could then continue in perpetuity.

A SPECIAL STEADY STATE: THE REGULATED FOREST

As seen in the previous chapter, a regulated forest has the same area in each age class. The cut always removes the oldest age class, and reforestation follows immediately. A regulated forest is thus in a steady state, because the growing stock and the cut remain constant in perpetuity. It is possible to constrain a management plan to lead to a regulated forest after a given amount of time. For example, to obtain a regulated forest with a rotation of 20 years by the end of the third decade, one would require that by that date the first two age classes each occupy half of the forest and that the other two be absent:

$$A_{14} = A_{24} = \frac{500}{2} = 250 \text{ ha} \quad \text{and} \quad A_{34} = A_{44} = 0 \text{ ha}$$

A Dynamic Model of the Even-aged Forest

However, although regulation is a sufficient condition for a steady state, it is not a necessary condition. For example, the state and harvest values

$$A_{13} = 425 \text{ ha}, \quad X_{13} = 350 \text{ ha}$$
$$A_{23} = 75 \text{ ha}, \quad X_{23} = 75 \text{ ha}$$

do not depict a regulated forest, but they do describe a steady state, because the stock we would observe at the beginning of decade 4 would be the same as at the beginning of decade 3, since:

$$A_{14} = X_{13} + X_{23} = 425 \text{ ha} \quad \text{and} \quad A_{24} = A_{13} - X_{13} = 75 \text{ ha}$$

In fact, the number of possible steady states is generally infinite. This means that managers have the flexibility to choose the steady state most suitable for their purpose without restricting themselves to a regulated forest. A drawback of the regulated forest is that, given the initial conditions, it may not be possible to reach a desired regulated forest within a specific amount of time, though it still may be possible to reach a different steady state.

6.5 OBJECTIVE FUNCTION

The growth equations and the steady-state equations define completely the transformation of the forest from its initial condition to the terminal state. In general, the equations have an infinite number of solutions. However, with linear programming, one can find a solution that maximizes a particular function of the decision variables. Here we shall consider two kinds of objectives: timber production and landscape diversity.

TIMBER PRODUCTION

Let us assume that the volume per unit area in each age class of our example forest is given by the yield function in Figure 6.1. For example, regardless of when it occurs, 1-ha cut from age class 1 yields 50 m³ of timber, a 1-ha cut from age class 2 yields 250 m³, and so on.

Thus, the volume cut from the entire forest during the 30 years of the conversion period is:

$$\begin{aligned} Z = {} & 50X_{11} + 250X_{21} + 500X_{31} + 600X_{41} \\ & + 50X_{12} + 250X_{22} + 500X_{32} + 600X_{42} \\ & + 50X_{13} + 250X_{23} + 500X_{33} + 600X_{43} \quad (\text{m}^3) \end{aligned}$$

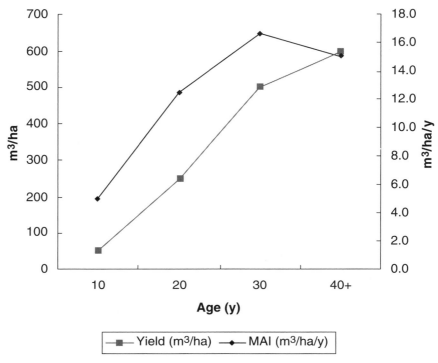

FIGURE 6.1 Yield and mean annual increment (MAI) of short-leaf pine on land of site index 18 m. (Data derived from Smalley and Bailey, 1974.)

where the first, second and third lines express the volume produced during the first, second, and third decades, respectively.

Z is the appropriate objective function if the objective is to maximize the total volume produced during the conversion of the forest to a steady state. Alternatively, if the management objective is to create the steady-state forest of highest productivity, the appropriate objective function would be:

$$Z' = 50X_{13} + 250X_{23} + 500X_{33} + 600X_{43} \quad (m^3)$$

because this is the expression of the constant volume that the steady-state forest would produce, decade after decade. Although the objective function Z' is part of the objective Z, maximizing either objective will generally give quite different results. In the next chapter we investigate how to write the objective function that gives appropriate economic weights to the production during and after conversion to the steady state.

LANDSCAPE DIVERSITY AND THE MAXIMIN CRITERION

Maximizing the minimum value in a set of values is a useful objective in many decision settings. To pursue our example, assume that the managers want a management plan that would keep the forest landscape as diverse as possible. Thinking of the forest as a mosaic of age classes, one way to promote diversity would be to ensure that there is some area in every age class. A plausible diversity objective would then be to maximize the smallest area in any age class at the beginning of each decade considered except the first. That is, with the notations of our model:

$$\max[\min(A_{12}, A_{22}, \ldots, A_{44})]$$

The initial conditions $A_{11}, A_{12}, \ldots, A_{14}$ are not part of the objective function because they are fixed. To express this objective in a form usable in linear programming, let A_{min} be a new variable designating the smallest area in any age class. Then, by definition, A_{min} must satisfy these constraints:

$$A_{12} \geq A_{min}$$
$$A_{22} \geq A_{min}$$
$$\vdots$$
$$A_{44} \geq A_{min}$$

And since the smallest area in any age class must be as large as possible, the objective function is:

$$\max A_{min}$$

This optimization is done by varying simultaneously the decision variables $X_{11}, X_{21}, \ldots, X_{43}$ and the smallest area in any age class and decade, A_{min}.

6.6 SPREADSHEET FORMULATION AND SOLUTION

Figure 6.2 shows the spreadsheet for the short-leaf pine forest of our example. The entries in bold are input data; the other entries are the results of formulas based on those data and on the decision variables. The input data in this

	A	B	C	D	E	F	G
1	MAX PRODUCTION						
2	Age		Decade				
3	class		1	2	3	4	
4			Stock (ha)				
5		1	100	200	300	300	
6		2	200	100	200	200	
7		3	50	200	0	0	
8		4	150	0	0	0	
9							
10			Cut (ha)				Yield (m³/ha)
11		1	0	0	100		50
12		2	0	100	200		250
13		3	50	200	0		500
14		4	150	0	0		600
15			Cut (m³)			max(Total)	
16			115000	125000	55000	295000	
17							
18				Key cell formulas			
19	Cell		Formula			copied to	
20	D5		=SUM(C11:C14)			D5:F5	
21	D6		=C5-C11			D6:F7	
22	D8		=C7-C13+C8-C14			D8:F8	
23	C16		=SUMPRODUCT($G11:$G14,C11:C14)			C16:E16	
24	F16		=SUM(C16:E16)				

FIGURE 6.2 Spreadsheet model for converting a short-leaf pine forest to a steady state in 30 years while maximizing production.

spreadsheet are the initial forest condition and the yields per unit area of the various age classes.

The cells in Figure 6.2 under the heading "Stock" show the state of the forest at the beginning of each decade, that is, the A_{ij} state variables. Cells C5:C8 contain the initial condition: 100 ha in age class 1, 200 ha in age class 2, 50 ha in age class 3, and 150 ha in age class 4. The cells under the heading "Cut" show the harvest and reforestation activities during each decade, that is, the decision variables X_{ij}. The cells G11:G14 contain the yield per hectare.

Cells D5:F8 contain the growth equations in Table 6.1, giving the stock of the forest at the beginning of each decade as a function of the stock 10 years earlier and of the decisions during the previous decade. For example, the formula in cell D5 corresponds to the equation $A_{12} = X_{11} + X_{21} + X_{31} + X_{41}$; the formula in cell D6 corresponds to $A_{21} = A_{11} - X_{11}$; and the formula in cell D8 corresponds to

$A_{42} = A_{31} - X_{31} + A_{41} - X_{41}$. Copying cells D5:D8 into cells E5:E8 and F5:F8 then gives the stock at the beginning of decades 3 and 4.

The formula in cell C16 corresponds to the equation giving the volume of timber produced during the first decade as a function of the decision variables: $50X_{11} + 250X_{21} + 500X_{31} + 600X_{41}$ (m³). Cell F16 contains the formula of the volume produced during the first 30 years, corresponding to the equation:

$$Z = 50X_{11} + 250X_{21} + 500X_{31} + 600X_{41}$$
$$+ 50X_{12} + 250X_{22} + 500X_{32} + 600X_{42}$$
$$+ 50X_{13} + 250X_{23} + 500X_{33} + 600X_{43} \quad (m^3)$$

MAXIMIZING TIMBER PRODUCTION

Figure 6.3 shows the Solver parameters to maximize production during 30 years while ending with a steady-state forest. The solver seeks the largest production, in target cell F16, by changing cells C11:E14, the decision variables defining the area cut and reforested from each age class in each decade: X_{11}, X_{21},..., X_{43}. The optimization is done subject to the following constraints:

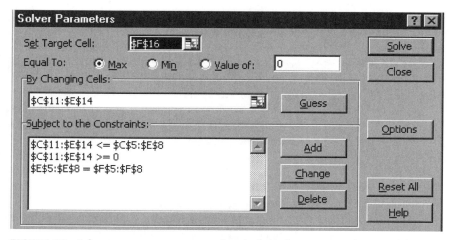

FIGURE 6.3 Solver parameters to convert a short-leaf pine forest to a steady state while maximizing production.

C11:E14 <= C5:E8, meaning that the area cut in each age class and decade must be less than the area available, corresponding to these equations:

$$X_{11} \leq A_{11}$$
$$X_{12} \leq A_{12}$$
$$\vdots$$
$$X_{43} \leq A_{43}$$

C11:E14 >= 0, because the area cut and reforested in each age class and period cannot be negative, corresponding to these equations:

$$X_{11}, X_{12}, \ldots, X_{43} \geq 0$$

E5:E8 = F5:F8, to ensure that the forest is in a steady state at the end of the third decade, corresponding to these equations:

$$A_{13} = A_{14}$$
$$A_{23} = A_{24}$$
$$A_{33} = A_{34}$$
$$A_{43} = A_{44}$$

The best solution for this problem is in Figure 6.2. It shows the best pattern of harvest and reforestation, the corresponding production, and the evolution of the forest state over time. For example, the plan calls for cutting and reforesting 50 ha and 150 ha from age classes 3 and 4 during the first decade, producing 115,000 m^3 of timber. This leads to a forest with 200 ha in age class 1, 100 ha in age class 2, 200 ha in age class 3, and 0 ha in age class 4 by the end of the decade. The total maximum production over the 30 years is 295,000 m^3. At the end of the plan, the forest would be in a steady state. The steady-state forest has 300 ha in age class 1 and 200 ha in age class 2. This steady state could be maintained in perpetuity by cutting and reforesting 100 ha from age class 1 and all of age class 2 every decade. Note that this steady state is not a regulated forest, nor does it maximize the productivity of the forest in the long run, because it cuts the timber when it is 20 years old, while the maximum mean annual increment occurs at age 30 (see Figure 6.1).

A solution that maximized the long-term productivity of the forest could be found with the same spreadsheet by changing the target cell from F16 to E16, that is, by maximizing the production of the steady-state forest. We leave this as an exercise (see Problem 6.5 at the end of this chapter).

A Dynamic Model of the Even-aged Forest

	A	B	C	D	E	F	G
1	MAX DIVERSITY OF FOREST LANDSCAPE						
2	Age		Decade				
3	class		1	2	3	4	
4			Stock (ha)				
5		1	100	100	100	100	
6		2	200	100	100	100	
7		3	50	200	100	100	max(Amin)
8		4	150	100	200	200	100
9							
10			Cut (ha)				Yield (m³/ha)
11		1	0	0	0		50
12		2	0	0	0		250
13		3	50	100	0		500
14		4	50	0	100		600
15			Cut (m³)			Total	
16			55000	50000	60000	165000	
17							
18				Key cell formulas			
19	Cell		Formula			copied to	
20	D5		=SUM(C11:C14)			D5:F5	
21	D6		=C5−C11			D6:F7	
22	D8		=C7−C13+C8−C14			D8:F8	
23	C16		=SUMPRODUCT($G11:$G14,C11:C14)			C16:E16	
24	F16		=SUM(C16:E16)				

FIGURE 6.4 Spreadsheet model for converting a short-leaf pine forest to a steady state in 30 years while maximizing landscape diversity.

Maximizing Landscape Diversity

Figure 6.4 shows a spreadsheet set up for the same short-leaf pine forest but with the objective of maximizing the landscape diversity of the forest throughout the conversion. The initial conditions and the yield data are the same as in Figure 6.2. The conversion to a steady state is still done in 30 years. The growth equations and the steady-state equations are unchanged. The only difference is in the objective function.

The cell being optimized is now G8, which contains the smallest area in any age class and decade, the variable A_{min} in the notation used earlier. The solver tries to make this variable as large as possible. Because A_{min} is also a variable, cell G8 has been added to the list of adjustable cells in the Solver parameters (Figure 6.5).

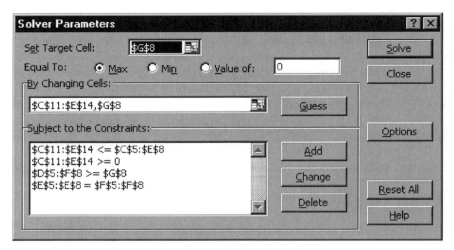

FIGURE 6.5 Solver parameters to convert a short-leaf pine forest to a steady state while maximizing landscape diversity.

Furthermore, the following constraints have been added to the solver parameters: D5:F8 >= G8, corresponding to these inequalities:

$$A_{12} \geq A_{min}$$

$$A_{22} \geq A_{min}$$

$$\vdots$$

$$A_{44} \geq A_{min}$$

which ensure that A_{min} is the smallest area in any age class in every decade.

Figure 6.4 shows the best solution for this problem. In contrast with the solution that maximized production shown in Figure 6.2, there would never be an age class missing from the forest. In fact, the smallest area in stock in any age class and decade is 100 ha. The conversion plan would end up with a steady-state forest that again is not exactly regulated. From the beginning of the fourth decade onward, there would be 100 ha in all age classes but the oldest, which would contain 200 ha. The harvest, 100 ha, would always occur in this oldest age class. The total production during the 30 years of conversion would be 145,000 m³. This is less than half what would be obtained by the plan that maximized production without regard to the consequences for the diversity of the forest landscape. It is up to the stakeholders to decide whether the gain in diversity is worth that much loss in production or whether a compromise solution is

A Dynamic Model of the Even-aged Forest

warranted. This issue of the opportunity cost of alternative policies in discussed further in Chapter 7.

6.7 GENERAL FORMULATION

The model developed in this chapter applies to the following general situation. We describe the state of an even-aged forest at successive periodic inventories by the area in each of m age classes. The time unit to define age classes is the length of the period between inventories, u. A_{ij} is the area in age class i at the jth inventory.

The forest is managed for p periods, each u years long. At the end of this conversion period we want to reach a sustainable forest structure and harvest. The management plan is defined by the decision variables X_{ij}, the area cut and reforested from age class i in period j. The growth equations describe the state of the forest at the start of each successive period as a function of the initial condition and of the prior decisions.

GROWTH EQUATIONS

$$A_{11} = A_{11}^0$$
$$A_{21} = A_{21}^0$$
$$\vdots$$
$$A_{m1} = A_{m1}^0$$

$$A_{1j} = \sum_{i=1}^{m} X_{i,j-1} \qquad j = 2,\ldots,p+1$$

$$A_{ij} = A_{i-1,j-1} - X_{i-1,j-1} \qquad i = 2,\ldots,m-1;\ j = 2,\ldots,p+1$$

$$A_{mj} = (A_{m-1,j-1} - X_{m-1,j-1}) + (A_{m,j-1} - X_{m,j-1}) \qquad j = 2,\ldots,p+1$$

where A_i^0 is the initial area in age class i. To be feasible, the state variables and decision variables must be such that the area harvested and reforested from each age class in each period is nonnegative and less than the corresponding stock:

$$X_{ij} \geq 0 \qquad i = 1,\ldots,m;\ j = 1,\ldots,p$$
$$X_{ij} \leq A_{ij} \qquad i = 1,\ldots,m;\ j = 1,\ldots,p$$

STEADY-STATE CONSTRAINTS

These constraints ensure that the forest left at the end of the plan is sustainable, by forcing the stock to be constant in all age classes, during the last two inventories:

$$A_{ip} = A_{i,p+1} \qquad i = 1, \ldots, m$$

which implies that the harvest obtained in period p may continue indefinitely.

OBJECTIVE FUNCTION

If the objective is to maximize the total timber production, the objective function would be:

$$\max Z = \sum_{i=1}^{m} \sum_{j=1}^{p} v_i X_{ij}$$

where v_i is the volume per unit area in age class i.

If, instead, the objective is to maximize the landscape diversity of the forest, the objective function would be:

$$\max A_{min}$$

subject to:

$$A_{ij} \geq A_{min} \qquad i = 1, \ldots, m; j = 2, \ldots, p+1$$

where A_{min} is a new variable designating the smallest area in any age class and period.

Verify that the special cases described in Section 6.6 fit within this general formulation. Of course, no one needs to remember these formulas. All that is needed is to understand the spreadsheets in Figures 6.2 and 6.4 and to be able to adapt them to different initial conditions, conversion periods, end states, and objective functions.

6.8 SUMMARY AND CONCLUSION

In this chapter we have developed a model of the even-aged forest that can describe fully the condition of the various age classes at any point in time, given initial conditions and the intervening harvests and reforestation. Conditions were added to guarantee that the forest at the end of the plan had a sustainable structure and harvest. Sustainability was ensured by steady-state constraints, such that the stock in each age class remains constant beyond the last period of the plan. We observed that a steady-state forest could look quite different from the classical regulated forest. Regulation is sufficient to ensure sustainability, but it is not necessary. The growth equations and the steady-state equations became the constraints of a linear program for which a large number of solutions were possible. Particular solutions were obtained with objective functions that maximized either timber production or landscape diversity. The highest landscape diversity was defined as the mosaic of age classes that maximized the smallest area in any age class and period. This MaxiMin criterion was formulated as a new objective function within the linear programming framework. In the next chapter we shall see how the model can also be used to design harvest schedules with economic objectives, subject to additional environmental and managerial constraints.

PROBLEMS

6.1 Section 6.3 shows a method for representing the area in any age class in any period for an even-aged forest. Table 6.1 illustrates this method for four decades of growth of a forest that has initially four age classes. Extend the table to show the growth of this same forest for two more decades.

6.2 Construct a table similar to Table 6.1 for an even-aged forest with five age classes. Write the equations to predict the forest state at the beginning of period 3. Assume that the initial forest state consists of 50 ha in age class 1, 100 ha in age class 2, 0 ha in age class 3, 200 ha in age class 4, and 40 ha in age class 5.

6.3 Consider the forest described by the growth equations in Table 6.1. Assume that the desired forest state at the beginning of the fourth period is a regulated forest with a rotation of 30 years. Using the state and decision variables in Table 6.1, write the end-state constraints needed to achieve this conversion.

6.4 Consider the short-leaf pine forest conversion problem for which a spreadsheet model is shown in Figure 6.2.
(a) Modify this model so that the end state of the forest would be not only a steady state but also a regulated forest with a rotation of 30 years. (*Note:* The area in all age classes must add up to exactly 500 ha for the solution to be feasible.)
(b) Solve this modified model, and discuss the effects of this additional constraint on the solution relative to that shown in Figure 6.2.

6.5 Consider the short-leaf pine forest conversion problem for which a spreadsheet model is shown in Figure 6.2.

(a) Modify this model so that the objective is to maximize the periodic production of the steady-state forest created by conversion, instead of the production during all of the conversion period.

(b) Solve this modified model, and discuss the effects of this change on the solution.

6.6 Consider the short-leaf pine forest conversion problem for which a spreadsheet model is shown in Figure 6.2.

(a) Modify this model to change the conversion period to 40 years instead of 30 years.

(b) Solve this modified model, and discuss the effects of changing the conversion period on the solution.

6.7 Consider the short-leaf pine forest conversion problem for which a spreadsheet model is shown in Figure 6.2. Assume that for ecological reasons you want to maintain 150 ha in the oldest age class forever.

(a) What constraint should you add to the model to ensure this result?

(b) Modify the spreadsheet in Figure 6.2 accordingly, and compute the new management plan. Discuss the effects of this constraint on the solution.

6.8 Consider the short-leaf pine forest conversion problem in which landscape diversity is maximized, for which a spreadsheet model is shown in Figure 6.4. Assume you wished to maximize the diversity of the steady-state forest only at the end of the conversion and were not concerned about diversity during the conversion. Specifically, assume you wished to maximize the smallest area in any age class in the steady-state forest.

(a) How would you modify the objective function and the constraints to reflect this new objective?

(b) Modify the spreadsheet in Figure 6.4 accordingly, and compute the new management plan. Discuss the effects of this new objective on the solution.

6.9 Consider the short-leaf pine forest conversion problem in which landscape diversity is maximized, for which a spreadsheet model is shown in Figure 6.4. Assume that for esthetic reasons your objective is to maximize the area in the oldest age class by the end of the conversion period.

(a) How would you modify the objective function and the constraints to reflect this new objective?

(b) Modify the spreadsheet in Figure 6.4 accordingly, and compute the new management plan. Discuss the effects of this new objective on the solution.

ANNOTATED REFERENCES

Clutter, J.L., J.C. Fortson, L.V. Pienaar, G.H. Brister, and R.L. Baily. 1983. *Timber Management: A Quantitative Approach*. Wiley, New York. 333 pp. (Chapter 10's "XYZ Timber Co." model has a structure similar to that used in this chapter.)

Connaughton, K. 2001. Sustainability: The key forest policy issue of the new millennium? *Journal of Forestry* 99(2):7. (Introduction to a special issue of the *Journal of Forestry* designed as a discussion guide on sustainability.)

Davis, L.S., K.N. Johnson, P.S. Bettinger, and T.E. Howard. 2001. *Forest Management: To Sustain Ecological, Economic, and Social Values.* McGraw-Hill, New York. 804 pp. (Chapter 10 discusses "Model I and Model II" structures, the latter being similar to that used in this chapter.)

Dykstra, D.P. 1984. *Mathematical Programming for Natural Resource Management.* McGraw-Hill, New York. 318 pp. (Chapter 6 presents a linear programming model with a structure similar to that used in this chapter.)

Iverson, D.C., and R.M. Alston. 1986. *The Genesis of FORPLAN: A Historical and Analytical Review of Forest Service Planning Models.* USDA Forest Service, Intermountain Research Station, Report INT-214. 30 pp. (A review of the evolution of linear programming models from the model structure used in Chapters 4 and 5 to that used in this chapter.)

Johnson, K.N., and H.L. Scheurman. 1977. Techniques for prescribing optimal timber harvest and investment under different objectives—Discussion and synthesis. *Forest Science* Monograph No. 18. 31 pp. (Discusses the theoretical basis for different linear programming model structures for forest resource management. In their terminology, the model used in this chapter is of the "Model II" type, because a new harvesting unit is created out of the acres regenerated in each period.)

Johnson, K.N., T.W. Stuart, and S.A. Crim. 1986. FORPLAN version 2—An overview. USDA Forest Service, Land Management Planning Systems, Fort Collins, CO, 85 pp. (A linear programming-based decision-support system used by the U.S. Forest Service to prepare management plans for the national forests.)

Nautiyal, J.C., and P.H. Pearse. 1967. Optimizing the conversion to sustained yield—A programming approach. *Forest Science* 13(2):131–139. (Origin of the forest dynamic equations used in this chapter.)

Smalley, G.W., and R.L. Bailey. 1974. *Yield Tables and Stand Structure for Short-Leaf Pine Plantations in Tennessee, Alabama, and Georgia Highlands.* U.S. Forest Service, Research Paper SO-97. (Source of yield data used in this chapter.)

Sustainable Forestry Working Group. 1998. *The Business of Sustainable Forestry: Case Studies.* The John D. and Catherine T. MacArthur Foundation, Chicago. (Collection of case studies on "sustainable" forest enterprises.)

Toman, M.A., and P.M.S. Ashton. 1996. Sustainable forest ecosystems and management: A review article. *Forest Science* 42(3):366–377. (Review of both economic and ecological perspectives on sustainability.)

CHAPTER 7

Economic Objectives and Environmental Policies for Even-Aged Forests

7.1 INTRODUCTION

In the preceding chapter we used a dynamic model to forecast the growth of an even-aged forest. Given the initial state of the forest and a specific harvesting and reforestation schedule, we could predict what the forest would look like at any future date. The model also ensured that at the end of the management plan we would have a steady-state forest that could be sustained forever. We applied that model to calculate harvest and reforestation schedules to maximize either timber production or landscape diversity.

The purpose of this chapter is to extend the model in Chapter 6 to define the economic value of a forest for timber production and the effects of environmental constraints on this value. Forests produce many goods and services other than timber, but timber is one of the few with a well-defined price, set by markets. Forest value for timber production is therefore amenable to estimation, a task that we tackle in this chapter.

This concept of forest value is useful even if the management objectives for a forest have nothing to do with timber production, for it informs us about the opportunity cost of the policies designed to favor the production of other goods and services. If a national forest has a value of, say, $1 billion for timber production and the forest is turned into a national park with no timber production, this implies that the services and amenities provided by the park are worth at least $1 billion. As we shall see, there are many, more mundane, situations where this concept of opportunity cost is helpful to compare management alternatives with mixed economic and environmental objectives.

The beginning of the chapter deals with one of the simplest cases for estimation of forest value: where we start with bare land and want to estimate the value of that land for timber production. Then we shall use the dynamic model of the even-aged forest developed in the last chapter to calculate the value of a forest with many initial age classes, converted to a steady state within a specified period. The method will give us a way to find the management that would bring about the highest forest value and to predict the effect of constraints such as even-flow policies on forest value. At the end we shall use this same model to investigate the effect of different environmental goals on forest value and thus to measure the trade-off between economic and environmental objectives.

7.2 LAND EXPECTATION VALUE AND ECONOMIC ROTATION

What is the value of land if we use it to grow trees? The answer can be found with a simple, yet powerful, formula originally developed by Martin Faustmann in 1849 (Faustmann, 1995).

LAND EXPECTATION VALUE

Consider a piece of land on which we plan to grow trees. When the trees have reached a particular *rotation age*, R, they are harvested, yielding the volume per hectare v_R. The land is reforested immediately after harvest. This sequence of reforestation, growth, and harvest is repeated in perpetuity, with each rotation identical to the preceding rotation (Figure 7.1). Faustmann's insight was to

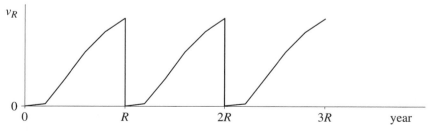

FIGURE 7.1 Stand growth and harvest in Faustmann's model.

Economic Objectives and Environmental Policies for Even-Aged Forests

realize that the value of the land is equal to the present value of the income from this infinite sequence of harvests.

Let w be the price of timber, per unit of volume, net of harvesting cost, and let c be the cost of reforestation per hectare. Then the initial cost of establishing the forest is c, and the income at the end of each successive rotation, net of the reforestation cost, is:

$$wv_R - c$$

Therefore, denoting the interest rate by r, the net present value of the income, starting from bare land, for an infinite sequence of successive rotations is:

$$\text{LEV} = -c + \frac{wv_R - c}{(1+r)^R} + \frac{wv_R - c}{(1+r)^{2R}} + \frac{wv_R - c}{(1+r)^{3R}} + \cdots$$

Or, in a more compact form (see Appendix A):

$$\text{LEV} = -c + \frac{wv_R - c}{(1+r)^R - 1}$$

This, then, is the land expectation value, that is, the value of bare land used for this kind of silviculture.

ECONOMIC ROTATION

Faustmann's formula gives the land expectation value for a particular rotation. It can also be used to find the *economic rotation*, that is, the rotation leading to the highest land expectation value for a given yield function, timber price, reforestation cost, and interest rate.

For example, Figure 7.2 shows a spreadsheet model to compute the land expectation value for land forested to short-leaf pine for rotations of 10, 20, 30, or 40 years. All entries in bold are input data; the other entries are the results of formulas. The yield data are the same as those used in the previous chapter (see Figure 6.1), the timber price is set at $50/m^3, the interest rate at 6% per year, and the reforestation cost at $500/ha.

Cells E5:E8 show that the highest net income per unit area is obtained with a rotation of 40 years. However, cells F5:F8 show that the land expectation value is highest for a rotation of only 20 years. With an interest rate of 6%, it is financially better to earn $12,000 every 20 years than $29,500 every 40 years.

Cells C5:C8 show the mean annual increment for different rotations. The highest productivity occurs for a rotation of 30 years and gives 16.7 m^3/ha/y. Yet it is better, from a purely financial point of view, to harvest the stand at 20 years of

	A	B	C	D	E	F
1	LAND EXPECTATION VALUE					
2	Stand	Stand	Mean annual	Gross	Net	
3	age	yield	increment	income	income	LEV
4	(y)	(m^3/ha/y)	(m^3/ha/y)	($/ha)	($/ha)	($/ha)
5	10	50	5.0	2500	2000	2029
6	20	250	12.5	12500	12000	4937
7	30	500	16.7	25000	24500	4665
8	40	600	15.0	30000	29500	2677
9	Financial data					
10	Timber price	($/$m^3$)		50		
11	Interest rate	(/y)		0.06		
12	Reforestation cost	($/ha)		500		
13						
14	Key cell formulas					
15	Cell	Formula			Copied to	
16	C5	=B5/A5			C6:C8	
17	D5	=$D$10*B5			D6:D8	
18	E5	=D5-$D$12			E6:E8	
19	F5	=E5/((1+D11)^A5-1)-$D$12			F6:F8	

FIGURE 7.2 Spreadsheet model to compute the land expectation value.

age. The higher the interest rate, the larger would be this difference between the economic rotation and the rotation with the highest mean annual increment.

Opportunity Cost and Alternative Land Uses

Faustmann's formula has many applications besides computing the economic rotation. It can be amended to include the value of nontimber goods and services provided by trees of a certain age, as long as we have prices for these goods and services. But even if prices are not available, the formula is useful to show the opportunity cost of decisions. For example, assume that the owners of the land considered in Figure 7.2 like big trees for esthetic reasons and thus choose a rotation of 40 years instead of 20 years. The opportunity cost of this choice, in terms of land expectation value, is $2,260/ha ($4,937 − $2,677). Choosing the longer rotation makes sense only if the esthetic value of the older trees is worth at least that much to the owners.

Not only is Faustmann's formula useful to compare forest management alternatives, it is also useful to compare totally different land uses. Suppose that the owner of the land considered in Figure 7.2 is a farmer who could grow corn on the same land instead of trees. Because corn is a yearly crop, $R = 1$ y, and the land expectation formula simplifies to:

$$\text{LEV} = -c + \frac{I - c}{r}$$

where I is the yearly crop income, net of harvesting cost, and c is the yearly planting cost. For example, assume $I = \$600/\text{ha}$, $c = \$200/\text{ha}$, and $r = 6\%/\text{y}$. Then the land expectation value for growing corn is $\$6,467/\text{ha}$. If the farmer decides to grow trees instead, the implication is that the tangible or intangible benefits of doing so must be at least $\$1,530/\text{ha}$ ($\$6,467 - \$4,937$).

7.3 ESTIMATING FOREST VALUE BY LINEAR PROGRAMMING

Faustmann's fundamental insight can be generalized to define the value of any forest, regardless of its initial condition, and the values of the constraints that apply to its management. In general, the forest value in terms of timber production is the present value of the income that the forest is capable of producing over an infinite time horizon, subject to various management goals and constraints. This future income is produced by the land and by any tree initially growing on it. Thus the forest value we seek includes land and trees.

In the remainder of this chapter we shall study how to compute forest value with linear programming. The basic forest model is the same as in Chapter 6, and we shall continue to use the same example. The initial condition of the forest is a mosaic of even-aged short-leaf pine stands: There are 100 ha in age class 1, that is, with trees 1–10 years old. There are 200 ha in age class 2, with trees 11–20 years old. There are 50 ha in age class 3, with trees 21–30 years old, and there are 150 ha in age class 4, with trees 31–40 years old. The forest is inventoried every decade, and the state variable A_{ij} is the area in age class i at the beginning of decade j.

The silviculture remains the same as in Chapter 6: immediate artificial regeneration after harvest. The decision variable X_{ij} refers to the area harvested and reforested in age class i during decade j. The yield function is that described in Figure 6.1. For example, regardless of when it occurs, 1 ha harvested from age class 1 yields 50 m³/ha. The conversion period is set initially at 30 years. Therefore, the equations that give the stock in each age class at the beginning of each decade are the same as in Table 6.1. The steady-state equations are also the same: The area in each age class at the beginning of the third decade must be the same as the area at the beginning of the fourth decade.

The difference lies in the objective function. Here it expresses the forest value. This forest value consists of the present value of the income up to the steady state (the first 30 years) plus the present value of the income after the steady state. Although the steady-state production continues forever, the fact that it is constant will allow a simple computation of its value, similar to Faustmann's formula.

PRESENT VALUE OF INCOME UP TO THE STEADY STATE

As seen in the previous chapter, the volume of timber produced in each decade for the first 30 years is:

$$Q_1 = 50\ X_{11} + 250 X_{21} + 500 X_{31} + 600 X_{41}$$
$$\text{(m}^3\text{)} \quad \text{(m}^3\text{/ha)} \quad \text{(ha)}$$

$$Q_2 = 50 X_{12} + 250 X_{22} + 500 X_{32} + 600 X_{42}$$

$$Q_3 = 50 X_{13} + 250 X_{23} + 500 X_{33} + 600 X_{43}$$

where Q_j is the volume of timber produced in decade j and X_{ij} is the area cut and reforested from age class i during the jth decade. Let the price of timber be $50/m^3$. The gross income in each decade, B_j, is:

$$B_j = 50\ Q_j$$
$$\text{(\$)} \quad \text{(\$/m}^3\text{)}\ \text{(m}^3\text{)}$$

Assume that the reforestation cost is $500/ha. Then the reforestation cost in each decade, C_j, is:

$$C_j = 500(X_{1j} + X_{2j} + X_{3j} + X_{4j})$$
$$\text{(\$)} \quad \text{(\$/ha)} \quad \text{(ha)}$$

Thus, the net income in each decade, N_j, is:

$$N_j = B_j - C_j$$
$$\text{(\$)} \quad \text{(\$)} \quad \text{(\$)}$$

Assume an interest rate of 6% per year. Then the present value (PV) of the net income from decade j is:

$$PV_j = \frac{N_j}{(1+0.06)^{10j}} \quad \text{(\$)}$$

Although the net revenue N_j is generated throughout decade j, for simplicity we count it as if it occurred at the end of the decade. Alternatively, the net revenue could be accounted for at the beginning of the decade or in the middle. This would only multiply the objective function by a constant and thus leave the best solution unchanged when the objective is to maximize present value.

The total present value of the net income up to the steady state is:

$$PV_c = PV_1 + PV_2 + PV_3 \quad (\$)$$

Total present value is a linear function of the decision variables, $X_{11}, X_{21}, \ldots, X_{43}$, because each periodic present value PV_j is a linear function of the net income N_j, which itself is a linear function of the benefit and cost, B_j and C_j, which are themselves linear functions of the decision variables.

PRESENT VALUE OF INCOME AFTER THE STEADY STATE

In our example, the initial forest is converted to a steady-state forest within 30 years. This means that the stock at the beginning of the third decade is the same as at the end, and the production during the third decade can continue forever.

Using the foregoing notation, the steady-state forest will produce a constant periodic income from the third decade onward equal to N_3 (\$). It can be shown (see Appendix A) that the net present value of this constant periodic income, if it started in the first decade, would be:

$$\frac{N_3}{(1+0.06)^{10}-1}$$

However, the forest will take 30 years to reach the steady state. Therefore, the net present value of the income after the steady state is:

$$PV_\infty = \frac{N_3}{[(1+0.06)^{10}-1](1+0.06)^{30}}$$

This is a linear function of the decision variables because the net income during the third decade, N_3, is a linear function of X_{13}, X_{23}, X_{33}, and X_{43}, which define the harvest and reforestation activities during the third decade.

FOREST VALUE

The forest value, FV, is the total net present value of the timber income that the forest would produce during and after the conversion to the steady state:

$$FV = PV_c + PV_\infty$$

This is the value of the initial forest, land and trees, from the point of view of timber production alone. Forest value measures what the owners may expect to earn from the forest by following a particular management regime, defined by the decision variables. The only constraint is that the management end with a steady state, that is, a sustainable regime.

The management regime leading to the highest forest value can be obtained by solving the following linear programming problem: Find the harvest and reforestation activities, $X_{11}, X_{21}, \ldots, X_{43}$, such that:

$$\max FV$$

subject to:

Growth equations (as in Table 6.1)
Steady-state:

$$A_{13} = A_{14}$$
$$A_{23} = A_{24}$$
$$A_{33} = A_{34}$$
$$A_{43} = A_{44}$$

Cut less than stock:

$$X_{11} \leq A_{11}$$
$$X_{12} \leq A_{12}$$
$$\vdots$$
$$X_{43} \leq A_{43}$$

Nonnegative cut:

$$X_{11}, X_{12}, \ldots, X_{43} \geq 0$$

7.4 SPREADSHEET OPTIMIZATION OF FOREST VALUE

Figure 7.3 shows the spreadsheet to compute the management plan leading to the highest forest value for the short-leaf pine forest studied in Chapter 6. The initial age classes of the forest and the yield data, in bold characters, are the same as in Figure 6.2. The new data are the timber price of $50/m^3, the reforestation cost of $500/ha, and the interest rate of 6% per year.

Economic Objectives and Environmental Policies for Even-Aged Forests 117

	A	B	C	D	E	F	G	H	I	J
1	MAX FOREST VALUE									
2	Age		Decade							
3	class		1	2	3	4				
4			Stock (ha)							
5	1		100	250	250	250				
6	2		200	100	250	250				
7	3		50	150	0	0				
8	4		150	0	0	0				
9							Yield	Price	Cost	Interest
10			Cut (ha)				(m³/ha)	($/m³)	($/ha)	(/y)
11	1		0	0	0		50	50	500	0.06
12	2		50	100	250		250			
13	3		50	150	0		500			
14	4		150	0	0		600			
15			Cut (m³)			Total				
16			127,500	100,000	62,500	290,000				
17			Net income ($)							
18			6,250,000	4,875,000	3,000,000					
19			Present value up to steady state ($)			After ($)	Total ($)			
20			3,489,967	1,520,048	522,330	660,469	6,192,815	max		
21										
22			Key cell formulas							
23	Cell		Formula				copied to			
24	D5		=SUM(C11:C14)				D5:F5			
25	D6		=C5-C11				D6:F7			
26	D8		=C7-C13+C8-C14				D8:F8			
27	C16		=SUMPRODUCT($G11:$G14,C11:C14)				C16:E16			
28	F16		=SUM(C16:E16)							
29	C18		=$H$11*C16-$I$11*SUM(C11:C14)				C18:E18			
30	C20		=C18/(1+$J$11)^(C3*10)				C20:E20			
31	F20		=(E18/((1+J11)^10-1))/(1+J11)^(E3*10)							
32	G20		=SUM(C20:F20)							

FIGURE 7.3 Spreadsheet model to maximize forest value.

The decision variables X_{ij} are in cells C11:E14. The growth equations in cells D5:F8 are the same as in Figure 6.2. The formulas to compute the periodic production are also the same. For example, cell C16 contains the formula of the volume produced during the first decade, corresponding to this equation:

$$Q_1 = 50X_{11} + 250X_{21} + 500X_{31} + 600X_{41} \quad (m^3)$$

Cells C18:E18 contain the formulas to compute the net income every decade. For example, cell C18 contains the formula of the net income during the first decade, corresponding to this equation:

$$N_1 = B_1 - C_1 = 50Q_1 - 500(X_{11} + X_{21} + X_{31} + X_{41}) \quad (\$)$$

Cells C20:E20 contain the formulas for the present value of the net period income. For example, cell C20 gives the present value of the income during the first decade, corresponding to this formula:

$$PV_1 = \frac{N_1}{(1+0.06)^{10}}$$

Cell F20 contains the formula for the net present value of the net income obtained after the steady state, corresponding to this equation:

$$PV_\infty = \frac{N_3}{[(1+0.06)^{10}-1](1+0.06)^{30}}$$

The forest value, then, is in cell G20, containing the sum of the present value up to the steady state and after the steady state, corresponding to this equation:

$$FV = PV_1 + PV_2 + PV_3 + PV_\infty$$

Figure 7.4 shows the Solver parameters to maximize this forest value. The target cell is G20. The decision variables are in cells C11:E14, and the constraints specify the steady state at the end of the third decade (E5:E8 = F5:F8), the cut in each decade and age class being less than the stock (C11:C14 <= C5:E8), and the nonnegativity of the cut (C11:E14 >= 0).

The best solution is shown in Figure 7.3. Compared to the solution that maximized production for the same initial conditions and conversion period (Figure 6.2), maximizing forest value would produce 5,000 m³ less timber

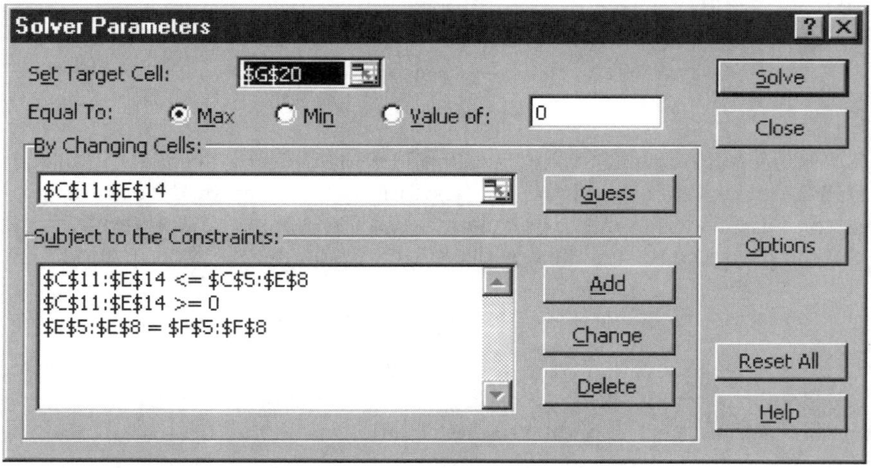

FIGURE 7.4 Solver parameters to maximize forest value.

during the 30 years. But it would produce 12,500 m³ more during the first decade. This is due to the fact that the present value formula gives more weight to early revenues than to late revenues. Still, the steady-state forest contributes $660,469, or about 11% of the total forest value.

In this example, the steady state after 30 years is a regulated forest. This steady-state forest has two age classes of 250 ha each. The oldest age class is cut and reforested every decade. Thus, the rotation in the steady-state forest is 20 years. This is also the economic rotation given by Faustmann's formula (Figure 7.2). The initial condition of the forest becomes less important as time goes by; only the land productivity determines long-term production, and its highest return is governed by Faustmann's formula.

7.5 EVEN-FLOW POLICY: COST AND BENEFITS

The model we have just developed can be extended to a wide range of applications where the economic effects of specific forest policies are of interest. As a first example, we apply the model to study the economic consequences of an even-flow policy. The *even-flow* or *nondeclining even-flow* principle states that forests should be managed in such a way that no more timber should be cut today than can be cut in the future.

The rationale for even-flow policies is to ensure stability of forest production. While the steady-state constraints ensure sustainability and stability in the long run, the even-flow principle seeks stability of production in the short run as well. The implicit belief is that an even flow of timber production will stabilize local economies. In addition, even-flow policies are expected to have environmental benefits because they tend to limit the size of the harvests, especially in old timber stands.

Within the example used so far, a strict even-flow policy means that the timber production must be the same in every decade, during the period considered explicitly in the plan, and after.

The steady-state constraints, which force the growing stock at the beginning of the third decade to be the same as at the beginning of the fourth decade, ensure that the production during the third decade, Q_3, can be sustainable in perpetuity. Thus, all that is needed to ensure strict even flow is that the production during the first and second decades also equal the production in the third decade:

$$Q_1 = Q_2 = Q_3$$

We can use the model developed so far to predict the effect of this strict even-flow policy on forest value. Figure 7.5 shows a spreadsheet model to maximize the forest value of the short-leaf pine forest in our example while

	A	B	C	D	E	F	G	H	I	J
1	MAX FOREST VALUE WITH EVEN-FLOW POLICY									
2	Age		Decade							
3	class		1	2	3	4				
4			Stock (ha)							
5		1	100	248	183	183				
6		2	200	100	183	183				
7		3	50	0	100	100				
8		4	150	152	33	33				
9							Yield	Price	Cost	Interest
10			Cut (ha)				(m³/ha)	($/m³)	($/ha)	(/y)
11		1	0	65	0		50	50	500	0.06
12		2	200	0	83		250			
13		3	48	0	67		500			
14		4	0	118	33		600			
15			Cut (m³)			Total				
16			74,163	74,163	74,163	222,488				
17			Net income ($)							
18			3,583,971	3,616,507	3,616,507					
19			Present value up to steady state ($)			After ($)	Total ($)			
20			2,001,271	1,127,644	629,671	796,197	4,554,782	max		
21										
22				Key cell formulas						
23	Cell		Formula				copied to			
24	D5		=SUM(C11:C14)				D5:F5			
25	D6		=C5-C11				D6:F7			
26	D8		=C7-C13+C8-C14				D8:F8			
27	C16		=SUMPRODUCT($G11:$G14,C11:C14)				C16:E16			
28	F16		=SUM(C16:E16)							
29	C18		=$H$11*C16-$I$11*SUM(C11:C14)				C18:E18			
30	C20		=C18/(1+$J$11)^(C3*10)				C20:E20			
31	F20		=(E18/((1+J11)^10-1))/(1+J11)^(E3*10)							
32	G20		=SUM(C20:F20)							

FIGURE 7.5 Spreadsheet model to maximize forest value with an even-flow policy.

following an even-flow policy. The initial state, the yield function, the state variables, the decision variables, and the growth equations are the same as in Figure 7.3.

The Solver parameters in Figure 7.6 show the usual steady-state constraints, the cut-less-than-stock constraints, and the nonnegative cut constraints. The new constraints specify that production is constant in every decade:

$$C16 = E16, \quad \text{corresponding to} \quad Q_1 = Q_3$$

$$D16 = E16, \quad \text{corresponding to} \quad Q_2 = Q_3$$

The solution that maximizes forest value under this strict even-flow policy is shown in Figure 7.5. It shows that strict even-flow could be achieved by producing 74,163 m³ every decade. The forest left at the end of the third decade

Economic Objectives and Environmental Policies for Even-Aged Forests

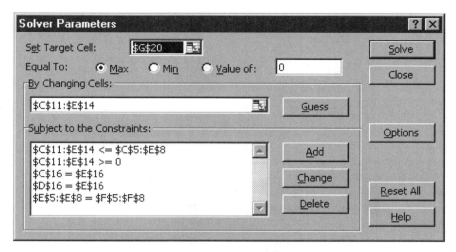

FIGURE 7.6 Solver parameters to maximize forest value with an even-flow policy.

could continue to produce this periodic harvest forever. The best forest value is about $4.5 million, 23% less than the forest value without the even-flow constraints. Thus, the even-flow policy has a substantial cost in terms of foregone timber production.

However, a strict even-flow policy also has benefits. Even-flow constraints eliminated the rapid decline in production in the second and third decades observed in Figure 7.3. At the scale of a state or country, an even-flow policy might smooth out the boom-and-bust cycles that have been so pervasive in forest history. It is to reduce those fluctuations and to foster economic stability that the even-flow principle has been advocated widely.

Another advantage of the even-flow policy is that it leads to a steady-state forest much more diverse than that obtained without the even-flow constraints. As seen in Figure 7.5, all four age classes are present in the steady state with the even-flow policy, compared to only age classes 1 and 2 without it (Figure 7.3). This diversity is valuable for esthetic reasons, and to provide a variety of habitats for a wide range of fauna and flora.

7.6 MIXED ECONOMIC AND ENVIRONMENTAL OBJECTIVES

As just seen, a policy that forces constant periodic timber production, such as the even-flow policy, may have indirect environmental benefits, such as conserving old trees and improving biological diversity. However, specific ecological objectives can be pursued more directly with similar models by making appropriate changes in the objective function and/or constraints. Meanwhile, having forest value as the objective function or as a constraint gives a useful measure of the trade-off between economic and ecological objectives.

Economically Efficient and Diverse Landscape

In Chapter 6 we studied a model to convert an even-aged forest to a steady state while maximizing landscape diversity. This was done by defining a new variable, A_{min}, equal to the smallest area in any age class in any period. The problem was then to find the sequence of harvesting and reforestation activities that maximized A_{min} while leading to a steady-state forest.

Another way of approaching this problem would be to treat A_{min} not as a variable to maximize, but as an input parameter, A_{min}^*. One would then seek the economically efficient management plan that would keep the area in any age class and period at or above A_{min}^*. By "economically efficient," we mean the plan that maximizes forest value, subject to the landscape diversity constraints.

Figure 7.7 shows the spreadsheet formulation of this problem for our example forest. The only change with respect to Figure 7.3 is in cell H6, which contains the value of $A_{min}^* = 100$ ha. So the solution must be such that there are at least 100 ha in any age class at the beginning of each decade.

Figure 7.8 shows the Solver parameters. The objective function, in G20, is the forest value, defined as earlier. In addition to the constraints that keep the growth less than the cut and the cut nonnegative, there are constraints forcing the stock in each age class to be at least 100 ha (D5:E8 >= H6).

Best Solution and Opportunity Cost

The best solution for this example is shown in Figure 7.7. As required, there are at least 100 ha in every age class at the beginning of each decade. Incidentally, this policy has also led to periodic timber production that varies little from decade to decade.

With the constraint on landscape diversity, the highest forest value is about $4.7 million. Comparing this result to the forest value of about $6.2 million obtained without landscape diversity constraint (Figure 7.3) shows that the opportunity cost of achieving landscape diversity is $1.5 million of foregone timber revenues. Thus, the environmental gains of landscape diversity should be worth at least $1.5 million to justify this policy; otherwise the lower bound, A_{min}^*, should be smaller than 100 ha.

Economic Objectives and Environmental Policies for Even-Aged Forests   123

	A	B	C	D	E	F	G	H	I	J
1	MAX FOREST VALUE WITH LANDSCAPE DIVERSITY									
2	Age		Decade							
3	class		1	2	3	4				
4			Stock (ha)					Lower bound		
5	1		100	150	150	150		on forest area:		
6	2		200	100	150	150		100	(ha)	
7	3		50	150	100	100				
8	4		150	100	100	100				
9								Yield	Price	Cost	Interest
10			Cut (ha)					(m^3/ha)	($\$/m^3$)	($\$$/ha)	(/y)
11	1		0	0	0			50	50	500	0.06
12	2		50	0	50			250			
13	3		0	50	0			500			
14	4		100	100	100			600			
15			Cut (m^3)			Total				
16			72,500	85,000	72,500	230,000				
17			Net income ($\$$)							
18			3,550,000	4,175,000	3,550,000					
19			Present value up to steady state ($\$$)			After ($\$$)	Total ($\$$)			
20			1,982,301	1,301,785	618,091	781,555	4,683,732	max		
21										
22			Key cell formulas							
23	Cell		Formula				copied to			
24	D5		=SUM(C11:C14)				D5:F5			
25	D6		=C5-C11				D6:F7			
26	D8		=C7-C13+C8-C14				D8:F8			
27	C16		=SUMPRODUCT($G11:$G14,C11:C14)				C16:E16			
28	F16		=SUM(C16:E16)							
29	C18		=$H$11*C16-$I$11*SUM(C11:C14)				C18:E18			
30	C20		=C18/(1+$J$11)^(C3*10)				C20:E20			
31	F20		=(E18/((1+J11)^10-1))/(1+J11)^(E3*10)							
32	G20		=SUM(C20:F20)							

FIGURE 7.7 Spreadsheet model to maximize forest value with landscape diversity constraints.

Alternatively, if the opportunity cost is less than the environmental gain, A^*_{min} could be increased. However, there are definite limits to the possible values of A^*_{min}. For example, verify that with the initial state assumed in Figure 7.7 it is not feasible to manage the forest to keep at least 150 ha in stock in every age class and every period.

7.7 GENERAL FORMULATION

The model studied in this chapter uses the same representation of forest growth as that in Chapter 6. We describe the state of an even-aged forest at successive periodic inventories by the area in each of m age classes. The time unit to classify

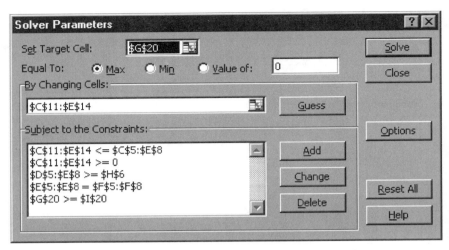

FIGURE 7.8 Solver parameters to maximize forest value with a landscape diversity constraint.

age classes is the length of the period between inventories, u. A_{ij} is the area in age class i at the jth inventory.

The forest is managed for p periods. At the end of this conversion period we want to leave a sustainable forest. The management plan is defined by the decision variables X_{ij}, the area cut and reforested from age class i in period j.

The growth equations, not repeated here, describe the state of the forest at the start of each successive period as a function of the initial condition and of the prior decisions.

As in Chapter 6, the area cut and reforested must be nonnegative, and the area cut must be less than the stock in each age class and period. The steady-state constraints ensure that the forest left at the end of the plan is sustainable.

The objective function used throughout this chapter is the forest value, FV. This is the net present value of the income that the forest would produce over an infinite horizon. It consists of the present value of the income up to the steady state, PV_c, and of the present value after the steady state, PV_∞:

$$FV = PV_c + PV_\infty$$

where:

$$PV_c = \sum_{j=1}^{p} \frac{N_j}{(1+r)^{ju}}$$

and

$$PV_\infty = \frac{N_p}{[(1+r)^u - 1](1+r)^{pu}}$$

where $N_j = R_j - C_j$ is the net income, R_j is the gross revenue, and C_j is the reforestation cost during period j:

$$R_j = wQ_j = w\sum_{i=1}^{m} v_i X_{ij} \quad \text{and} \quad C_j = c\sum_{i=1}^{m} X_{ij}$$

where w is the unit price of timber, Q_j is the timber harvest in period j, v_i is the volume per unit area in age class i, and c is the reforestation cost per unit area.

The forest value FV is a linear function of the decision variables, X_{ij}; thus it can be used as the objective function in a linear program to seek the management plan that converts a particular initial forest to a steady state within a specified amount of time while maximizing the forest value.

This maximum forest value can be viewed as the value of the initial stock and of the land for a forest managed purely for timber production, with the only constraint being to leave a sustainable regime at the end of the management plan.

Other constraints may be imposed to control timber production. For example, under the even-flow policy, the production of the forest should be constant before and after the steady state. This is expressed by these constraints:

$$Q_1 = Q_2 = \cdots = Q_p$$

Environmental goals and their effects on forest value can be investigated by adding other constraints. For example, landscape diversity of the forest can be ensured by forcing the area in stock in every age class and period to be at least equal to a threshold value, A^*_{min}, with the following set of constraints:

$$A_{ij} \geq A^*_{min} \quad i = 1,\ldots,m; \quad j = 2,\ldots,p+1$$

7.8 SUMMARY AND CONCLUSION

The dynamic model of even-aged forest management developed in Chapter 6 has been expanded in this chapter to investigate the economic consequences of different management regimes and objectives. According to Faustmann's theory, the value of bare forest land, or land expectation value, is equal to the net present value of the income that the land is capable of producing over an infinite horizon. The economic rotation for an even-aged stand is the rotation that maximizes the land expectation value. Faustmann's formula is useful to compare forest management regimes and to compare land uses.

Similarly, the value of a complex even-aged forest, including land and trees, is, from the point of view of timber production, the present value of the income that the forest will produce in the future. For an even-aged forest consisting of a mosaic of age classes, the management leading to the highest forest value can be found by linear programming. The objective function is the sum of the present value of the income up to the steady state and the present value of the income after the steady state.

Under a strict even-flow policy, periodic production is held constant before and after reaching the steady state. The highest forest value with an even flow may be much lower than the forest value obtained without these constraints. However, even flow can yield environmental benefits because it tends to result in older and more diverse forests.

Various environmental objectives can be investigated with the model developed in this chapter, by changing the constraints or the objective function. In particular, the forest value defined in this chapter may be a constraint, or it may be the objective function. Setting it as the objective function does not necessarily mean that timber production has the highest priority. Recall that linear programming finds the best solution within the feasible region defined by all the constraints. Even if forest value is optimized, the constraints reflecting other goals may reduce forest value drastically.

The main benefit of computing forest value, regardless of the forest management goals, is to assess the opportunity cost of decisions in terms of the value of timber income foregone, a value that is well defined by markets. Once a decision is made, the implication is that the value of the tangible and intangible benefits is at least equal to the opportunity cost in terms of decreased timber production.

PROBLEMS

7.1 Set up your own spreadsheet model for computing land expectation value like the one shown in Figure 7.2. Using the same data, verify that your results are the same as in Figure 7.2.

(a) Reduce the interest rate from 6% to 3%, and then increase it to 10%. How does this change the land expectation value for a given rotation?

(b) How does this change the economic rotation?

(c) For an interest rate of 3%, how does decreasing the price of timber from $50/m^3 to $25/m^3 change the land expectation value for a given rotation?

(d) For an interest rate of 3%, how does decreasing the price of timber from $50/m^3 to $25/m^3 change the economic rotation?

(e) For an interest rate of 3% and a timber price of $50/m³, how does decreasing the reforestation cost from $500/ha to $250/ha affect the land expectation value for a given rotation?

(f) How does this change the economic rotation?

7.2 Set up your own spreadsheet model for computing forest value like the one shown in Figure 7.3. Using the same data, verify that your results are the same as in Figure 7.3.

(a) With the data in Figure 7.3, compute the coefficients of the harvest variables (X_{ij}) in the objective function. The objective function, the forest value, is the sum of the present value up to and after the steady state.

(b) Verify that substituting the solution values of the harvest variables into the objective function obtained in part (a) gives the forest value.

7.3 With the spreadsheet model to maximize forest value (Figure 7.3), change the interest rate from 6% per year to 3% per year. How does this affect the forest value, the production per decade, and the structure of the steady-state forest?

7.4 Consider the forest represented by the spreadsheet model in Figure 7.3. Suppose that for environmental reasons you want to transform this forest into a steady-state forest, with the largest possible area in the oldest age class.

(a) Modify the model to maximize the area in the oldest age class in the steady-state forest.

(b) How does this change in objective affect the forest value, the periodic production, and the structure of the steady-state forest?

(c) What is the opportunity cost of this policy?

7.5 Consider the spreadsheet model for maximizing forest value with an even-flow policy (Figure 7.5). Because the strict even-flow policy has a high opportunity cost, you consider an alternative policy that would produce a forest value of at least $5 million, lead to a steady state in 30 years, and minimize the total volume cut. Modify the spreadsheet model to reflect this new policy by:

(a) Removing the even-flow constraints.

(b) Adding a constraint that the forest value must be at least $5 million.

(c) Changing the objective function to minimize the total volume cut during 30 years. Discuss how these changes affect the forest value and the periodic production in the best solution.

7.6 Consider the forest represented by the spreadsheet model in Figure 7.3. Assume that you are concerned with the possible disruption harvesting would cause to wildlife species associated with each age class.

(a) Modify the objective of the model to find the management plan that minimizes the largest area cut in every period and age class. (*Hint:* Create a new variable, C_{max}, corresponding to the largest area cut in every period and age class, and set the objective function to minimize C_{max}. Add constraints to ensure that C_{max} is the largest area in every period and age class.

(b) How does this new policy change the forest value, the periodic production, and the structure of the steady-state forest?

(c) What is the opportunity cost of this policy?

(d) Why is there any harvesting at all in the solution to this modified model?

7.7 Assume that you are the supervisor of a national forest ranger district in the Pacific Northwest. Your task is to develop a long-term management plan for the district. Your objective is to maintain a diverse forest landscape in order to protect the forest against epidemics and fires and to provide habitat for varied wildlife and for esthetic reasons. The property currently consists of a mosaic of even-aged stands of Douglas fir. The stands are classified in age classes of 20 years each. Many of the stands are old growth (more than 140 years old). In your management, harvesting is immediately followed by planting the land with 1-year-old seedlings. The current and anticipated price of timber is $50/m^3, the cost of reforestation is $500/ha, and the interest rate is 3% per year. Anticipated per-acre yields depend only on stand age, as shown in this table along with the forest's initial state:

Initial State and Expected Yields for a Douglas Fir Forest

Age class	Age (y)	Yield (m³/ha)	Initial area (ha)
1	1–20	3	500
2	21–40	27	500
3	41–60	57	500
4	61–80	73	500
5	81–100	87	1,000
6	101–120	97	1,000
7	121–140	103	2,000
8	141+	110	4,000

(a) Modify the spreadsheet in Figure 7.3 to predict the forest state up to the beginning of period 6 (that is, after 100 years). Check the growth equations by showing how this forest would grow if you cut and reforested one-fourth of the forest every 20 years, always cutting from the oldest age classes.

(b) Use this modified spreadsheet to find the management plan that maximizes forest value while leading to a sustainable forest.

(c) What is the forest structure in period 5? How does it score in terms of landscape diversity?

(d) Modify this spreadsheet to determine the plan that maximizes landscape diversity, in the sense of maximizing the smallest area in any age class and period. The plan should also end with a sustainable forest.

(e) What is the opportunity cost of the plan that maximizes landscape diversity?

(f) Modify the spreadsheet again to determine a compromise plan that would maximize landscape diversity without allowing forest value to be less than half of the unconstrained maximum forest value found in part (b).

(g) Compare this compromise plan with those that maximize diversity or forest value.

ANNOTATED REFERENCES

Armstrong, G.W., J.A. Beck, Jr., and B.L. Phillips. 1984. Relaxing even-flow constraints to avoid infeasibility with the Timber Resources Allocation Method (RAM). *Canadian Journal of Forest Research* 14:860–863. (Discusses the implications of constraints linking harvest flows between planning periods.)

Binkley, C.S. 1984. Allowable cut effects without even flow constraints. *Canadian Journal of Forest Research* 14:317–320. (Discusses the implications of constraints linking harvest flows between planning periods.)

Conrad, J.M. 1999. *Resource Economics*. Cambridge University Press, Cambridge, UK. 213 pp. (Chapter 4 discusses determination of the Faustmann rotation.)

Countryman, D.W. 1989. Investment analysis of upland oak stands with sugar maple understories: Management for oak vs. conversion to sugar maple in Iowa and Missouri. *Northern Journal of Applied Forestry* 6:165–169. (Application of LEV calculations for alternative forest management regimes.)

Davis, L.S., K.N. Johnson, P.S. Bettinger, and T.E. Howard. 2001. *Forest Management: To Sustain Ecological, Economic, and Social Values*. McGraw-Hill, New York. 804 pp. (Chapter 10 discusses different kinds of harvest flow constraints.)

Faustmann, M. 1995. Calculation of the value which forest land and immature stands possess for forestry. *Journal of Forest Economics* 1(1):7–44. (Translation of an 1849 article establishing the land expectation value principle used in this chapter).

Hann, D.W., and J.D. Brodie. 1980. *Even-Aged Management: Basic Managerial Questions and Available or Potential Techniques for Answering Them*. U.S. Forest Service General Technical Report INT-83. Intermountain Forest and Range Experiment Station, Ogden, UT. 29 pp. (Review of the different approaches that have been or could be used to schedule harvests for even-aged forests.)

Holland, D.N., R.J. Lilieholm, D.W. Roberts, and J.K. Gilless. 1994. Economic trade-offs of managing forests for timber production and vegetative diversity. *Canadian Journal of Forest Research* 24:1260–1265. (Describes a linear programming model incorporating diversity stand species, basal area, and vertical crown diversity indexes.)

Öhman, K. 2000. Creating continuous areas of old forest in long-term forest planning. *Canadian Journal of Forest Research* 30:1817–1823. (Describes a harvest-scheduling model designed to optimize the unfragmented areas of old-growth forests.)

O'Toole, R. 1988. *Reforming the Forest Service*. Island Press, Washington, DC. 247 pp. (Critique of USDA Forest Service policies, including even-flow and related policies).

Pickens, J.B., B.M. Kent, and P.G. Ashton. 1990. The declining even-flow effect and the process of national forest planning. Forest Science 36(3):665–679. (Discusses some of the effects of the highly constrained linear programming models used by the U.S. Forest Service for forest resource management planning.)

USDA Forest Service. 1999. Spectrum users guide. Available at: http://www.fs.fed.us/imi/planning_center/download_center.html (Linear programming model successor of FORPLAN, to optimize land allocation and harvest scheduling.)

CHAPTER 8

Managing the Uneven-Aged Forest with Linear Programming

8.1 INTRODUCTION

In an uneven-aged (or selection) forest, many trees of different age and size coexist on small tracts of land. In contrast with the even-aged forest, distinct areas of homogeneous age-classes cannot be distinguished. However, the ideal uneven-aged forest, where trees of all ages appear on the same acre, is rare. Trees may be grouped in patches of similar age, but these patches are usually too small to be administered like the even-aged compartments that we dealt with in previous chapters.

Large contiguous tracts are never clear-cut in the uneven-aged forest. Rather, one selects single trees or group of trees within stands. Consequently, unlike even-aged stands, uneven-aged stands have no beginning and no end. There are always trees left on each hectare of the uneven-aged forest, even immediately after harvest. For this reason, uneven-aged management is sometimes referred to as *continuous-cover forestry*. The French term for the uneven-aged forest is *la futaie jardinée*, or "the garden forest," reflecting the care in management and the high value that mountain communities in the Jura and in the Alps place on their forests, not only as sources of income, but also for protection against avalanches and erosion.

In uneven-aged silviculture, regeneration is usually, though not necessarily, natural. It comes from the stock of saplings in the understory emerging through the openings left by cutting the large trees. Therefore, this form of management works best with trees that are shade tolerant, for example, maples, hemlocks, cedars, spruces, and firs. Nevertheless, many forests of ponderosa pine in the West of the United States are uneven aged, despite the species needs for enough light for good regeneration. In this case, instead of the pure form of selection cutting, trees are cut in small patches, leading to an overall structure that is essentially uneven aged for management purposes.

Uneven-aged management leads to a forest with a more natural aspect than its even-aged counterpart. For that reason, it is very attractive for forests managed for multiple uses, including recreation. It is often the only form of cutting that is acceptable for small, private woodlots. In that case, a good-looking forest not only pleases the owners, but often enhances the value of their property.

Unfortunately, uneven-aged management is often believed to be inferior from a purely economic point of view. This is certainly not the case for a woodlot in which timber production is only a secondary object. But even for pure timber management, the case against uneven-aged management is doubtful. Starting a new crop of good trees is often the most costly operation in forestry. This cost is minimal in uneven-aged systems with natural regeneration.

On the other hand, the costs of harvesting, per unit of volume, are generally greater in a selection forest. There are two reasons for this. First, more area must be covered to extract a given volume than is the case with clear-cutting. This means higher costs for roads and for movement of machinery and people. Second, the felling of trees and the hauling of logs are more delicate operations in a selection forest. Care must be taken not to damage the trees that are left, especially young saplings that will constitute the future crops. This is a labor-intensive process that only skilled labor can do and that is difficult to mechanize. For this reason, uneven-aged management is most appropriate for the production of large trees yielding expensive, high-quality timbers for which the cost of harvesting represents only a small part of the value of the final product.

Perhaps because they are more complex than even-aged systems, uneven-aged forest systems have not been studied as much. Relatively few models of selection forests exist, and little is known about the real economics of these forests for timber production. The object of this chapter is to study such a model and to use it to investigate problems of interest to forest managers. These problems include the length of the cutting cycle, that is, the interval between successive cuttings on a given tract of land, and the intensity of the cut, that is, the number and size of trees to be removed, if any, given the owner's economic and environmental objectives.

8.2 A GROWTH MODEL OF THE UNEVEN-AGED FOREST STAND

The model presented here deals with an uneven-aged stand. A stand is an area of forest that can be treated as a unit because it has uniform land quality, topography, and species composition. Typically, a stand is no less than 1 ha and no more than 20 ha in area. The state of a stand is described by the size distribution of trees on an average hectare. Usually, this distribution is determined from a few sample plots rather than by exhaustive counting of all the trees.

The first three columns of Figure 8.1 show the size distribution for a sugar maple stand in Wisconsin that we shall use throughout this chapter. For simplicity, only three size (diameter) classes were used. This may be enough for some purposes, for example, if the trees are classified as poles, small saw timbers, or large saw timbers. For other purposes, there may be a dozen or more size classes. The trees may also be distinguished by species or species categories, such as shade-intolerant vs. shade-tolerant trees.

The data in cells C3:C5 show the typical "inverse J"-shape size distribution found in uneven-aged stands: The number of trees per unit area decreases progressively as the size of the trees increases. Cells E3:E5 show the basal area of an average tree in each size class, which is the area of the cross section of the tree, measured at breast height. Cells F3:F6 show the basal area per hectare of the trees in each size class, as well as the total basal area per hectare. The basal area

	A	B	C	D	E	F
1	TREE DISTRIBUTION					
2	Diameter class	Diameter range (cm)	Number of trees (/ha)	Average diameter (cm)	Basal area of average tree (m^2)	Total basal area (m^2/ha)
3	1	10-19.9	840	15	0.02	14.8
4	2	20-34.9	234	27	0.06	13.4
5	3	35+	14	40	0.13	1.8
6	Total		1088			30.0
7						
8			*Key cell formulas*			
9	*Cell*	*Formula*			*Copied to*	
10	E3	=PI()*(D^2)/40000			E3:E5	
11	F3	=E3*C3			F3:F5	
12	F6	=SUM(F3:F5)				
13	C6	=SUM(C3:C5)				

FIGURE 8.1 Spreadsheet to compute the basal area distribution of a stand of trees.

is a useful measure of the forest density per unit of land. It plays an important part in the model studied here, and it is relatively easy to measure in the field.

The tree size distribution in Figure 8.1 represents the stand state at a particular point in time, the time when the stand has been inventoried. Over time, the stand state changes because some trees die, some are cut, and new trees appear in the smallest size class.

To represent the general stand *state* at a given point in time, t, we use three variables, y_{1t}, y_{2t}, and y_{3t}, where y_{it} is the number of trees per acre in size class i at time t.

The stand growth model is a set of equations that predicts the state of the stand at time $t + 1$, given its current state. The time from t to $t + 1$ is a fixed unit of one or more years. There is one equation for each size class:

$$\begin{aligned} y_{1,t+1} &= a_1 y_{1t} + R_t \\ y_{2,t+1} &= b_1 y_{1t} + a_2 y_{2t} \\ y_{3,t+1} &= \phantom{b_1 y_{1t} + {}} b_2 y_{2t} + a_3 y_{3t} \end{aligned} \tag{8.1}$$

where the variable R_t in the first equation stands for the *recruitment* (or *ingrowth*), the number of young trees that enter the smallest size class during the interval t to $t + 1$.

Each parameter a_i is the fraction of live trees in size class i at time t that are still alive and in the same size class at time $t + 1$. Each parameter b_i is the fraction of live trees in size class i at time t that are still alive and have grown into size class $i + 1$ at time $t + 1$.

Consequently, the fraction of trees in age class i at time t that are dead at time $t + 1$ is $1 - a_i - b_i$, because a tree can only remain in the same class, grow into a larger class, or die. The time unit used is short enough that no tree can skip one size class.

Table 8.1 shows examples of values of the parameters a_i and b_i for the stand summarized in Figure 8.1. The parameters are based on observations from permanent plots in sugar maple stands in Wisconsin. In Table 8.1, $a_1 = 0.80$ and $b_1 = 0.04$ mean that 80% of the trees in the smallest size class were in the same class 5 years later, while 4% of the trees in the smallest size class grew into the larger class in 5 years. The remaining 16% died.

TABLE 8.1 Proportion of Trees Staying in the Same Size Class, Growing into the Next Size Class, or Dying Within 5 Years

Size class, i	Proportion staying, a_i	Proportion growing up, b_i	Proportion dying, $1 - a_i - b_i$
1	0.80	0.04	0.16
2	0.90	0.02	0.08
3	0.90	0.00	0.10

Managing the Uneven-Aged Forest with Linear Programming

Assuming that the rates of transition in Table 8.1 are constant and independent of the stand state, and substituting them in the growth Equations (8.1) leads to:

$$\begin{aligned} y_{1,t+1} &= 0.80 y_{1t} + R_t \\ y_{2,t+1} &= 0.04 y_{1t} + 0.90 y_{2t} \\ y_{3,t+1} &= \phantom{0.04 y_{1t} +} 0.02 y_{2t} + 0.90 y_{3t} \end{aligned} \qquad (8.2)$$

To complete the model, we need an expression of the recruitment, R_t. The simplest option is to assume that it is constant. However, biometric studies have shown that although recruitment is very erratic, it is influenced by the stand state. For the sugar maple forests in our example, Buongiorno and Michie (1980) found that recruitment was affected mostly by the stand basal area and by the number of trees per hectare. More precisely, the average relation between recruitment, basal area, and number of trees was:

$$\underset{(\text{tree/ha/5 y})}{R_t} = 109 - 9.7 \underset{(\text{m}^2/\text{ha})}{B_t} + 0.3 \underset{(\text{tree/ha})}{N_t} \qquad (8.3)$$

where B_t is the stand basal area at t, N_t is the number of trees, and the recruitment is measured over a 5-year period. The equation shows that, other things being equal, recruitment tends to be lower on stands of high basal area and larger on stands with many trees.

In this form, recruitment is a function only of the state variables y_{1t}, y_{2t}, and y_{3t}, because:

$$N_t = y_{1t} + y_{2t} + y_{3t} \qquad (8.4)$$

and

$$\underset{(\text{m}^2/\text{ha})}{B_t} = 0.02 \underset{(\text{m}^2/\text{tree})}{y_{1t}} + 0.06 y_{2t} + 0.13 y_{3t} \qquad (8.5)$$

where each coefficient is the basal area of the average tree in the corresponding size class (see Figure 8.1). Substituting Equations (8.4) and (8.5) in Equation (8.3) then gives the expression of recruitment as a linear function of the state variables:

$$R_t = 109 + 0.12 y_{1t} - 0.29 y_{2t} - 0.96 y_{3t}$$

Substituting this expression for recruitment in the first of Equations (8.2) gives the final growth model:

$$\begin{aligned} y_{1,t+1} &= 0.92 y_{1t} - 0.29 y_{2t} - 0.96 y_{3t} + 109 \\ y_{2,t+1} &= 0.04 y_{1t} + 0.90 y_{2t} \\ y_{3,t+1} &= \phantom{0.04 y_{1t} +} 0.02 y_{2t} + 0.90 y_{3t} \end{aligned} \qquad (8.6)$$

This basic growth model involves only variables describing the state of the stand at times t and $t + 1$. We shall use it in the next section to describe the growth of a stand without human or catastrophic disturbance. Then we shall use the model to determine the best cutting regime given different management objectives.

8.3 PREDICTING THE GROWTH OF AN UNMANAGED STAND

STAND DYNAMICS

Let $y_{1,0}$, $y_{2,0}$, and $y_{3,0}$ be a particular initial stand condition for the uneven-aged stand for which we have just developed the growth equations. This is the stand state at time $t = 0$. We would like to predict its future state with the growth and death processes embedded in the model parameters, assuming no other disturbance. To do this we can apply the basic growth model of Equations (8.6) iteratively.

For example, to predict the undisturbed growth of the sugar maple forest displayed in Figure 8.1 we set the initial conditions at:

$$y_{1,0} = 840, \quad y_{2,0} = 234, \quad y_{3,0} = 14 \quad \text{(trees/ha)}$$

Substituting these initial conditions into the growth equations gives the stand state after 5 years:

$$y_{11} = 0.92 \times 840 - 0.29 \times 234 - 0.96 \times 14 + 109 = 801 \text{ (trees/ha)}$$

$$y_{21} = 0.04 \times 840 - 0.90 \times 234 \qquad\qquad = 244 \text{ (trees/ha)}$$

$$y_{31} = \qquad\qquad 0.02 \times 234 + 0.90 \times 14 \quad = 17 \text{ (trees/ha)}$$

Thus, the model predicts that after 5 years there would be fewer trees in size class 1 and more in size classes 2 and 3. Substituting these values of y_{11}, y_{21}, and y_{31} back into the growth equations gives the stand state after 10 years, y_{12}, y_{22}, and y_{32}. We can proceed in this manner for as long as we want.

Figure 8.2 shows how the recursive growth equations can be programmed in a spreadsheet. Regardless of how the computations are done, the approach is general and can be applied to a model with as many size classes as necessary. Figure 8.2 shows predictions of number of trees per acre, starting with the stand described in Figure 8.1. That stand was logged heavily in the recent past. As a result, there were initially many trees in the smallest size class. Were the stand to grow undisturbed for 75 years, the number of trees in the smallest size class would decline, the number in the middle size class would remain about constant, and that in the largest size class would increase.

	A	B	C	D	E	F	G	H
1	STAND DYNAMICS WITHOUT HARVEST							
2		Stock (trees/ha)						
3	Year	Size1	Size2	Size3				
4	0	840	234	14				
5	5	801	244	17				
6	10	758	252	20				
7	15	714	257	23				
8	20	669	260	26				
9	25	624	261	29				
10	30	580	259	31				
11	35	537	257	33				
12	40	497	253	35				
13	45	459	247	37				
14	50	425	241	38				
15	55	394	234	39				
16	60	366	226	40				
17	65	342	218	40				
18	70	322	210	41				
19	75	305	202	41				
20								
21				Key cell formulas				
22	Cell	Formula					Copied to	
23	A5	=A4+5					A6:A19	
24	B5	=0.92*B4-0.29*C4-0.96*D4+109					B6:B19	
25	C5	=0.04*B4+0.90*C4					C6:C19	
26	D5	=0.02*C4+0.90*D4					D6:D19	

FIGURE 8.2 Spreadsheet to predict the growth of an undisturbed forest stand.

STEADY STATE

The spreadsheet in Figure 8.2 projects the number of trees in each size class 75 years into the future. Pursuing the calculations further would show number of trees and basal areas that oscillate with very long periods, but at decreasing amplitude and ultimately converging toward a steady state in which the stand remains unchanged forever. The steady state is independent of the initial stand condition (see Problems 8.2 and 8.3 at the end of this chapter). Biologically, this steady state corresponds to the undisturbed climax forest, where the growth would just replace mortality and the average stand structure per unit area would change little over time.

There is a more direct way to determine the steady-state forest. By definition, a steady state means that, regardless of when the stand is observed, it always has the same number of trees in each size class. That is, in our example:

$$y_{i,t+1} = y_{it} = y_i \quad \text{for } i = 1, 2, 3 \text{ and for all } t$$

Substituting for $y_{i,t+1}$ and y_{it} with their unknown steady-state value, y_i, in the growth model of Equations (8.6) gives:

$$y_1 = 0.92 y_1 - 0.29 y_2 - 0.96 y_3 + 109$$
$$y_2 = 0.04 y_1 + 0.90 y_2 \quad (8.7)$$
$$y_3 = 0.02 y_2 + 0.90 y_3$$

This is a system of three equations in three unknowns that can be solved by substitution. The third equation yields $y_2 = 5 y_3$ and the second yields $y_1 = 2.5 y_2$, which implies that $y_1 = 12.5 y_3$. This shows that in the steady-state stand, there are 12.5 times as many trees in the smallest size class as in the largest, and 2.5 times as many as in the intermediate. Then, replacing y_1 and y_2 in the first equation by their expression in terms of y_3 gives $y_3 = 32.0$, which in turn implies $y_2 = 159.8$ and $y_1 = 399.6$ trees/ha.

The steady-state distribution has the classical inverse J shape of uneven-aged stands. However, compared with the initial forest, it has fewer trees in the two smallest size classes and more in the largest. This is plausible since the stand we started with is currently being managed and has its largest trees removed periodically.

A more powerful and informative way to calculate the steady state is by linear programming. The object is to find the number of trees at time t: y_{1t}, y_{2t}, y_{3t}, subject to:

Growth equations:

$$y_{1,t+1} = 0.92 y_{1t} - 0.29 y_{2t} - 0.96 y_{3t} + 109$$
$$y_{2,t+1} = 0.04 y_{1t} + 0.90 y_{2t}$$
$$y_{3,t+1} = \phantom{0.04 y_{1t} +} 0.02 y_{2t} + 0.90 y_{3t}$$

Steady-state constraints:

$$y_{1,t+1} = y_{1t}$$
$$y_{2,t+1} = y_{2t} \quad (8.8)$$
$$y_{3,t+1} = y_{3t}$$

Managing the Uneven-Aged Forest with Linear Programming 139

Nonnegativity constraints:

$$y_{it} \geq 0 \quad \text{for } i = 1, 2, 3 \quad (8.9)$$

(Note that, together, constraints (8.8) and (8.9) also ensure that $y_{i,t+1} \geq 0$.)

Any function of the state variables will do as the objective function because the system of Equations (8.6) and (8.8) has a unique solution, which we found earlier by substitution (for that reason, the nonnegativity constraints are not strictly necessary). For example, we could set the objective as maximizing the number of trees in the largest size class (max y_{3t}) or minimizing the number of trees in the medium size class (min y_{2t}) and get the same unique solution for the steady state.

Figure 8.3 shows a spreadsheet to compute the steady state for this example. Cells D5:D7 contain the growth formulas predicting the stand state at time $t + 1$ given its state at time t. Figure 8.4 shows the corresponding Solver parameters. Here, the arbitrary target cell contains the number of trees in the medium age class, which is minimized. The changing cells are the number of trees in each size class at time t.

The first set of constraints forces the state at time t to be the same as the state at time $t + 1$. The second set of constraints requires that the number of trees in each size class be nonnegative.

The Solver solution is shown in Figure 8.3. This solution is unique and is independent of the objective function. In effect, the Solver role in this case is to find the solution of a system of linear equations. Regardless of the initial condition of the stand, the growth and mortality processes embodied in the growth equations will lead to this steady state after a sufficiently long period of time. Figure 8.2 suggests that in this example the steady state would be nearly reached after a century.

	A	B	C	D	E
1	STEADY STATE WITHOUT HARVEST				
2			Stock		
3	Size	y_t		y_{t+1}	
4	class	(trees/ha)		(trees/ha)	
5	1	399.6		399.6	
6	2	159.8		159.8	min
7	3	32.0		32.0	
8					
9			Key cell formulas		
10	Cell	Formula			
11	D5	=0.92*B5-0.29*B6-0.96*B7+109			
12	D6	=0.04*B5+0.90*B6			
13	D7	=0.02*B6+0.90*B7			

FIGURE 8.3 Spreadsheet to compute the steady state of an undisturbed uneven-aged stand.

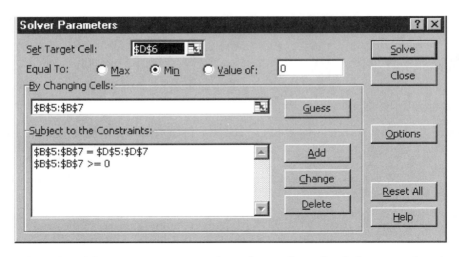

FIGURE 8.4 Solver parameters to compute the steady state of an undisturbed uneven-aged stand.

8.4 GROWTH MODEL FOR A MANAGED STAND

We shall now adapt the model just developed to predict the growth of a stand with periodic harvesting. This will be done in two steps, first establishing the relationships that govern the growth of a managed uneven-aged stand, and then determining the equations that define the steady state for such a stand.

STAND DYNAMICS WITH HARVEST

The number of trees cut per unit area in each size class describes the harvest at a certain point in time, t. In our example the harvest is designated by three variables, h_{1t}, h_{2t}, and h_{3t}, where h_{it} is the number of trees cut from size class i at time t. The number of trees left after the cut in each size class i is thus $y_{it} - h_{it}$. These remaining trees develop according to the growth Equations (8.6). Consequently, the growth of an uneven-aged stand that is cut periodically is described by Equations (8.6), where each y_{it} is replaced by $y_{it} - h_{it}$:

$$y_{1,t+1} = 0.92(y_{1t} - h_{1t}) - 0.29(y_{2t} - h_{2t}) - 0.96(y_{3t} - h_{3t}) + 109$$
$$y_{2,t+1} = 0.04(y_{1t} - h_{1t}) + 0.90(y_{2t} - h_{2t}) \qquad (8.10)$$
$$y_{3,t+1} = \phantom{0.04(y_{1t} - h_{1t}) +} 0.02(y_{2t} - h_{2t}) + 0.90(y_{3t} - h_{3t})$$

Managing the Uneven-Aged Forest with Linear Programming

	A	B	C	D	E	F	G	H	I	J	K	L
1	STAND DYNAMICS WITH HARVEST											
2		Stock (trees/ha)				Cut (trees/ha)				Proportion cut		
3	Year	Size1	Size2	Size3		Size1	Size2	Size3		0.20	0.15	0.1
4	0	840	234	14		168	35	1				
5	5	657	206	15		131	31	2				
6	10	529	179	16		106	27	2				
7	15	441	154	16		88	23	2				
8	20	382	132	16		76	20	2				
9	25	344	113	15		69	17	1				
10	30	322	97	14		64	15	1				
11	35	310	85	13		62	13	1				
12	40	305	75	12		61	11	1				
13	45	305	67	11		61	10	1				
14	50	307	61	10		61	9	1				
15	55	312	56	9		62	8	1				
16	60	317	53	8		63	8	1				
17	65	322	51	8		64	8	1				
18	70	327	49	7		65	7	1				
19	75	331	48	7		66	7	1				
20												
21					Key cell formulas							
22	Cell	Formula									Copied to	
23	A5	=A4+5									A5:A19	
24	B5	=0.92*(B4-F4)-0.29*(C4-G4)-0.96*(D4-H4)+109									B5:B19	
25	C5	=0.04*(B4-F4)+0.90*(C4-G4)									C5:C19	
26	D5	=0.02*(C4-G4)+0.90*(D4-H4)									D5:D19	
27	F4	$J$3*B4									F4:F19	
28	G4	$K$3*C4									G4:G19	
29	H4	$L$3*D4									H4:H19	

FIGURE 8.5 Spreadsheet to predict uneven-aged stand growth with a fixed-proportion periodic harvest.

This system of recursive equations describes the evolution of the stand under any sequence of harvests, regardless of their timing and level, as long as $h_{it} \leq y_{it}$.

For example, Figure 8.5 shows a spreadsheet to predict the growth of the stand, with a fixed-proportion periodic harvest. Specifically, the cut occurs every 5 years, starting at year 0. The cutting rule is always to take 20% of the smallest trees, 15% of the mid-size trees, and 10% of the largest trees.

As a result of this management regime, the forest has reached a near steady state after 75 years (note that the stock as well as the harvest in each size class change little between 65 and 75 years). In fact we shall see later that there is an infinite number of management regimes capable of maintaining a steady state.

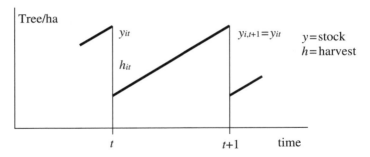

FIGURE 8.6 In the steady state, growth replaces harvest within a cutting cycle.

In the next section we shall study how to define a managed steady state and how to choose one that is best for specific management objectives.

STEADY STATE FOR A MANAGED STAND

We shall assume for now that the interval between harvests, or the cutting cycle, has the same length as the time unit in the growth equations. In our example, the interval between t and $t+1$ in the growth equations is 5 years, and thus the cutting cycle is also five years long.

A managed uneven-aged stand is in a steady state if the trees harvested in each size class are just replaced by the growth of the stand between harvests (Figure 8.6). As observed earlier, the growth of a managed stand between harvests is governed by Equations (8.10).

The steady state means that the stand state must be the same at the end as at the beginning of the cutting cycle:

$$y_{i,t+1} = y_{it} \quad \text{for } i = 1, 2, 3 \tag{8.11}$$

Furthermore, the cut must be less than the stock:

$$h_{it} \le y_{it} \quad \text{for } i = 1, 2, 3 \tag{8.12}$$

And the cut must be nonnegative:

$$h_{it} \ge 0 \quad \text{for } i = 1, 2, 3 \tag{8.13}$$

(Note that, together, Constraints (8.13) and (8.12) ensure that, $y_{it} \ge 0$; together with Constraint (8.11) this also ensures that $y_{i,t+1} \ge 0$.)

There is an infinite number of solutions to the system defined by Equations and Inequalities (8.10) to (8.13) and therefore an infinite number of feasible steady states. This gives the manager the opportunity to choose the steady state that is best for a particular purpose.

8.5 OPTIMIZING UNEVEN-AGED STANDS

A wide range of management objectives can be explored with the model we have just set up. The first example we will consider seeks the management that maximizes the production per unit of time, parallel to the search for the maximum mean annual increment in even-aged management. The second will find the management leading to the climax forest with the largest possible number of "big trees," for environmental reasons, especially esthetics.

MAXIMIZING SUSTAINABLE PRODUCTION

A classical goal of sustained-yield management is to maximize the volume produced per unit of time while maintaining the forest in a steady state. In our example this means that the stand is restored every 5 years to the state it was in 5 years earlier, and as a result the volume cut every 5 years is constant. Table 8.2 shows the volume and value of the average tree in each size class. Given this per-tree volume, the equation for the volume cut from the stand every 5 years is:

$$Z_Q = \underset{(m^3/\text{tree})}{0.20} \; \underset{(\text{tree/ha})}{h_{1t}} + 1.00 h_{2t} + 3.00 h_{3t}$$
$$\scriptstyle (m^3/ha)$$

The object is to find the harvest, h_{1t}, h_{2t}, and h_{3t}, and the corresponding growing stock, y_{1t}, y_{2t}, and y_{3t}, that maximize production while satisfying Constraints (8.10) to (8.13).

TABLE 8.2 Volume and Value of the Average Tree, by Size Class

Size class	Volume per tree (m³)	Value per tree ($)
1	0.20	0.30
2	1.00	8.00
3	3.00	20.00

	A	B	C	D	E
1	MAX PERIODIC PRODUCTION				
2			Stock		
3	Size	y_t		y_{t+1}	
4	class	(trees/ha)		(trees/ha)	
5	1	1362.5		1362.5	
6	2	54.5		54.5	
7	3	0.0		0.0	
8			Harvest		
9	Size	h_t	Volume		
10	class	(trees/ha)	(m³/tree)		
11	1	0	0.20		
12	2	54.5	1.00		
13	3	0	3.00		
14		Production		54.5	(m³/ha/5y)
15			max		
16					
17			Key cell formulas		
18	Cell	Formula			
19	D5	=0.92*(B5-B11)-0.29*(B6-B12)-0.96*(B7-B13)+109			
20	D6	=0.04*(B5-B11)+0.90*(B6-B12)			
21	D7	=0.02*(B6-B12)+0.90*(B7-B13)			
22	C14	=SUMPRODUCT(C11:C13,B11:B13)			

FIGURE 8.7 Spreadsheet model to maximize the periodic production of an uneven-aged stand in steady state.

Figure 8.7 shows a spreadsheet model to solve this problem. Cells D5:D7 contain the formulas for the growth equations predicting the stand state at time $t + 1$, given the stand state and harvest at t. Cell C14 contains the formula for the periodic timber production, Z_Q, as a function of the harvest variables and of the volume per tree in cells C11:C13.

Figure 8.8 shows the corresponding Solver parameters. The target cell is the periodic production. The changing cells contain the harvest and stock variables, h_{it} and y_{it}. The constraints specify that the harvest must be less than the stock, the harvest must be nonnegative, and the stock at time t must be equal to the stock at time $t + 1$ to ensure a steady state.

The best solution, shown in Figure 8.7, is a stand in which no tree ever grows into the largest size class. Furthermore, all the trees in the second size class are removed every 5 years. The growth of the residual trees in the first size class and the recruitment of new trees are enough to restore the stand to its preharvest state after 5 years. This system gives the highest periodic production, about 54 m³/ha/5y.

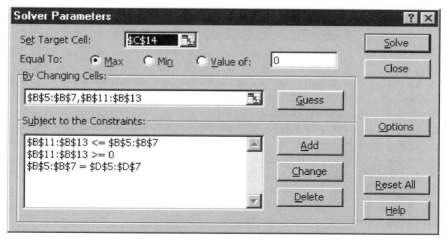

FIGURE 8.8 Solver parameters to maximize the periodic production of an uneven-aged stand.

SEEKING ENVIRONMENTAL QUALITY

Timber production is rarely the only objective of forest management. In fact, some owners have little interest in timber production. They may own a forest purely for recreational and esthetic reasons. Public forests are generally managed for many purposes, including watershed protection, recreation, wildlife habitat, and biological diversity.

Thus, the management of forests often reflects a multitude of goals. The various goals may be expressed with an objective function, which must be unique, and with one or more constraints. To pursue our example, assume that the owners have a special interest in maintaining a beautiful forest. In their eyes, beauty is associated with the presence of many big trees: the more, the better. As far as timber production is concerned, they would be satisfied with half of the maximum production computed earlier, that is, about 25 m³/ha/5 y.

The model just developed can help find the best management for this set of goals. The decision variables are the same as to maximize production. Growth Equations and Constraints (8.10) to (8.13) are the same. However, periodic production is now subject to a constraint:

$$0.20 \underset{(m^3/tree)}{h_{1t}} + 1.00 h_{2t} + 3.00 h_{3t} \geq \underset{(m^3/ha)}{25}$$

Furthermore, the new objective function is the number of trees in the largest size class, y_{3t}, which should be as large as possible.

	A	B	C	D	E
1	MAX BIG TREES				
2			Stock		
3	Size	y_t		y_{t+1}	
4	class	(trees/ha)		(trees/ha)	
5	1	841.3		841.3	
6	2	111.5		111.5	
7	3	17.3	max	17.3	
8			Harvest		
9	Size	h_t	Volume		
10	class	(trees/ha)	(m³/tree)		
11	1	0	**0.20**		
12	2	25.0	**1.00**		
13	3	0	**3.00**		
14		Production		25.0	(m³/ha/5y)
15		Lower bound		25	(m³/ha/5y)
16					
17			*Key cell formulas*		
18	Cell	Formula			
19	D5	=0.92*(B5-B11)-0.29*(B6-B12)-0.96*(B7-B13)+109			
20	D6	=0.04*(B5-B11)+0.90*(B6-B12)			
21	D7	=0.02*(B6-B12)+0.90*(B7-B13)			
22	C14	=SUMPRODUCT(C11:C13,B11:B13)			

FIGURE 8.9 Spreadsheet to maximize the number of large trees with a production constraint.

Figure 8.9 shows the spreadsheet set up to solve this problem. It is similar to the one in Figure 8.7, except for the addition of the lower bound on production in cell C15. The Solver parameters in Figure 8.10 have been changed so that the target cell is now B7, which contains the number of trees in the largest size class. The constraint C14 >= C15 has been added to force the periodic production to be at least 25 m³/ha/5 y.

The results in Figure 8.9 show that the maximum sustainable number of trees in the largest size class is about 17 trees/ha. This is done by thinning the medium size class, removing about 25 trees/ha every 5 years.

Relaxing the timber production constraint would allow for more large trees. Exactly how many is revealed by the sensitivity option of the Solver in Figure 8.11. The shadow price for the production constraint in cell C14 shows that decreasing the timber production constraint from 25 to 24 m³/ha/5 y would increase the number of trees in the largest size class by 0.6 trees/ha. In the next chapter we shall learn how to put a monetary value on this kind of trade-off.

Managing the Uneven-Aged Forest with Linear Programming

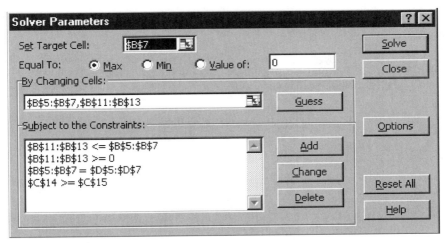

FIGURE 8.10 Solver parameters to maximize the number of large trees with a production constraint.

Constraints

Cell	Name	Final Value	Shadow Price	Constraint R.H. Side
$B$5	(trees/ha)	841.3	0.3	0
$B$6	(trees/ha)	111.5	0.6	0
$B$7	(trees/ha)	17.3	7.2	0
$B$11	(trees/ha)	0	0	0
$B$12	(trees/ha)	25.0	0.0	0
$B$13	(trees/ha)	0	0	0
$C$14	Production (m³/tree)	25.0	-0.6	25

FIGURE 8.11 Solver sensitivity option data showing the shadow price of the production constraint.

8.6 GENERAL FORMULATION

The example used in this chapter is a particular case of a general model of uneven-aged stand management. For conciseness, matrix notation is used throughout (see Appendix B for a brief introduction to matrix algebra).

GROWTH MODEL

The column vector $\mathbf{y}_t$ designates the state of an uneven-aged stand at time t, where each vector element is the average number of trees per hectare in each of n size classes. Similarly, the column vector $\mathbf{h}_t$ designates the harvest. Stand growth is represented by these recursive equations:

$$y_{t+1} = G(y_t - h_t) + c \qquad (8.14)$$

where G and c are matrices of constant parameter. The time interval t to $t+1$ is u years. In the example used in this chapter, $n = 3$, and:

$$y_t = \begin{bmatrix} y_{1t} \\ y_{2t} \\ y_{3t} \end{bmatrix}, \quad h_t = \begin{bmatrix} h_{1t} \\ h_{2t} \\ h_{3t} \end{bmatrix}, \quad G = \begin{bmatrix} 0.92 & -0.29 & -0.96 \\ 0.04 & 0.90 & 0 \\ 0 & 0.02 & 0.90 \end{bmatrix}, \quad c = \begin{bmatrix} 109 \\ 0 \\ 0 \end{bmatrix}$$

Recursive Equation (8.14) allows prediction of the stand state at any point in the future, starting from a particular initial state y_0 and subject to a specific sequence of harvests: $h_0, h_1, \ldots, h_T$:

$$y_1 = G(y_0 - h_0) + c$$
$$y_2 = G(y_1 - h_1) + c$$
$$\vdots$$
$$y_T = G(y_{T-1} - h_{T-1}) + c$$

UNDISTURBED STEADY STATE

When there is no harvest and $h_t = 0$ for all t, the stand will converge to a steady state as t increases. The steady state is independent of the initial conditions and depends only on the growth parameters. This undisturbed steady state is found by solving this system of linear equations:

$$y_{t+1} = Gy_t + c \qquad (8.15)$$

with this steady state condition:

$$y_{t+1} = y_t \qquad (8.16)$$

This system has a unique solution, which can be found by linear programming, with Equations (8.15), (8.16), and $y_t \geq 0$ as constraints and with an arbitrary objective function, such as:

$$\min_{y_t} y_{1t}$$

where y_{1t} is the number of trees in the smallest size class.

MAXIMIZING PERIODIC PRODUCTION

The sustainable regime that maximizes production per unit of time is found by solving a linear program with the following constraints:

Growth equations:

$$y_{t+1} = G(y_t - h_t) + c$$

Steady-state constraints:

$$y_{t+1} = y_t$$

Cut less than stock, and nonnegative:

$$h_t \le y_t$$
$$h_t \ge 0 \qquad (8.17)$$

Objective function is the constant periodic production:

$$\max_{h_t, y_t} Z_Q = vh_t$$

where v is a row vector of dimension n, the elements of which are the volumes per tree in each of the n size classes.

ENVIRONMENTAL OBJECTIVES

A variety of environmental objectives can be studied with this model. For example, we may seek the regime that obtains the highest sustainable number of large trees, subject to a lower bound on periodic production. The objective function is then:

$$\max_{h_t, y_t} y_{nt}$$

where y_{nt} is the number of trees in the largest size class. This objective function is maximized subject to Constraints (8.14), (8.16), and (8.17) and this additional constraint:

$$vh_t \ge Z_Q^*$$

where Z_Q^* is the lower bound on production.

8.7 SUMMARY AND CONCLUSION

In this chapter we studied a growth model of uneven-aged forest stands and applied it to different management issues. The stand state is defined by the number of trees per unit area in different size classes. The growth model consists of a set of recursive equations predicting the state of the stand as a function of its past state and of the last harvest. The model parameters are the probabilities that live trees in particular size classes will stay in that size class or move up a size class in a specified interval. The model is completed with a recruitment equation that predicts the rate at which trees enter the smallest size class. Recruitment is modeled as a function of the stand state.

This model was used to predict first the growth of an undisturbed stand and then the growth with a particular sequence of harvests. In the absence of disturbances, the model predicts the convergence of the stand to a natural steady state, or climax, where the growth and death processes just balance and the stand state remains constant over time. This undisturbed steady state can be found by solving the growth equations with the condition that the stand state remain the same over time. The solution can also be found by linear programming with an arbitrary objective function, subject to the growth equations and steady-state conditions.

With human intervention, in the form of harvest, a steady state is defined as a system where the stand growth during the cutting cycle just replaces the harvest. Thus, the stand state at the end of the cycle is the same as at the beginning, and the harvest can continue forever. There is an infinite number of possible steady states. Those that best meet specific objectives can be found by linear programming.

In particular, we specified linear programs to find sustainable management regimes (harvest and corresponding growing stock) that maximized the production per unit of time. This emphasis on timber productivity would have environmental drawbacks for some owners because it leaves few if any of the largest trees. Another model gave priority to environmental and esthetic considerations by finding a regime that maximized the sustainable number of large trees while producing a specified amount of timber. These are just two examples of management alternatives. The model structure has the potential to deal with many more scenarios. In the next chapters we will look in particular at the economic dimension of uneven-aged management, with a view to measure the trade-off between economic and environmental goals.

PROBLEMS

8.1 The table here shows the fractions of trees staying in the same size class or growing into a larger class over a 5-year period for a sugar maple stand.

Managing the Uneven-Aged Forest with Linear Programming

Proportion of Trees Staying in the Same Size Class or Growing into the Next Size Class Within 5 Years (sugar maple)

Size class	Proportion staying in same class	Proportion growing into next class
1	0.79	0.02
2	0.88	0.01
3	0.85	0.00

(a) Assuming that the definitions of the size classes are the same as those used in the example in Section 8.2, do the given data describe a faster or a slower growing stand than the one described in Table 8.1?

(b) What proportion of the trees in each size class dies every 5 years?

8.2 Set up the spreadsheet model in Figure 8.2 on your own spreadsheet, and verify the results. Then extend the stand projection to 500 years.

(a) How does the number of trees in each size class change over time?

(b) How long does it take for the stand to approach a steady state?

(c) Compare this approximate steady state with the exact steady state in Figure 8.3.

8.3 In the spreadsheet model in Figure 8.2, change the initial stand state to $y_{1,0} = 300$ trees/ha, $y_{2,0} = 100$ trees/ha, and $y_{3,0} = 20$ trees/ha.

(a) What is the effect on the number of trees, by size class, over time?

(b) What is the effect on the steady state?

8.4 Consider the following equations of recruitment into the smallest size class for an uneven-aged stand:

$$R_t = 100 - 10B_t + 0.2N_t$$

where all terms are defined as in Equation (8.3).

(a) For a stand of 30 m²/ha of basal area and 1200 trees/ha, what is the recruitment rate?

(b) Assume two stands with 20 m²/ha basal area, but one with 1200 trees/ha and another with 1500 trees/ha. Which stand has the higher recruitment rate?

(c) Assume two stands with 1200 trees/ha, but one with 20 m²/ha basal area and another with 15 m²/ha. Which stand has the higher recruitment rate?

8.5 Equation (8.1) describes the state of an uneven-aged stand at time $t + 1$ in terms of state of the stand at time t and recruitment. Expand this system of growth equations to include a fourth size class.

8.6 The owners of a vacation home have asked for your advice regarding the management of the northern oaks grown on their property. Because the home is located in the center of the property, they are interested in pursuing uneven-aged silviculture to minimize the esthetic impact of harvesting activities. They have already obtained some growth data for northern oak stands growing on similar sites. The data are shown in the table. Use these data to construct a set of equations similar to Equations (8.6) to predict the growth of the oak stand. Use the recruitment equation $R_t = 80 - 10B_t + 0.1N_t$.

Proportion of Trees Staying in the Same Size Class or Growing into the Next Size Class Within 5 Years (oak)

Size class	Proportion staying in same class	Proportion growing into next class	Basal area per tree (m²)
1	0.78	0.04	0.02
2	0.80	0.06	0.07
3	0.80	0.04	0.14
4	0.90	0.00	0.25

8.7 Consider an uneven-aged stand of trees that grows according to the following equations:

$$y_{1,t+1} = 0.61y_{1t} - 0.40y_{2t} - 1.4y_{3t} - 2.4y_{4t} + 100$$

$$y_{2,t+1} = 0.04y_{1t} + 0.80y_{2t}$$

$$y_{3,t+1} = \phantom{0.04y_{1t} +} 0.06y_{2t} + 0.80y_{3t}$$

$$y_{4,t+1} = \phantom{0.04y_{1t} + 0.06y_{2t} +} 0.04y_{3t} + 0.90y_{4t}$$

where the interval between times t and $t + 1$ is 5 years.
(a) Write the steady-state conditions for this stand similar to Equations (8.7).
(b) Set up a spreadsheet model similar to that in Figure 8.3 to find this steady state by linear programming. Verify that the steady state is unique by changing the objective function.

8.8 Consider two different initial states for stands growing according to the equations in Problem 8.7.

Initial state 1: $y_{1,0} = 200, y_{2,0} = 30, y_{3,0} = 0, y_{4,0} = 0$
Initial state 2: $y_{1,0} = 20, y_{2,0} = 10, y_{3,0} = 5, y_{4,0} = 1$

State 1 describes a stand that has been cut over, leaving a residual stand of small trees. State 2 describes a stand with relatively few trees in the smallest size class as a result of mortality from browsing deer. Set up a spreadsheet model similar to that in Figure 8.2 to predict the growth of the stand given different initial states. Can you detect any convergence toward the steady state identified in Problem 8.7?

8.9 Consider an uneven-aged stand whose growth can be described by the equations given in Problem 8.7.

(a) Modify these growth equations to describe the growth of a stand from which harvesting can occur in any size class every 5 years.

(b) Assume the harvesting policy for the stand will be to harvest all the trees in the largest size class and nothing from the two smaller size classes. Set up a spreadsheet similar to that in Figure 8.5 to predict the growth of the stand over the next 100 years, given this policy. What do you observe?

8.10 Consider an uneven-aged stand whose growth can be described by the equations given in Problem 8.7. Wood volumes per tree are given in the table. Set up a spreadsheet model similar to that in Figure 8.9 to find the steady-state regime that would maximize the number of stock trees in the third and fourth size classes and produce at least 10 m^3/ha/5 y.

Volume of the Average Tree, by Size Class

Size class	Volume per tree (m^3)
1	0.1
2	0.5
3	1.5
4	3.0

ANNOTATED REFERENCES

Avery, T.E., and H.E. Burkhart. 2002. *Forest Measurements*. McGraw-Hill, New York. 456 pp. (Chapter 17 discusses the stand growth projection of the sort used in the matrix models in this chapter.)

Buongiorno, J., and B.R. Michie. 1980. A matrix model of uneven-aged forest management. *Forest Science* 26(4):609–625. (Source of the matrix model presented in this chapter.)

Davis, L.S., K.N. Johnson, P.S. Bettinger, and T.E. Howard. 2001. *Forest Management: To Sustain Ecological, Economic, and Social Values*. McGraw-Hill, New York. 804 pp. (Chapter 11 discusses a "Model II" linear programming model for uneven-aged management.)

Favrichon, V. 1998. Modeling the dynamics and species composition of a tropical mixed-species uneven-aged natural forest: Effects of alternative cutting regimes. *Forest Science* 44(1):113–124. (Describes a matrix growth model to compare the dynamics of logged and unlogged stands in French Guiana.)

Getz, W., and R.G. Haight. 1989. *Population Harvesting: Demographic Models of Fish, Forest, and Animal Resources*. Princeton University Press, Princeton, NJ. 391 pp. (Theory and application of matrix models like that used in this chapter.)

Hann, D.W., and B.B. Bare. 1979. *Uneven-Aged Forest Management: State of the Art (or Science?)*. U.S. Forest Service General Technical Report INT-50. Intermountain Forest and Range Experiment Station, Ogden, UT. 18 pp. (Review of the philosophy and history of uneven-aged management and modeling efforts.)

Leak, W.B., and J.H. Gottsacker. 1985. New approaches to uneven-aged management in New England. *Northern Journal of Applied Forestry* 2:28–31. (Discusses the importance of simplicity in measurements and models for practical implementation of uneven-aged management.)

Mendoza, G.A., and A. Setyarso. 1986. A transition matrix forest growth model for evaluating alternative harvesting schemes in Indonesia. *Forest Ecology and Management* 15:219–228. (Uses a matrix model similar to that in this chapter to evaluate alternative cutting regimes for dipterocarp forests.)

Mengel, D.L., and J.P. Roise. 1990. A diameter-class matrix model for southeastern U.S. coastal plain bottomland hardwood stands. *Southern Journal of Applied Forestry* 14:189–195. (Extends the matrix model used in this chapter to allow for density-dependent transition probabilities.)

Nyland, R.D. 2002. *Silviculture: Concepts and Applications*. McGraw-Hill, New York. 682 pp. (Chapter 2 compares even- and uneven-aged management systems.)

Osho, J.S.A. 1991. Matrix model for tree population projection in a tropical rain forest of southwestern Nigeria. *Ecological Modeling* 59:247–255. (Uses a matrix model similar to that in this chapter.)

Rorres, C. 1978. A linear programming approach to the optimal sustainable harvesting of a forest. *Journal of Environmental Management* 6(3):245–254. (Makes theoretical extensions to the model proposed by Usher.)

Schulte, B., J. Buongiorno, C.R. Lin, and K. Skog. 1998. *Southpro: A Computer Program for Managing Uneven-Aged Loblolly Pine Stands*. USDA Forest Service, Forest Products Laboratory, Gen. Tech. Rep. FPL-GTR-112. 47 pp. (Extends the matrix model used in this chapter to consider both species groups and state-dependent parameters.)

Solomon, D.S., R.A. Hosmer, and H.T. Hayslett, Jr. 1986. A two-stage matrix model for predicting growth of forest stands in the northeast. *Canadian Journal of Forest Research* 16:521–528. (Extends the matrix model used in this chapter to consider several species.)

Usher, M.B. 1966. A matrix approach to the management of renewable resources, with special reference to selection forests. *Journal of Applied Ecology* 3(2):355–367. (Uses a matrix model similar to that in this chapter to evaluate management options for scots pine.)

Vanclay, J.K. 1995. Growth models for tropical forests: A synthesis of models and methods. *Forest Science* 41(1):7–42. (Discusses some of the strengths and weaknesses of matrix models, noting their usefulness when data are limited.)

CHAPTER 9

Economic and Environmental Management of Uneven-Aged Forests

9.1 INTRODUCTION

In the preceding chapter we studied a dynamic model for forecasting the growth of uneven-aged forest stands. Given the initial state of the stand and a specific schedule of periodic harvests, we could predict how the stand would look at any future date, with or without human intervention. This model also gave us a way to define sustainable management regimes. Among all possible sustainable regimes, we used linear programming to identify those that best met objectives such as maximizing productivity or esthetics.

The purpose of this chapter is to extend this model to define the economic value of uneven-aged forest stands for timber production and the effects of environmental constraints on this value. Economic objectives are important for many forest owners. Some own forests mostly for financial reasons, and they are interested in managing them to maximize timber income. Other owners have nonfinancial reasons for managing or not managing their forest, such as soil protection, esthetics, and biological diversity.

Even for owners with no direct financial objective, knowledge of the economic potential of the forest is useful to judge the opportunity cost of management decisions. Even in a publicly owned forest that is being devoted exclusively to recreation, it is worth knowing how much timber income is lost due to this choice. Knowledge of the full opportunity cost of alternative land uses is essential for making good decisions.

A large part of this chapter deals with steady-state conditions. That is, we seek ideal combinations of harvest and growing stock that are sustainable and that best meet different management goals. Steady-state regimes can be viewed as ideal conditions that management should strive for. At the end of the chapter we shall turn to the issue of how to convert a given stand to such a steady state.

We shall begin by studying purely economic management regimes, adapting Faustmann's theory studied in Chapter 7 to the case of valuing an uneven-aged stand and thus finding the land expectation value in uneven-aged silviculture. We shall then apply the model to investigate the effect of different environmental goals on land expectation value and thus measure the trade-off between economic and environmental objectives in uneven-aged management.

The last part of the chapter deals with the issue of converting a given initial stand to the steady state that best meets various combinations of economic and environmental objectives.

9.2 ECONOMIC STEADY STATE FOR UNEVEN-AGED STANDS

The object here is to determine the structure of sustainable uneven-aged stands and the corresponding harvest that maximize the return to the land. We shall proceed in two steps. First finding the best stock and harvest for a given cutting cycle, and then determining the effect of changing the cycle.

LAND EXPECTATION VALUE FOR UNEVEN-AGED STANDS

When we studied Faustmann's formula for even-aged stands in Chapter 7 we defined the land expectation value (LEV) as the present value of the perpetual income stream produced by periodic crops of trees, beginning with bare land.

We can apply the same concept to an uneven-aged stand in a steady state. The forest value of an even-aged stand (land plus trees) is the present value of the income that it will produce in perpetuity. The value of the land only (the fixed input), or the land expectation value, is the present value of the income produced by the land and trees, minus the initial value of the trees (a variable input). The object is to choose the stock of trees in the uneven-aged stand and the corresponding harvest to maximize the land expectation value.

Economic and Environmental Management of Uneven-Aged Forests

This optimization is done within constraints that define the biological growth of the stand under consideration, the steady state, and ancillary conditions.

CONSTRAINTS

Biological Growth

Throughout this chapter we shall continue to use as an example the stand of sugar maple trees studied in Chapter 8. The stands evolves over time according to these growth equations:

$$y_{1,t+1} = 0.92(y_{1t} - h_{1t}) - 0.29(y_{2t} - h_{2t}) - 0.96(y_{3t} - h_{3t}) + 109$$
$$y_{2,t+1} = 0.04(y_{1t} - h_{1t}) + 0.90(y_{2t} - h_{2t}) \qquad (9.1)$$
$$y_{3,t+1} = \qquad\qquad 0.02(y_{2t} - h_{2t}) + 0.90(y_{3t} - h_{3t})$$

where y_{it} is the number of trees in size class i at time t, h_{it} is the number of trees cut from size class i at time t, and the interval between times t and $t+1$ is 5 years.

Steady State

For the stand to be in a steady state, the number of trees at time $t+1$ must be the same as the number at times t, in every size class:

$$y_{i,t+1} = y_{it} \quad \text{for } i = 1, 2, 3 \qquad (9.2)$$

This ensures that the stand growth from time t to time $t+1$ just replaces the harvest and, therefore, that the harvest h_{it} could be repeated at $t+1, t+2, t+3,...$ and so on forever.

Feasible Harvest

Equations (9.1) and (9.2) have an infinite number of solutions. However, a meaningful solution must be such that the number of trees harvested in every size class is less than the stock:

$$h_{it} \leq y_{it} \quad \text{for } i = 1, 2, 3 \qquad (9.3)$$

Nonnegativity

Last, the number of trees harvested in every size class cannot be negative:

$$h_{it} \geq 0 \quad \text{for } i = 1, 2, 3 \qquad (9.4)$$

Even with Constraints (9.3) and (9.4), Equations (9.1) and (9.2) still have an infinite number of solutions. There is an infinite number of possible steady states, some better than others for specific ends. Here, we seek the steady state that maximizes the return to the land, what we called the land expectation value in Chapter 7.

ECONOMIC OBJECTIVE FUNCTION

The land expectation value of an uneven-aged stand in steady state derives from the stream of constant periodic returns that the stand will produce indefinitely. For the sugar maple stand in our example, and using the data on tree value in Table 8.2, the equation for the value of the periodic harvest in steady state is:

$$V_h = 0.30\ h_{1t} + 8h_{2t} + 20h_{3t}$$
$$\text{(\$/ha)}\quad \text{(\$/tr)}\ \text{(tr/ha)}$$

This value is realized at time t and every 5 years thereafter, and is sustainable forever. Thus, the present value of all the harvests starting at time t, assuming an interest rate of 5% per year, is:

$$P_h = V_h + \frac{V_h}{(1+0.05)^5 - 1}\quad (\$/\text{ha})$$

However, this income results from the land and from the stock of trees that uneven-aged management keeps on it. Thus P_h is the value of the land and the trees that are on it, in other words, the forest value at time t. The value of the stock of trees alone is:

$$V_s = 0.30\ y_{1t} + 8y_{2t} + 20y_{3t}$$
$$\text{(\$/ha)}\quad \text{(\$/tr)}\ \text{(tr/ha)}$$

Thus, the value of the land alone (the land expectation value) is:

$$\text{LEV} = P_h - V_s\quad (\$/\text{ha})$$

The formula shows that the owner affects the returns to the fixed input, the land, by simultaneously choosing the harvest and the inventory of growing stock.

The land expectation value is a linear function of the harvest and stock variables, because P_h is a linear function of V_h, and V_h is itself a linear function of h_{1t}, h_{2t}, and h_{3t}. As a result, we may seek the management of highest land expectation value by linear programming.

Economic and Environmental Management of Uneven-Aged Forests 159

LINEAR PROGRAMMING SOLUTION

The decision variables are the harvest variables and the stock variables: h_{1t}, h_{2t}, h_{3t}, y_{1t}, y_{2t}, and y_{3t}. The objective function is LEV. The constraints are the growth Equations (9.1), the steady state Constraints (9.2), the harvest-less-than-stock Equations (9.3), and the nonnegativity-of-the-harvest Equations (9.4).

Figure 9.1 shows a spreadsheet to compute the highest land expectation value for the uneven-aged stand used throughout this chapter. As in Chapter 8, cells D5:D7 contain the formulas for the growth equations, predicting the stand state at time $t+1$ given the stand state and the harvest at t. Cells C11:C13 contain the tree value data. With these data, the formula in cell C14 computes the harvest value, V_h, and the formula in cell C15 computes the forest value, P_h. The formula in cell C16 gives the stock value, V_s.

	A	B	C	D	E
1	MAX LAND EXPECTATION VALUE, 5 YEAR CYCLE				
2			Stock		
3	Size	y_t		y_{t+1}	
4	class	(trees/ha)		(trees/ha)	
5	1	1362.5		1362.5	
6	2	54.5		54.5	
7	3	0.0		0.0	
8			Harvest		
9	Size	h_t	Value		
10	class	(trees/ha)	($/tree)		
11	1	0.0	0.30		
12	2	54.5	8.00		
13	3	0.0	20.00		
14		Harvest value:	436	($/ha)	
15		Forest value:	2014	($/ha)	
16		Stock value:	845	($/ha)	
17	max	LEV:	1169	($/ha)	
18					
19			Key cell formulas		
20	Cell	Formula			
21	D5	=0.92*(B5-B11)-0.29*(B6-B12)-0.96*(B7-B13)+109			
22	D6	=0.04*(B5-B11)+0.90*(B6-B12)			
23	D7	=0.02*(B6-B12)+0.90*(B7-B13)			
24	C14	=SUMPRODUCT(C11:C13,B11:B13)			
25	C15	=C14+C14/((1.05)^5-1)			
26	C16	=SUMPRODUCT(C11:C13,B5:B7)			
27	C17	=C15-C16			

FIGURE 9.1 Spreadsheet to maximize land expectation value.

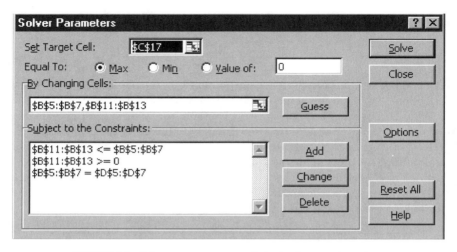

FIGURE 9.2 Solver parameters to maximize land expectation value.

The land expectation value in cell C17 is the target cell in the solver parameters shown in Figure 9.2. The adjustable cells contain the number of trees in stock in each size class and the number of trees cut. The first line of constraints ensures that the cut is less than the stock, the second line that the cut is non-negative, and the third line that the stock is the same at time t as at time $t + 1$ so that the stand is in a steady state.

The results in Figure 9.2 show that the best management decision is to harvest all of the middle-sized trees every 5 years. The forest value, land plus trees, of a stand in this steady state is $2,014/ha. The value of the growing stock is $845/ha, and the land expectation value is $1,169/ha. Comparing the solutions in Figures 9.1 and 8.7 shows that the management that gives the highest land expectation value is the same as that which maximizes periodic production, although this is not always the case.

Thus, $1,169/ha is our estimate of the best value of the land used in this type of uneven-aged management. Together with other data, this land expectation value can help us choose between management alternatives and between different land uses, such as forestry, agriculture, and wilderness.

ECONOMIC CUTTING CYCLE

So far, we have assumed that the interval between harvests is the same as the time unit of the growth equations. In our example, both were equal to 5 years. The choice of cutting cycle, however, is itself a managerial decision. A longer cutting cycle may be desirable for environmental reasons, to decrease the

Economic and Environmental Management of Uneven-Aged Forests 161

frequency of disturbances. Longer cutting cycles may even be superior in economic terms. There is usually a fixed cost attached to a harvest entry, namely, a cost per unit area, that is independent of the amount harvested. The existence of a fixed cost may lengthen the best economic cutting cycle.

To study the effect of different cutting cycles, it is convenient to write the growth Equations (9.1) in matrix notation (see Appendix B):

$$\mathbf{y}_{t+1} = \mathbf{G}(\mathbf{y}_t - \mathbf{h}_t) + \mathbf{c} \quad (9.5)$$

where $\mathbf{y}_t$ is the state of the stand at time t, $\mathbf{h}_t$ is the harvest at time t, and $\mathbf{G}$ and $\mathbf{c}$ are the constant parameters in Equations (9.1).

Assume a 10-year cutting cycle instead of 5 years. This means that there is a harvest at time t, but no harvest at time $t + 1$. Therefore, the state of the stand at time $t + 2$ is:

$$\mathbf{y}_{t+2} = \mathbf{G}\mathbf{y}_{t+1} + \mathbf{c} \quad (9.6)$$

The steady state constraints require that the stand state be the same at the beginning and at the end of the cutting cycle; i.e.:

$$\mathbf{y}_{t+2} = \mathbf{y}_t \quad (9.7)$$

In addition, the harvest must be less than the stock:

$$\mathbf{h}_t \leq \mathbf{y}_t \quad (9.8)$$

and the harvest cannot be negative:

$$\mathbf{h}_t \geq 0 \quad (9.9)$$

The object is to find the harvest $\mathbf{h}_t$ and the growing stock $\mathbf{y}_t$ that satisfy Constraints (9.5) to (9.9) while maximizing the land expectation value. In the objective function, the expressions of the value of the periodic harvest V_h and of the stock V_s remain the same as for the 5-year cutting cycle:

$$V_h = \mathbf{w}\mathbf{h}_t \quad \text{and} \quad V_s = \mathbf{w}\mathbf{y}_t$$

where $\mathbf{w} = (0.30, 8, 20)$ is the vector of tree values for each size class given in Table 8.2. However, the expression of the land expectation value changes, because the harvest occurs every 10 years instead of every 5:

$$\text{LEV} = P_h - V_s = V_h + \frac{V_h}{(1 + 0.05)^{10} - 1} - V_s \quad (\$/\text{ha})$$

	A	B	C	D	E	F
1	MAX LAND EXPECTATION VALUE, 10 YR CYCLE					
2			Stock			
3	Size	y_t		y_{t+1}		y_{t+2}
4	class	(trees/ha)		(trees/ha)		(trees/ha)
5	1	1266.8		1274.5		1266.8
6	2	96.6		50.7		96.6
7	3	1.0		0.0		1.0
8			Harvest			
9	Size	h_t	Value	h_{t+1}		
10	class	(trees/ha)	($/tree)	(trees/ha)		
11	1	0.0	**0.30**	0		
12	2	96.6	**8.00**	0		
13	3	1.0	**20.00**	0		
14		Harvest value:		793	($/ha)	
15		Forest value:		2054	($/ha)	
16		Stock value:		1173	($/ha)	
17	max	LEV:		881	($/ha)	
18						
19			*Key cell formulas*			
20	Cell	Formula				copy to
21	D5	=0.92*(B5-B11)-0.29*(B6-B12)-0.96*(B7-B13)+109				F5
22	D6	=0.04*(B5-B11)+0.90*(B6-B12)				F6
23	D7	=0.02*(B6-B12)+0.90*(B7-B13)				F7
24	C14	=SUMPRODUCT(C11:C13,B11:B13)				
25	C15	=C14+C14/((1.05)^10-1)				
26	C16	=SUMPRODUCT(C11:C13,B5:B7)				
27	C17	=C15-C16				

FIGURE 9.3 Spreadsheet to maximize land expectation value with a 10-year cutting cycle.

Figure 9.3 shows a spreadsheet to compute the management regime that maximizes land expectation value with a cutting cycle of 10 years for the example forest. Most of the data and formulas are the same as those for a 5-year cutting cycle, shown in Figure 9.1. Added are the formulas in cells F5:F7, copied from D5:D7, to predict the stand state at time $t + 2$, given the state at time $t + 1$, and the harvest at time $t + 1$ set at 0. The formula for the forest value, P_h, in cell C15 has been changed to reflect the fact that the harvest occurs every 10 years.

The Solver parameters in Figure 9.4 have the same target cell as before, corresponding to LEV, and the same adjustable cells, corresponding to $\mathbf{h}_t$ and $\mathbf{y}_t$. The first two rows of constraints are also the same, corresponding to Constraints (9.8) and (9.9). The steady-state constraints B5:B7 = F5:F7 are different, corresponding to Equation (9.7), to reflect the 10-year cutting cycle.

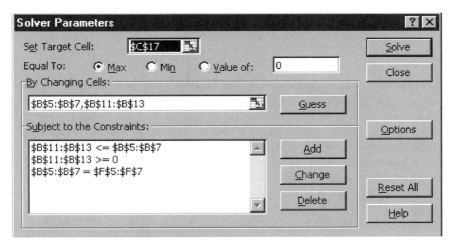

FIGURE 9.4 Solver parameters to maximize land expectation value with a 10-year cutting cycle.

The results in Figure 9.3 show that the best management with a 10-year cycle consists of cutting all of the trees in the medium and largest size classes. The value of the periodic harvest ($793/ha) is much higher than with a 5-year cutting cycle ($436/ha). The forest value, however, is only slightly higher ($2,054/ha against $2,014/ha), because the higher income occurs at longer intervals. The value of the growing stock with the 10-year cycle is higher ($1,173/ha against $845/ha) because we keep more and larger trees in the stand. As a result, the land expectation value is higher for the 5-year cutting cycle ($1,169/ha against $881/ha). From a purely economic point of view, the shorter cutting cycle is superior in this example.

ROLE OF FIXED COST

The result we have just obtained is general. With little or no fixed cost, a shorter cutting cycle is better from a purely economic standpoint. However, the presence of a fixed cost may change this result. The decision to harvest a stand usually entails costs that are independent of the amount harvested. For example, private owners may need to hire consultants to mark and administer a timber sale. Assume the fee, F, for such services is set per unit area ($/ha). Then the value of the periodic harvest, net of the fixed cost, is $V_h - F$, and the value of the growing stock is $V_s - F$. Therefore, the land expectation value with a fixed cost, LEV_F, is:

$$\text{LEV}_F = (V_h - F) + \frac{V_h - F}{(1+r)^D - 1} - (V_s - F) \quad (\$/\text{ha})$$

where r is the interest rate and D is the cutting cycle. This simplifies to:

$$LEV_F = V_h + \frac{V_h - F}{(1+r)^D - 1} - V_s = LEV - \frac{F}{(1+r)^D - 1} \quad (\$/ha)$$

Thus, the land expectation value with a fixed cost is equal to the land expectation value without a fixed cost, minus a constant that depends only on the fixed cost, the interest rate, and the cutting cycle. As a result, for a given cutting cycle, the harvest that maximizes land expectation value is the same with and without fixed cost. However, the land expectation value is different.

To pursue our example further, let the fixed cost be $200/ha, independent of the harvest. Then the land expectation value for a 5-year cutting cycle is:

$$1{,}169 - \frac{200}{(1+0.05)^5 - 1} = 445 \quad (\$/ha)$$

while the land expectation value for a 10-year cycle is:

$$881 - \frac{200}{(1+0.05)^{10} - 1} = 563 \quad (\$/ha)$$

In this scenario, the 10-year cutting cycle is better. Can you determine the level of fixed cost at which the owners should switch from a 5-year cutting cycle to a 10-year cycle?

9.3 ENVIRONMENTAL OBJECTIVES

A common objective of modern uneven-aged management is to create and maintain diverse stands. In fact, advocacy for uneven-aged silviculture is based largely on the view that it leads to diverse stands that look more natural than even-aged single-species stands. The diversity stems in part from the fact that uneven-aged management maintains trees of very different sizes on a given patch of land.

Furthermore, the natural regeneration most often used in uneven-aged management leads to the presence of diverse tree species in a stand. Both diversity of tree size, sometimes referred to as *structural diversity*, and the species diversity are useful in promoting general forest biodiversity.

MANAGING FOR TREE DIVERSITY IN FOREST STANDS

The model used throughout this chapter dealt with the size of trees only. For example, the state variable y_{it} and the harvest variable h_{it} designate the number of trees in size class i at time t and the number of trees cut from that size class.

Economic and Environmental Management of Uneven-Aged Forests

However, the model is more general. The subscript i may refer to the trees in a particular size class and species group. All the principles and methods presented so far remain the same; the only change is in the detail of the data and in the complication of the notations. For simplicity we shall continue to concern ourselves with tree size only. Keep in mind, however, that when we deal with tree size (structural) diversity, we are in fact laying down methods to deal with tree species diversity as well and for combinations of size and species diversity.

A MaxiMin Criterion of Diversity

An operational criterion for structural diversity is to ensure that a stand has trees in every size class. A way to achieve high diversity is then to seek the management regime that keeps the smallest number of trees in any size class as large as possible. This is the same MaxiMin principle we used to obtain landscapes of high diversity in even-aged forests (see Chapter 7).

Linear programming can be used to find the management regime leading to a sustainable (i.e., steady-state) stand of maximum diversity, in that sense. To do this we define a new decision variable, N_{min}, the smallest number of trees in any size class.

The linear programming problem then consists of finding the number of trees to cut in each size class, h_{1t}, h_{2t}, and h_{3t}, and the number of trees to keep in each size class, y_{1t}, y_{2t}, and y_{3t}, that maximize N_{min}, subject to growth Equations (9.1), steady-state constraints (9.2), harvest-less-than-stock-Constraints (9.3), and nonnegativity-Constraints (9.4). New constraints:

$$y_{it} \geq N_{min} \quad \text{for } i = 1, 2, 3 \qquad (9.10)$$

ensure that N_{min} is the smallest number of trees in any size class. The objective is to make this number as large as possible.

Spreadsheet Solution

Figure 9.5 shows a spreadsheet model to find the management that maximizes tree diversity with a 5-year cutting cycle. The only change with respect to the spreadsheet in Figure 9.1 is the addition of the variable N_{min} in cell C5.

The corresponding Solver parameters in Figure 9.6 indicate that C5 is the new target cell and that C5 is also an adjustable cell. The constraints are the same as in Figure 9.2, except for the addition of the constraints corresponding to Inequality (9.5).

The best solution in Figure 9.5 shows that cutting no trees at all maximizes stand diversity, in the sense of obtaining the highest number of trees in the size class with the least number of trees. Letting the natural growth and death

	A	B	C	D	E
1	MAX TREE DIVERSITY				
2			Stock		
3	Size	y_t	N_{min}	y_{t+1}	
4	class	(trees/ha)	(trees/ha)	(trees/ha)	
5	1	399.6	32.0	399.6	
6	2	159.8	max	159.8	
7	3	32.0		32.0	
8			Harvest		
9	Size	h_t	Value		
10	class	(trees/ha)	($/tree)		
11	1	0.0	0.30		
12	2	0.0	8.00		
13	3	0.0	20.00		
14		Harvest value:	0	($/ha)	
15		Forest value:	0	($/ha)	
16		Stock value:	2038	($/ha)	
17		LEV:	-2038	($/ha)	
18					
19			Key cell formulas		
20	Cell	Formula			
21	D5	=0.92*(B5-B11)-0.29*(B6-B12)-0.96*(B7-B13)+109			
22	D6	=0.04*(B5-B11)+0.90*(B6-B12)			
23	D7	=0.02*(B6-B12)+0.90*(B7-B13)			
24	C14	=SUMPRODUCT(C11:C13,B11:B13)			
25	C15	=C14+C14/((1.05)^5-1)			
26	C16	=SUMPRODUCT(C11:C13,B5:B7)			
27	C17	=C15-C16			

FIGURE 9.5 Spreadsheet model to maximize tree diversity in uneven-aged stand.

processes take their course leads to the climax stand of highest diversity. This may not be a general result, but similar results have also been obtained with much more detailed models and other measures of diversity (see the Annotated References at the end of this chapter).

OPPORTUNITY COST OF ENVIRONMENTAL GOALS

The results in Figure 9.5 show that the land expectation value for the management of highest tree diversity is −$2,038/ha. This value is negative due to the fact that the stand of highest diversity produces no income yet requires the owners to carry a stock with a liquidation value of $2,038/ha.

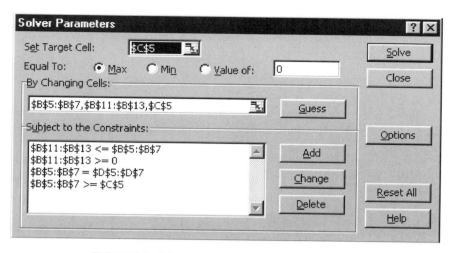

FIGURE 9.6 Solver parameters to maximize tree diversity.

But this is not the full opportunity cost of following the diversity-maximizing strategy. The data in Figure 9.1 show that by managing their uneven-aged stand without any consideration of tree diversity, the owners could obtain a maximum land expectation value of $1,169/ha. Therefore, the full opportunity cost of the diversity strategy is: 1,169 − (−2,038) = $3,207/ha.

The implication is that if the owners of this stand, be they public or private, choose to manage (in effect to leave alone) their forest, the goods, services, and amenities produced by the undisturbed forest are worth at least $3,207/ha. If this opportunity cost seems too high, a compromise needs to be struck between the economic and environmental objectives.

ECONOMICALLY EFFICIENT ENVIRONMENTAL MANAGEMENT

One way of shaping a management policy that balances economic and environmental objectives is to consider both types simultaneously. We know, however, that the objective function in any optimization problem must be unique. In our example, this means that we must choose to optimize either the land expectation value or the stand diversity and treat the other goal as a constraint.

One possibility is to seek the most efficient way, in an economic sense, of achieving a prespecified environmental goal. For example, assume that purely economic management is unacceptable to the owners of our example forest because it leaves no tree in the largest size class. On the other hand, the opportunity cost of the maximum-diversity management is too high. The owners feel that their diversity objective would be satisfied by a management that would keep at least 16 trees/ha in the largest size class, 80 trees/ha in the medium size class, and 200 trees/ha in the smallest size class (Figure 9.5 shows that this is about half the number of trees in the stand structure that maximizes diversity).

The economically efficient management to achieve this environmental goal can be found by solving the following linear programming problem: Find the harvest, h_{1t}, h_{2t}, and h_{3t}, and the growing stock, y_{1t}, y_{2t}, and y_{3t}, that maximize LEV, subject to growth Equations (9.1), steady-state Constraints (9.2), the harvest-less-than-stock Constraints (9.3), nonnegativity Constraints (9.4), and three new constraints:

$$y_{1t} \geq 200 \quad \text{(trees/ha)}$$
$$y_{2t} \geq 80 \quad \text{(trees/ha)} \qquad (9.11)$$
$$y_{2t} \geq 16 \quad \text{(trees/ha)}$$

Figure 9.7 shows a spreadsheet model to find this compromise solution with a 5-year cutting cycle. The spreadsheet is similar to the one in Figure 9.1 to maximize land expectation value. The only new data are the lower bounds on the number of trees in stock in every size class, in cells C5:C7. The corresponding Solver parameters in Figure 9.8 are the same as those to maximize land expectation value in Figure 9.2. The only additional constraint is B5:B7 >= C5:C7, corresponding to the Inequalities (9.11).

The best solution in Figure 9.7 shows that one should harvest about 27 trees in the medium size class every 5 years. This would maintain the desired 16 trees/ha in the largest size class and more than the desired number in the other two size classes. The land expectation value for this management is $–436/ha. This is still a net economic loss, but much smaller than the $2,038/ha loss that would result from maximizing diversity.

9.4 CONVERTING A STAND TO THE DESIRED STEADY STATE

The models we have explored so far gave a picture of the ideal steady-state uneven-aged forest stand that could maintain itself forever while fulfilling specific economic or environmental objectives. However, upon examining uneven-aged forest, even one with many stands, it is rare to find the stands in

	A	B	C	D	E
1	EFFICIENT TREE DIVERSITY				
2			Stock		
3	Size	y_t	Lower bound	y_{t+1}	
4	class	(trees/ha)	(trees/ha)	(trees/ha)	
5	1	880.5	**200**	880.5	
6	2	107.2	**80**	107.2	
7	3	16.0	**16**	16.0	
8			Harvest		
9	Size	h_t	Value		
10	class	(trees/ha)	($/tree)		
11	1	0.0	**0.30**		
12	2	27.2	**8.00**		
13	3	0.0	**20.00**		
14		Harvest value:		218	($/ha)
15		Forest value:		1006	($/ha)
16		Stock value:		1442	($/ha)
17	max	LEV:		-436	($/ha)
18					
19			*Key cell formulas*		
20	Cell	Formula			
21	D5	=0.92*(B5-B11)-0.29*(B6-B12)-0.96*(B7-B13)+109			
22	D6	=0.04*(B5-B11)+0.90*(B6-B12)			
23	D7	=0.02*(B6-B12)+0.90*(B7-B13)			
24	C14	=SUMPRODUCT(C11:C13,B11:B13)			
25	C15	=C14+C14/((1.05)^5-1)			
26	C16	=SUMPRODUCT(C11:C13,B5:B7)			
27	C17	=C15-C16			

FIGURE 9.7 Spreadsheet model to maximize land expectation value in a diverse stand.

this desirable state. In most cases the actual size distribution of trees differs from the desirable distribution. How, then, should one proceed to convert the stand to the desirable state?

A simple approach to conversion is to use the best steady-state distribution as a cutting guide, cutting only the trees in excess of the desired number in the steady state. For example, we found with the model in Figure 9.7 that the specified diversity objectives would be met efficiently, in the steady state, with a stand that had, after harvest, 880.5 trees/ha in the smallest size class, 80 trees/ha (107.2 − 27.2) in the medium size class, and 16 trees/ha in the largest size class. Using this distribution as the cutting guide, the trees removed every 5 years are

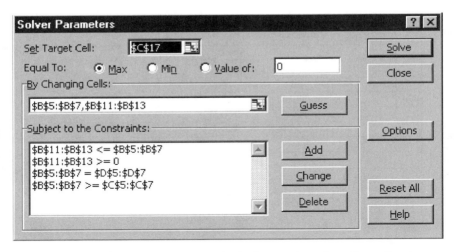

FIGURE 9.8 Solver parameters to maximize land expectation value with tree diversity.

defined by these equations:

$$h_{1t} = \max(0, y_{1t} - 880.5)$$
$$h_{2t} = \max(0, y_{2t} - 80.0) \quad (9.12)$$
$$h_{3t} = \max(0, y_{3t} - 16)$$

Figure 9.9 shows a spreadsheet to predict the consequences of this management rule over a period of 75 years. It is a modification of the spreadsheet used earlier to predict stand dynamics with harvest (see Figure 8.5). The difference lies in the specification of the harvest. The desired growing stock after harvest is in cells J4:L4. The formulas to calculate the harvest are based on Equations (9.12). For example, the harvest at time 0 in the largest size class is 0, because the desired growing stock is 16 trees/ha but the initial stand has only 14 trees/ha. Instead, the harvest in the second size class is 154 trees/ha, because the desired stock is 80 trees/ha and the actual stock is 234 trees/ha.

Figure 9.9 shows that in this instance, the cutting rule based on the steady-state solution leads quickly to a state that is similar to the desired state. Indeed, this method always approaches the desired state. It reaches it asymptotically if the desired state is truly a steady state. The speed of convergence to the desired state depends on the difference between the initial stand state and the desired state. In particular, if the initial stand has more trees than desired in every size class, then conversion is achieved immediately by the first harvest. The greater the deficit of trees in any size class relative to the desired level, the longer the conversion will take.

Economic and Environmental Management of Uneven-Aged Forests 171

	A	B	C	D	E	F	G	H	I	J	K	L
1	CONVERSION TO STEADY STATE											
2		Stock (trees/ha)				Cut (trees/ha)				Desired stock		
3	Year	Size1	Size2	Size3		Size1	Size2	Size3		after cut (trees/ha)		
4	0	840	234	14		0	154	0		880.5	80.0	16.0
5	5	845	106	14		0	26	0				
6	10	850	106	14		0	26	0				
7	15	854	106	15		0	26	0				
8	20	857	106	15		0	26	0				
9	25	860	106	15		0	26	0				
10	30	863	106	15		0	26	0				
11	35	866	107	15		0	27	0				
12	40	868	107	15		0	27	0				
13	45	870	107	15		0	27	0				
14	50	871	107	15		0	27	0				
15	55	873	107	15		0	27	0				
16	60	874	107	15		0	27	0				
17	65	875	107	15		0	27	0				
18	70	876	107	16		0	27	0				
19	75	877	107	16		0	27	0				
20												
21				Key cell formulas								
22	Cell	Formula								Copied to		
23	A5	=A4+5								A5:A19		
24	B5	=0.92*(B4-F4)-0.29*(C4-G4)-0.96*(D4-H4)+109								B5:B19		
25	C5	=0.04*(B4-F4)+0.90*(C4-G4)								C5:C19		
26	D5	=0.02*(C4-G4)+0.90*(D4-H4)								D5:D19		
27	F4	=MAX(0,B4-J$4)								F4:H19		

FIGURE 9.9 Spreadsheet model to convert an uneven-aged stand to an economically efficient and diverse stand structure.

9.5 GENERAL FORMULATION

The growth of an uneven-age stand, over a cutting cycle from time t to time $t + k$, where k is an integer, is described by these equations:

$$\begin{aligned}
\mathbf{y}_{t+1} &= \mathbf{G}(\mathbf{y}_t - \mathbf{h}_t) + \mathbf{c} \\
\mathbf{y}_{t+2} &= \mathbf{G}\mathbf{y}_{t+1} + \mathbf{c} \\
&\vdots \\
\mathbf{y}_{t+k} &= \mathbf{G}\mathbf{y}_{t+k-1} + \mathbf{c}
\end{aligned} \qquad (9.13)$$

The steady-state constraints require that the stand state be the same at the beginning as at the end of the cycle:

$$y_{t+k} = y_t \qquad (9.14)$$

In addition, the harvest must be less than the stock, and the harvest must be nonnegative:

$$h_t \leq y_t$$
$$h_t \geq 0 \qquad (9.15)$$

A solution of Equations and Constraints (9.13) to (9.15) describes a sustainable management regime. The choice of a particular management regime depends on the economic and environmental objectives of the forest owners.

Economic Management

The economic sustainable management consists of the harvest h_t and growing stock y_t that lead to the highest land expectation value LEV:

$$\max \text{LEV} = \mathbf{w}h_t - F + \frac{\mathbf{w}h_t - F}{(1+r)^{ku} - 1} - (\mathbf{w}y_t - F) \qquad (9.16)$$

where $\mathbf{w}$ is a column vector of value per tree, r is the interest rate per year, u is the time from t to $t+1$, in years, and F is the fixed cost per unit area. LEV is a linear function of h_t and y_t, and its highest value subject to Constraints (9.13) to (9.15) can be found by linear programming. For a given cutting cycle, k, the economic management regime is independent of the fixed cost. When fixed costs are high, longer cutting cycles tend to give higher land expectation values.

Environmental Management

A variety of environmental goals can be expressed with the objective function and/or with constraints on the sustainable harvest h_t and growing stock y_t. In particular, the management leading to the stand of highest tree diversity may be found by maximizing the least number of trees in any tree class, designated by a new variable, N_{min}, subject to Restrictions (9.8) to (9.10) and the additional constraint:

$$y_t \geq N_{min}$$

The difference between the land expectation value under this management and the highest possible land expectation value measures the opportunity cost of the environmental objective.

Mixed Objectives

Different combinations of economic and environmental objectives can also be represented with the model. For example, the economically efficient way of maintaining a stand with a specific number of trees in each size class can be found by solving a linear program with the land expectation value of Equation (9.16) as the objective function, subject to growth Equations and Constraints (9.13) to (9.15) and the additional constraints:

$$\mathbf{y}_t \geq \mathbf{y}^L$$

where $\mathbf{y}^L$ is a vector of the desired number of trees in each size class to be maintained in the stand. Some choices of $\mathbf{y}^L$ may be unfeasible.

Conversion to Desired State

A stand in a particular initial state $\mathbf{y}_0$ can be converted to a chosen steady state $\mathbf{y}^*$, $\mathbf{h}^*$ by applying the following harvesting rule:

$$\mathbf{h}_t = \max(0, \mathbf{y}_t - (\mathbf{y}^* - \mathbf{h}^*)) \quad \text{for } t = 0, \ldots, T$$

As T increases, $\mathbf{y}_t$ and $\mathbf{h}_t$ converge toward $\mathbf{y}^*$ and $\mathbf{h}^*$.

9.6 SUMMARY AND CONCLUSION

In this chapter, we have studied how to model some aspects of the economic and environmental management of uneven-aged forest stands. Throughout, we described the biological growth of the stands with the model developed in Chapter 8. The analysis emphasized sustainable management regimes that keep the forest stand in a steady state. In this context, the economic objective is to find the sustainable growing stock and harvest that maximize the land expectation value.

The land expectation value is the present value of the perpetual periodic income from the forest, minus the value of the initial growing stock. The land expectation value is influenced by the presence of fixed costs, i.e., costs per unit area independent of the level of harvest. High fixed costs make longer cutting cycles more economical, other things being equal.

Knowledge of the economic management regime, leading to the highest land expectation value, is relevant even when the management objective is noneconomical, because it clarifies the opportunity cost of alternative management policies.

Purely environmental objectives were also explored with the model. In particular, we used the maxmin criterion to investigate the management that led to the highest stand diversity, in the sense of maximizing the least number of trees in any size class. The difference between the land expectation value obtained with this management regime and the one resulting in the highest unrestricted land expectation value defines the opportunity cost of the environmental objective.

To manage the opportunity cost, we presented a method that would reach a specific environmental goal, such as a particular stand structure, while maximizing the land expectation value. Another, symmetric approach would be to optimize a particular environmental objective, such as stand diversity, subject to guaranteeing a specific land expectation value.

After choosing a desirable steady-state regime, stands in any initial state can be transformed to this desirable steady state with the simple rule of periodically harvesting the trees in excess of the desired distribution and letting the rest grow. Following this rule, any initial stand would converge to the desired state. The conversion would be immediate if the number of trees in the initial stand exceeded the desired number in all size classes.

PROBLEMS

9.1 Set up your own spreadsheet model to maximize the land expectation value of an uneven-aged forest stand like the one shown in Figure 9.1. Using the same data, verify that your results are the same as in Figure 9.1.

(a) Find the highest land expectation value for interest rates of 5%, 10%, and 15%, other things remaining the same. What is the effect of different interest rates on the maximum land expectation value?

(b) What is the effect of interest rates of 5%, 10%, and 15% on the best harvest?

(c) What is the effect of interest rates of 5%, 10%, and 15% on the best growing stock?

9.2 With the spreadsheet model to maximize the land expectation value of an uneven-aged stand (Figure 9.1), do the following:

(a) Solve the model with the original unit values for the trees in each size class, then with the original unit values reduced by half, and then with the original unit values doubled. Keep the interest rate at 5%. What is the relationship between the land expectation value and the unit values for trees?

(b) What is the effect of halving or doubling the unit value of the trees on the best harvest?

(c) What is the effect of halving or doubling the unit value of the trees on the best growing stock?

9.3 Equations (9.5) to (9.9) describe biological growth, steady state, feasible harvest, and nonnegativity constraints for a model to maximize the land expectation value of an uneven-aged forest with a ten-year cutting cycle.

(a) Rewrite the constraints to reflect a 15-year cutting cycle.

(b) Rewrite the constraints to reflect a 20-year cutting cycle.

(c) Rewrite the objective function for this model [shown just after Equation (9.9)] to reflect a 15-year cutting cycle.

(d) Rewrite the objective function for this model to reflect a 20-year cutting cycle.

9.4 (a) Set up your own spreadsheet model to maximize the land expectation value of an uneven-aged forest stand like the one shown in Figure 9.3. Using the same data, verify that your results are the same as in Figure 9.3.

(b) Modify this spreadsheet to reflect a 15-year cutting cycle by:
  i. Adding formulas to grow the stand from time $t + 2$ to time $t + 3$
  ii. Setting the harvest at time $t + 2$ at 0
  iii. Changing the cutting cycle in the formula of forest value
  iv. Changing the Solver parameters to reflect the steady state between times t and $t + 3$

(c) Compare the best harvest, growing stock, forest value, stock value, and land expectation value for cutting cycles of 5 years (Figure 9.1), 10 years (Figure 9.3), and 15 years, other things being equal. Considering only these three options, what is the economic cutting cycle?

(d) Recalculate the land expectation value for cutting cycles of 5, 10, and 15 years, assuming a fixed cost of $200/ha. Considering only these three options, what is the economic cutting cycle?

9.5 (a) Set up your own spreadsheet model to maximize tree diversity in an uneven-aged forest stand like the one shown in Figure 9.5. Using the same data, verify that your results are the same as in Figure 9.5.

(b) Add to the model a constraint to produce a constant harvest every 5 years worth at least $400. How does this change the harvest, growing stock, and opportunity cost of maximizing tree diversity? (*Note:* The land expectation value without any consideration of tree diversity is show in Figure 9.1.)

9.6 (a) Modify the spreadsheet model to maximize tree diversity in an uneven-aged forest stand (Figure 9.5) to maximize land expectation value while keeping at least 40 trees/ha in the largest size class without bound in the other two. Is there a feasible solution?

(b) Use this model to determine the highest number of trees you could maintain in the largest size class. What is the land expectation value for this management?

(c) Use this model to maximize the land expectation value while maintaining at least 30 trees/ha in the largest size class.

(d) Change the lower bound on the number of trees in the largest size class from 30 trees/ha to 20 trees/ha, 10 tree/ha, and 0 tree/ha. Make a table of the land expectation value as a function of the lower bound on the number of trees in the largest size class. Interpret the results.

9.7 (a) Set up your own spreadsheet model to convert an uneven-aged stand to an economically efficient and diverse stand structure like the one shown in Figure 9.9. Using the same data, verify that your results are the same as in Figure 9.9.

(b) Assume two different initial stand states:

Initial state 1: $y_{1,0} = 900$, $y_{2,0} = 250$, $y_{3,0} = 20$
Initial state 2: $y_{1,0} = 100$, $y_{2,0} = 50$, $y_{3,0} = 0$

Assuming the same desired stock and cutting rule as in Figure 9.9, how do these different initial conditions affect the growth of the stand over 75 years?

9.8 Modify the spreadsheet model to convert an uneven-aged stand to an economically efficient and diverse stand structure (Figure 9.9) to do the following:

(a) Compute the present value of the harvest every 5 years. Assume that the interest rate is 5% per year and that the tree values are as in Figure 9.1.

(b) Compute the cumulative present value of all the harvests over 75 years. This represents most of stand value, that is, the value of the initial stand of trees and the land they are growing on under this management regime, because the present value of the harvests from year 80 onward is negligible.

(c) How does the stand value differ given the two different initial states specified in Problem 9.7?

ANNOTATED REFERENCES

Adams, D.M., and A.R. Ek. 1974. Optimizing the management of uneven-aged forest stands. *Canadian Journal of Forest Research* 4:274–287. (Early nonlinear programming analysis of uneven-aged management.)

Bare, B., and D. Opalach. 1988. Determining investment-efficient diameter distributions for uneven-aged northern hardwoods. *Forest Science* 34(1):243–249. ("Balanced" diameter distributions are not investment efficient and are not necessary to achieve sustained yield.)

Boscolo, M., and J. Buongiorno. 1997. Managing a tropical rainforest for timber, carbon storage and tree diversity. *Commonwealth Forestry Review* 76(4):246–254. (Uses models similar to those in this chapter to determine the opportunity cost of storing carbon, maintaining diverse species and structural diversity.)

Buongiorno, J., J.L. Peyron, F. Houllier, and M. Bruciamacchie. 1995. Growth and management of mixed-species, uneven-aged forests in the French Jura: Implications for economic returns and tree diversity. *Forest Science* 41(3):397–429. (Discusses the MaxiMin criterion as an index of diversity, and investigates the trade-offs between diversity and economics in a multispecies model.)

Chang, S.J. 1981. Determination of the optimal growing stock and cutting cycle for uneven-aged stand. *Forest Science* 26:609–625. (Uses marginal analysis-rather than optimization-to define LEV for uneven-aged management.)

Duerr, W.A., and W.E. Bond. 1952. Optimum stocking of a selection forest. *Journal of Forestry* 50(1):12–16. (Early economic analysis of uneven-aged forest management.)

Haight, R.G., J.D. Brodie, and D.M. Adams. 1985. Optimizing the sequence of diameter distributions and selection harvests for uneven-aged stand management. *Forest Science* 31(2):451–462. (Discusses optimization for different initial conditions.)

Ingram, C.D., and J. Buongiorno. 1996. Income and diversity trade-offs from management of mixed lowland dipterocarps in Malaysia. *Journal of Tropical Forest Science* 9(2):242–270. (Suggests the optimality of the cutting rule mentioned at the end of this chapter for the economic conversion of a stand to an economic steady state.)

Kant, Shashi. 1999. Sustainable management of uneven-aged private forests: A case study from Ontario, Canada. *Ecological Economics* 30:131–146. (Uses a matrix model similar to that in this chapter to consider the impacts of guiding interest rates, taxes, and subsidies on forest structure and environmental values.)

Lu, H.C., and J. Buongiorno. 1993. Long- and short-term effects of alternative cutting regimes on economic returns and ecological diversity of mixed-species forests. *Forest Ecology and Management* 58:173–192. (Uses a multispecies model similar to the one in this chapter.)

Martin, G.L. 1982. *Investment-Efficient Stocking Guides for All-Aged Northern Hardwoods Forests*. CALS Research Report R3129. University of Wisconsin, Madison. 12 pp. (Presents tables of optimal growing stock and harvest as a function of the interest rate, cutting cycle, and site quality. Computed with Adams and Ek's model.)

CHAPTER 10

Multiple Objectives Management with Goal Programming

10.1 INTRODUCTION

In all the applications of linear programming studied in the preceding chapters, we assumed that there was a single overriding management objective, such as maximizing forest value or maximizing landscape diversity. This objective was represented by the objective function. Other objectives, for example, maintaining an even flow of timber production and maintaining a specific number of large trees in a stand, were expressed by constraints.

This way of handling multiple management objectives may not be satisfactory, for several reasons. Representing goals by standard linear programming constraints is somewhat rigid. For example, managers who follow even-flow policies have some flexibility in the amount of timber they produce year after year. The harvest does not have to be exactly constant; rather, it should be "nearly" constant. A constraint that imposes strict constancy is not only unrealistic; it may also lead to infeasible solutions, while a slight relaxation might allow for a feasible solution. In large problems with many constraints, it is usually hard to identify the constraint(s) responsible for infeasibility.

Furthermore, representing some goals by constraints in effect gives them priority over the goal reflected in the objective function, because the objective function is optimized within the feasible region defined by the constraints. Deciding which goal should be selected as the objective function and which ones should be reflected by constraints is often arbitrary and difficult.

Goal programming attempts to correct these limitations while retaining the useful basic structure and numerical solution of linear programming. Goal programming provides a way of striving toward selected objectives simultaneously, treating them all in the same manner, although perhaps giving them different weights.

10.2 EXAMPLE: RIVER POLLUTION CONTROL REVISITED

To introduce the basic concepts of goal programming we will use the example of the Maine pulp mill given in Section 2.3. Recall that the board of directors of the cooperative owning the pulp mill had decided that they wanted to produce enough mechanical and chemical pulp to keep at least 300 workers employed and to generate $40,000 of gross revenue per day while minimizing the amount of pollution caused by the mill. This multiple-objective problem was formulated as a standard linear programming model, as follows: Find X_1 and X_2, the daily production levels of mechanical and chemical pulp such that:

$$\text{Min } Z = X_1 + 1.5X_2 \qquad \text{(daily pollution, BOD units)}$$

subject to:

$X_1 + X_2 \geq 300$ (workers employed)

$100X_1 + 200X_2 \geq 40{,}000$ ($ of daily revenues)

$X_1 \leq 300$ (tons/day of mechanical capacity)

$X_2 \leq 200$ (tons/day of chemical capacity)

$X_1, X_2 \geq 0$

This model is useful, but it has some drawbacks. First, it treats the pollution goal in a way that is entirely different from the employment and revenue goals. Pollution is treated as an "optimizing" goal, while employment and revenues are treated as "satisficing" goals. Unless there is a compelling reason for pollution to figure in the objective function, it could as well be represented by a constraint, while employment or revenue could be the objective function (to maximize). Remember, however, that there can be only one objective function.

Second, the level of the goals may be unrealistic. If employment or/and revenue targets are too high, the problem may have no feasible solution. This would easily be corrected in this small example, but unrealistic goals can cause much trouble in large problems. Goal-programming techniques can be used to reduce these limitations of linear programming.

10.3 GOAL-PROGRAMMING CONSTRAINTS

In goal programming, all or some of the management goals are expressed by *goal constraints*. Consider the employment goal expressed by the first constraint in the foregoing linear program. In goal programming, that constraint is written:

$$X_1 + X_2 + L^- - L^+ = 300 \quad \text{(workers)}$$

where L^- and L^+ are *goal variables*, both nonnegative like other linear programming variables, such that L^- is the amount by which employment falls short of the 300 worker goal and L^+ is the amount by which employment exceeds the goal. Depending on the values of L^- and L^+, three cases may occur:

$L^- = 0$ and $L^+ > 0$, in which case $X_1 + X_2 = 300 + L^+$, that is, $X_1 + X_2 > 300$ and employment exceeds the goal of 300 workers.

$L^- = 0$ and $L^+ = 0$, in which case $X_1 + X_2 = 300$ and employment just meets the goal.

$L^- > 0$ and $L^+ = 0$, in which case $X_1 + X_2 = 300 - L^-$, that is $X_1 + X_2 < 300$ and employment is less than the goal.

With this system, the employment goal cannot cause infeasibility. Even if the goal is unrealistically high, L^- fills the gap between $X_1 + X_2$ and the goal. On the other hand, employment may also exceed the goal, thanks to the variable L^+.

In a similar manner, we can write a goal constraint for the revenues goal:

$$100X_1 + 200X_2 + R^- - R^+ = 40{,}000 \quad (\$/\text{day})$$

where R^- is the amount by which daily revenues fall short of \$40,000 and R^+ is the amount by which they exceed \$40,000. Both goal variables are nonnegative.

To handle the pollution goal in the same way, a level must be set for the pollution goal, say 400 units of BOD per day. Then the corresponding goal constraint is:

$$X_1 + 1.5X_2 + P^- - P^+ = 400 \quad (\text{BOD/day})$$

where P^- is the number of units of BOD below the goal and P^+ is the number in excess.

The other constraints remain the same as in the linear program; they set the limits on the capacity of production of the plant. In summary, all the constraints in the pulp mill problem recast as a goal program are:

$$X_1 + X_2 + L^- - L^+ = 300 \quad \text{(workers)}$$

$$100X_1 + 200X_2 + R^- - R^+ = 40{,}000 \quad (\$/\text{day})$$

$$X_1 + 1.5 X_2 + P^- - P^+ = 400 \quad \text{(BOD/day)} \qquad (10.1)$$

$$X_1 \le 300 \quad \text{(tons/day)}$$

$$X_2 \le 200 \quad \text{(tons/day)}$$

The next step, by far the most difficult part of goal programming, is to specify the objective function.

10.4 GOAL-PROGRAMMING OBJECTIVE FUNCTION

The objective function of a goal-programming problem contains some or all of the goal variables. The general purpose of the objective function is to make the total deviation from all goals as small as possible.

WEIGHTING THE GOALS

Because the goal variables are in completely different units, it is generally unsatisfactory to minimize the ordinary sum of all deviations, which in our example is to write the objective function as:

$$\min Z = L^- + L^+ + R^- + R^+ + P^- + P^+$$

What we should minimize instead is a weighted sum of the deviations from all the goals:

$$\min Z = w_l^- L^- + w_l^+ L^+ + w_r^- R^- + w_r^+ R^+ + w_p^- P^- + w_p^+ P^+$$

where $w_l^-, \ldots, w_p^+$ are constant *weights*. These weights have two purposes: (1) to make all weighted deviations commensurate, and (2) to express the relative importance of each goal. For example, w_l^- and w_p^+ must be such that $w_l^- L^-$ and $w_p^+ P^+$ are in the same units. Also, the relative magnitudes of w_l^- and w_p^+ express how important it is to fall short of the employment objective by one worker, relative to exceeding the pollution objective by one unit of BOD per day.

Multiple Objectives Management with Goal Programming

Assigning appropriate weights to the goal variables is not a simple task. It involves considerable judgment as well as trial and error. A possible approach is to compute an initial solution based on a first set of goal levels and weights. Then, if the stakeholders do not find this solution satisfactory, the goals and/or the weights are changed and a new solution is computed. These iterations continue until the solution is acceptable to all stakeholders. At that point, the weights should be good indicators of the relative importance of each goal to the stakeholders.

SIMPLIFYING THE WEIGHTS

To simplify the choice of weights, it is best to reduce the objective function to the simplest expression consistent with the problem at hand. Often, only a few variables are needed. In our example, the cooperative is concerned about falling short of the employment goal but not about exceeding it. Therefore, L^- only needs to be in the objective function. Any positive value of L^+ is welcome. Similarly, the cooperative is concerned about underachieving with respect to the revenue objective, so R^- must be kept small and should be in the objective function, while R^+ need not be. Finally, the cooperative wants to keep pollution low, so P^+, but not P^-, should be in the objective function. In summary, the relevant expression of the objective function for our example is:

$$\min Z = w_l^- L^- + w_r^- R^- + w_p^+ P^+$$

The difficult task of choosing weights can also be simplified by working with *relative deviations* from the goals. Rewrite the objective function as:

$$\min Z = u_l^- \frac{L^-}{300} + u_r^- \frac{R^-}{40,000} + u_p^+ \frac{P^+}{400}$$

where the new weights, u_l^-, u_r^-, and u_p^+, now express the relative importance of deviating by one percentage point from the corresponding goals. For example, assume that the cooperative owning the pulp mill feels that it is indifferent toward a 1% deviation from any of the three goals. This is equivalent to setting $u_l^- = u_r^- = u_p^+ = 1$. The expression of the objective function is then:

$$\min Z = \frac{1}{300} L^- + \frac{1}{40,000} R^- + \frac{1}{400} P^+$$

where Z is dimensionless. In this equation, the coefficients are very small, especially that of R^-. This may lead to round-off problems in calculating a solution. To avoid this, we multiply all the coefficients by the same large number, say

10,000, which will not change the value of the variables in the best solution. The new value of the objective function is then:

$$\min Z = 33.3L^- + 0.25R^- + 25P^+ \qquad (10.2)$$

A greater concern for employment than for the other two goals could be translated by setting $w_l^- = 2$ and $w_r^- = w_p^+ = 1$, leading to this objective function:

$$\min Z = 66.6L^- + 0.25R^- + 25P^+$$

Working with relative deviations from goals has the advantage of eliminating the different units of measurement. However, it should be kept in mind that this scheme has a precise meaning in terms of the relative value of the goals. For example, let us determine the relative value of employment and revenues implied by objective function of Equation (10.2).

To do this, note that the change in the objective function is related to the change in each one of the goal variables by this equation:

$$dZ = 33.3dL^- + 0.25dR^- + 25dP^+$$

where the prefix d indicates a change in the corresponding variable. Keeping pollution constant, that is, $dP^+ = 0$, the changes in employment and revenues that keep the value of the objective function constant must satisfy this equation:

$$0 = 33.3dL^- + 0.25dR^-$$

that is:

$$\frac{dR^-}{dL^-} \approx -133 \quad (\$/\text{day/worker})$$

Thus, the implication of the objective function of Equation (10.2) is that the members of the pulping cooperative are willing to see revenues decline by \$133 per day if employment increases by 1 worker.

10.5 SPREADSHEET FORMULATION AND SOLUTION

The structure of the goal-programming problem expressed by Constraints (10.1) and Objective Function (10.2) is that of an ordinary linear program. Thus, it can be solved by the simplex method. The corresponding spreadsheet is shown in Figure 10.1. The entries in bold are the coefficients of the constraints, the goal levels, the capacity limits, and the weights of the objective function. Cells B4:I4 contain the production variables, X_1 and X_2, and the goal variables measuring the deviations from the goals. Cells J5:J9 contain the formula for the

Multiple Objectives Management with Goal Programming

	A	B	C	D	E	F	G	H	I	J	K	L	M
1	RIVER POLLUTION PROBLEM												
2		Production		Deviations from goals									
3		X_1	X_2	L^-	L^+	R^-	R^+	P^-	P^+				
4		200	100	0	0	0	0	50	0	Total			
5	Employment	1	1	1	-1					300	=	300	(workers)
6	Revenues	100	200			1	-1			40000	=	40000	($/d)
7	Pollution	1	1.5					1	-1	400	=	400	(BOD/d)
8	Mech capacity	1								200	<=	300	(t/d)
9	Chem capacity		1							100	<=	200	(t/d)
10	Objective			33.3		0.25				25.0	0	min	
11													
12						Key cell formulas							
13	Cell formula											Copied to	
14	J5	=SUMPRODUCT(B5:I5,B$4:I$4)										J5:J10	

FIGURE 10.1 Spreadsheet goal-programming model for the river pollution problem.

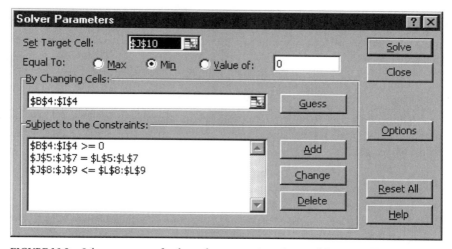

FIGURE 10.2 Solver parameters for the goal-programming solution of the river pollution problem.

left-hand side of Constraints (10.1), while cell J10 contains the formula of the objective function.

Figure 10.2 shows the Solver parameters for this problem. The target cell, J10, contains the weighted sum of the deviations from the goals. The Solver minimizes this objective by changing the production variables and the goal variables, B4:I4. The first set of constraints, B4:I4 >= 0, expresses the non-negativity of the decision variables. The second set of constraints, J5:J7 = L5:L7, refers to the employment goal, the revenues goal, and the pollution goal.

The third set of constraints, J8:J9 <= L8:L9, expresses the capacity limits on mechanical and chemical pulp production.

The best solution, shown in Figure 10.1, is:

$$X_1^* = 200 \text{ tons/day of mechanical pulp}$$

$$X_2^* = 100 \text{ tons/day of chemical pulp}$$

$$P^{-*} = 50 \text{ BOD/day}$$

All other variables are zero. That is, the employment and the revenue goals are met exactly, and the pollution goal is more than met, since pollution is less than the target. In a situation like this, where all goals are achieved, one may suspect that the original goals were too conservative. In essence, the system we are working with seems capable of doing better than what we asked. The solution seems *inefficient*.

Aware of this, the managers of the pulping cooperative decide to increase all targets boldly. They double the employment goal to 600 workers, increase the revenue goal by 50% to $60,000 per day, and reduce the pollution goal by 25% to 300 units of BOD per day. The new objective function is, using relative deviations from goals:

$$\min Z' = u_l^- \frac{L^-}{600} + u_r^- \frac{R^-}{60,000} + u_p^+ \frac{P^+}{300}$$

Assuming we still weight relative deviations from each goal identically so that the weights are 1 (but to avoid very small numbers we set them at 10,000 instead), we get:

$$\min Z' = 16.7 \, L^- + 0.17 \, R^- + 33.3 \, P^+$$

The revised spreadsheet model is in Figure 10.3. The only changes from Figure 10.1 are in the goal levels and the objective function weights. The Solver parameters remain the same as in Figure 10.2. The new best solution shown in Figure 10.3 is:

$$X_1^* = 300 \text{ tons/day of mechanical pulp}$$

$$X_2^* = 150 \text{ tons/day of chemical pulp}$$

$$L^{-*} = 150 \text{ workers}$$

$$P^{+*} = 225 \text{ BOD/day}$$

Therefore, this second trial falls short of the employment goal and pollutes more than we would like to while meeting the revenue goal exactly.

It may well be that the cooperative does not like this solution either. In that case, it can proceed through another iteration, changing weights in the objective

	A	B	C	D	E	F	G	H	I	J	K	L	M
1	RIVER POLLUTION PROBLEM, REVISED GOALS												
2		Production		Deviations from goals									
3		X_1	X_2	L^-	L^+	R^-	R^+	P^-	P^+				
4		300	150	150	0	0	0	0	225	Total			
5	Employment	1	1	1	-1					600	=	600	(workers)
6	Revenues	100	200			1	-1			60000	=	60000	($/d)
7	Pollution	1	1.5					1	-1	300	=	300	(BOD/d)
8	Mech capacity	1								300	<=	300	(t/d)
9	Chem capacity		1							150	<=	200	(t/d)
10	Objective			16.7		0.17			33.3	9997.5	min		
11													
12						Key cell formulas							
13	Cell formula											Copied to	
14	J5		=SUMPRODUCT(B5:I5,B$4:I$4)									J5:J10	

FIGURE 10.3 Spreadsheet for the river pollution problem with revised goals.

function or the level of the goals or both. It can also try to force a solution to satisfy certain restrictions. For example, assume that someone is adamant about the employment goal of 600 workers. This is equivalent to requiring L^- to be zero, which can be imposed simply by eliminating L^- from the model. In that case, the first constraint is equivalent to $X_1 + X_2 - L^+ = 600$; that is, $X_1 + X_2 \geq 600$. However, this is just a regular linear programming constraint, which may lead to infeasibilities. In fact, in this particular example it is not possible to reach an employment target of 600 workers. X_1 can be at most 300 tons/day, and X_2 can be at most 200 tons/day, implying a maximum employment of 500 workers.

In general, elimination of any of the goal variables may lead to an *infeasibility*. No infeasibility may arise from any one of the constraints in which both goal variables are present, however, because the goal variables always fill the gap between what is achieved and the set goal level.

10.6 OBJECTIVE FUNCTIONS WITH ORDINAL WEIGHTS

Up to now, the weights used in the objective function were cardinal numbers, measuring the relative value of each goal. There is another way of expressing the objective function in goal programming that uses *ordinal* instead of cardinal weights.

The procedure assumes that the decision makers are only able to rank the goals in order of importance, without specifying how much more important one goal is relative to another.

In our example, assume that the pulping cooperative has decided that employment has top priority, pollution has second priority, and revenues come last. The goal levels are the same as in Figure 10.3. As before, the cooperative managers are concerned about underachieving the employment and revenue goals and about exceeding the pollution goal. Consequently, the relevant objective function is:

$$\min Z = w_l^- L^- + w_r^- R^- + w_p^+ P^+$$

Goal programming with *ordinal weights* assumes that: $w_l^- \gg w_p^+ \gg w_r^-$; that is, the employment goal has a very large (in fact infinite) weight relative to the weight of the pollution goal, which itself has an infinite weight relative to the weight of the revenue goal.

The consequences of this weighting scheme are that all resources available must be used first to approach as close as possible to the employment goal. Then any remaining resources must be used to approach as close as possible to the pollution goal. Whatever is left is used to approach the revenue goal.

There are special algorithms to implement this concept. They solve a sequence of linear programs. In our example, the first linear program is:

$$\min L^-$$

subject to the constraints in Figure 10.3. The best solution is (see Problem 10.2):

$$X_1^* = 300 \text{ tons/day}$$
$$X_2^* = 200 \text{ tons/day}$$
$$L^{-*} = 100 \text{ workers}$$
$$P^{+*} = 300 \text{ BOD/day}$$
$$R^{+*} = \$10{,}000/\text{day}$$

The second linear program is:

$$\min P^+$$

subject to the constraints in Figure 10.3 and the additional constraints $L^- = 100$, $L^+ = 0$ workers, to keep employment as high as possible.

The solution of this second linear program is the same as the first one (see Problem 10.2). Thus, the pollution cannot be decreased if L^- is to be minimized before any other goal is considered. In this example, giving top priority to employment has, in effect, determined both revenues and pollution.

Ordinal ranking of goals is appealing because it seems, superficially at least, to do away with the difficult problem of specifying relative weights for the various goals. Nevertheless, it is questionable whether it leads to solutions that reflect the true values of the stakeholders. In forestry, as in all human endeavors, few goals are absolute. The statement that goal A has top priority over goal B is more a figure of speech than a strict guideline. It rarely means, as the algorithm presented earlier implies, that goal A must be satisfied to the maximum possible extent before goal B is considered at all. Most values are relative, and goal programming with cardinal weights is more likely to reflect these values. This leaves the problem of determining the relative weights. It is a difficult task, but one can hardly escape it.

10.7 GOAL PROGRAMMING IN EVEN-AGED FOREST MANAGEMENT

Goal-programming procedures can be very useful in adding flexibility to forest management models. As an example we will modify the dynamic model of even-aged management studied in Chapters 6 and 7. The model showed how to convert a forest from its initial state into a steady-state forest within a particular time period. This conversion was done while optimizing timber production, environmental objectives, forest value, or a combination of those objectives.

A potential application of this model is to find the management that would convert the forest to a state that stakeholders might judge desirable for a variety of economic and environmental reasons.

CONVERTING TO A DESIRABLE FOREST LANDSCAPE

As an example, consider the model of the even-aged short-leaf pine forest of Chapter 6, but with a less diverse initial state. At the beginning, there are trees in two age classes only: 400 ha with trees aged 1–10 years, and 100 ha with trees aged 11–20 years. There are no trees aged 21–30 years and no trees 31 years and older. One of our goals will be to make this forest more diverse.

In Chapter 6 we studied the recursive growth equations to predict the area by age class at the end of each decade as a function of the area at the beginning of the decade and of the harvest during the decade. Every area harvested was reforested immediately. Except for the initial state, the growth equations stay the same as those in Table 6.1.

Assume one of the management goals is to create a forest of high landscape diversity. While the initial state has only young stands, we would like to create a forest with a variety of age classes. This conversion should be done in 30 years. According to Shannon's index, the highest diversity would be achieved by a forest with equal area in every age class. In our example the total area of the forest is 500 ha, so the goal is to end up at the start of the fourth decade with four age classes, each one covering 125 ha. This is expressed by the new set of constraints:

$$A_{14} = A_{24} = A_{34} = A_{44} = 125 \text{ ha}$$

where A_{i4} is the area in stock in age class i at the start of the fourth decade. In addition, there will often be some constraint on production. Here, we assume that the stakeholders require a total annual production of 150,000 m^3 over 30 years.

It turns out that there is no feasible solution for this problem. That is, there is no sequence of harvest/reforestation activities that, starting with this initial forest, could produce both the desired distribution of age classes within 30 years and the desired production (see Problem 10.5).

A natural question to ask is: How close could we get to the desired distribution of age classes? This can be answered by goal programming. The four relevant goal constraints are:

$$A_{14} + A_1^- - A_1^+ = 125 \text{ ha}$$
$$A_{24} + A_2^- - A_2^+ = 125 \text{ ha} \tag{10.3}$$
$$A_{34} + A_3^- - A_3^+ = 125 \text{ ha}$$
$$A_{44} + A_4^- - A_4^+ = 125 \text{ ha}$$

where A_{i4} is the area that is actually in age class i at the start of the fourth decade. This area falls short of 125 acres by the amount designated by the goal variable A_i^-, and it exceeds it by the amount designated by the goal variable A_i^+. As usual, all the goal variables are nonnegative. The objective function, then, is to minimize the total deviation from the desired distribution of age classes at the start of the fourth decade:

$$\min Z = A_1^- + A_1^+ + A_2^- + A_2^+ + A_3^- + A_3^+ + A_4^- + A_4^+$$

Weights are not necessary because all variables are already in the same units (ha). Weights could be used, for example, if one were more concerned about a deviation in the older age classes than in the younger.

Spreadsheet Formulation and Solution

Figure 10.4 shows a spreadsheet version of this goal-programming model. It extends the spreadsheet in Figure 6.2. The initial stock of the forest is different, but the growth formulas in cells D5:F8 and the formulas for the periodic and total volume cut are the same.

Multiple Objectives Management with Goal Programming

	A	B	C	D	E	F	G	H	I	J	K
1	APPROACHING A DIVERSE LANDSCAPE										
2	age		Decade				Deviation:				
3	class		1	2	3	4	A^-	A^+	Total		Goal
4			Stock (ha)								
5		1	400	125	125	175	0	50	125	=	125
6		2	100	275	125	125	0	0	125	=	125
7		3	0	100	250	125	0	0	125	=	125
8		4	0	0	0	75	50	0	125	=	125
9								100	min		
10			Cut (ha)				m³/ha				
11		1	125	0	0		50				
12		2	0	25	0		250				
13		3	0	100	175		500				
14		4	0	0	0		600				
15			Cut(m³)			Total					
16			6250	56250	87500	150000	>=	150000			
17											
18				Key cell formulas							
19	Cell		Formula						copied to		
20	D5		=SUM(C11:C14)						D5:F5		
21	D6		=C5−C11						D6:F7		
22	D8		=C7−C13+C8−C14						D8:F8		
23	C16		=SUMPRODUCT($G11:$G14,C11:C14)						C16:E16		
24	F16		=SUM(C16:E16)								
25	H9		=SUM(G5:H8)								
26	I5		=F5+G5−H5						I5:I8		

FIGURE 10.4 Spreadsheet to convert an even-aged forest to a diverse landscape.

Cells K5:K8 contain the goal levels: the area in each age class at the start of the fourth decade. Cells G5:H8 contain the goal variables: the deviations between the actual areas and the goal areas. Cells I5:I8 contain the formulas for the left-hand side of Goal Constraints (10.3). The formula for the objective function is in cell H9. Cell F16 contains the formula for total production, and cell H16 contains its lower bound: 150,000 m³.

The corresponding Solver parameters are in Figure 10.5. The target cell, H9, is set to minimize the sum of the deviations from the goal area by age class. The changing cells are the areas cut by age class, in cells C11:E14, and the goal variables in cells G5:H8. The constraints specify that the cut must be less than the stock, the cut must be nonnegative, total production must exceed 150,000 m³, the goal variables must be nonnegative, and Goal Constraints (10.3) must hold.

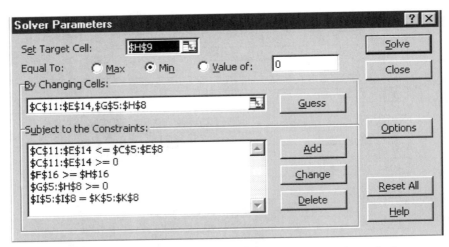

FIGURE 10.5 Solver parameters to convert an even-aged forest to a diverse landscape.

Figure 10.4 shows the best solution. Due to the production constraint, the desired forest landscape could not be obtained exactly. After 30 years, the forest would have 50 ha more than the goal in the youngest age class and 50 ha less than the goal in the oldest, for a total least deviation of 100 ha with respect to all the area goals.

10.8 GOAL PROGRAMMING IN UNEVEN-AGED STAND MANAGEMENT

Goal-programming methods may also be used to advantage in the context of uneven-aged stand management. In Chapters 8 and 9 we studied a linear programming model of uneven-aged management, paying special attention to sustainable regimes. Sustainability was guaranteed by steady-state constraints, whereby the harvest was just replaced by the stand growth over the cutting cycle. Within all possible steady states, we sought those that best met different economic or environmental goals, expressed either by the objective function or by constraints. In so doing it is not unusual to set constraints that are mutually exclusive, thus leaving a problem with no solution. The ability of goal programming to ensure feasibility is then advantageous. Not only are we sure to get a solution, but we can also find out how close we can get to the desired goal levels.

APPROACHING A DESIRABLE STAND STRUCTURE

In the examples of Chapters 8 and 9, the biological growth of an uneven-aged stand was described by a set of Growth Equations (8.10), predicting the stand state after 5 years, given its current state and harvest. Stand state and harvest

Multiple Objectives Management with Goal Programming

were described by the number of trees standing and harvested in each of three size classes. In the steady-state analysis, both the current stock and the harvest for each size class were choice variables. The steady-state constraints specified that the current state must be the same at the beginning as at the end of the cycle. For simplicity, and without loss of generality, we will assume again that the cutting cycle is equal to the length of the time unit in the growth model, 5 years in this example.

Assume that the owners of a woodlot that grows according to Equations (8.10) are fond of the current forest structure. They would like to maintain it in perpetuity. The current stand structure has 500 trees/ha in the smallest size class, 100 trees/ha in the intermediate size, and 20 trees/ha in the largest size class. We already know from Chapter 8 that letting the stand grow undisturbed would not give this stand structure but instead result in a climax forest with about 400 trees/ha in the smallest size class, 160 trees/ha in the intermediate size class, and 32 trees/ha in the largest size class (see Figure 8.3). This is in itself a useful to know: Unless the forest is already in its climax state, leaving it alone will not maintain it in its current state; natural growth and mortality will alter it.

Interestingly, even managing the stand could not obtain the desired sustainable structure. You may verify this with the model in Figure 8.7, set up to maximize timber production in the steady state. Add these constraints:

$$y_{1t} = 500 \text{ trees/ha}$$
$$y_{2t} = 100 \text{ trees/ha} \quad (10.4)$$
$$y_{3t} = 20 \text{ trees/ha}$$

and verify that there is no feasible solution. This is another important lesson: Although, as seen in Chapter 8, the number of managed steady states is infinite, not all stand structures are possible. They are limited by the biological potential of the forest under consideration. In particular, with Growth Equations (8.10), the desired steady state is not sustainable.

What management would produce a sustainable stand as close as possible to the desired structure of Equations (10.4)? This question can be answered by goal programming. To do this, we introduce goal variables in Equations (10.4) to allow the actual number of trees to differ from the desired. The goal constraints become:

$$y_{1t} + y_1^- - y_1^+ = 500 \text{ trees/ha}$$
$$y_{2t} + y_2^- - y_2^+ = 100 \text{ trees/ha} \quad (10.5)$$
$$y_{3t} + y_3^- - y_3^+ = 20 \text{ trees/ha}$$

where the goal variable y_1^- is the number of trees less than the goal of 500 trees/ha in the smallest size class, y_1^+ is the number of trees more than the 500 trees/ha, and the other goal variables have a similar definition for the medium and large size classes. All goal variables are nonnegative.

The objective function should keep the values of the goal variables as small as possible. One possibility is to minimize the sum of all the deviations from the goals. This is meaningful because all the goal variables have the same unit, trees/ha. Let us assume, however, that the large trees matter more to the owners than the small trees and that we weight the deviations from each goal by the volume per tree in Table 8.2. Thus, the objective function is:

$$\min Z = 0.2(y_1^- + y_1^+) + 1.0(y_2^- + y_2^+) + 3.0(y_3^- + y_3^+)$$

Spreadsheet Formulation and Solution

Figure 10.6 shows a spreadsheet version of this goal-programming model. It is a modification of the spreadsheet used in Chapter 8 to compute the steady-state regime maximizing production (Figure 8.7). The stand growth formulas in cells D5:D7 are the same.

	A	B	C	D	E	F	G	H	I	J
1	APPROACHING A DESIRABLE TREE DISTRIBUTION									
2			Stock			Deviation				
3	Size	y_t		y_{t+1}		y^-	y^+			
4	class	(trees/ha)		(trees/ha)		(trees/ha)		Total		Goal
5	1	500.0		500.0		0.0	0.0	500.0	=	500
6	2	100.0		100.0		0.0	0.0	100.0	=	100
7	3	18.0		18.0		2.0	0.0	20.0	=	20
8			Harvest				5.9	min		
9	Size	h_t	Volume							
10	class	(trees/ha)	(m³/tree)							
11	1	27.8	0.20							
12	2	9.9	1.00							
13	3	0.0	3.00							
14										
15										
16			*Key cell formulas*							
17	Cell	Formula							Copied to	
18	D5	=0.92*(B5-B11)-0.29*(B6-B12)-0.96*(B7-B13)+109									
19	D6	=0.04*(B5-B11)+0.90*(B6-B12)									
20	D7	=0.02*(B6-B12)+0.90*(B7-B13)									
21	G8	=SUMPRODUCT(C11:C13,F5:F8+G5:G8)									
22	H5	=B5+F5-G5							H5:H7	

FIGURE 10.6 Spreadsheet model to approach a desirable and sustainable stand.

Multiple Objectives Management with Goal Programming

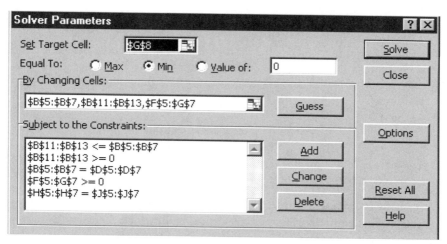

FIGURE 10.7   Solver parameters to approach a desirable and sustainable stand.

Cells J5:J7 contain the goal levels for the number of trees in each size class. Cells F5:G7 contain the goal variables: the deviations between the actual number of trees in each size class and the goal. Cells H5:H7 contain the formulas for the left-hand side of Goal Constraints (10.5). The formula for the objective function is in cell G8.

The corresponding Solver parameters are in Figure 10.7. The target cell, G8, is set to minimize the sum of the weighted deviations from the goal number of trees. The changing cells are the number of trees in stock, in cells B5:B7, the number of trees cut, in cells B11:B13, and the goal variables, in cells F5:G7. The constraints specify that the cut must be less than the stock, the cut must be nonnegative, the stock at the beginning of the cutting cycle must be the same as at the end, the goal variables must be nonnegative, and Goal Constraints (10.5) must hold.

Figure 10.6 shows the best solution. The sustainable stand structure closest to the one desired would have two trees/ha less than desired. The number of trees in the other two size classes would be as desired. This stand structure would be sustained by thinning the stand every 5 years, taking about 28 trees/ha from the smallest size class and about 10 trees/ha from the medium size class.

10.9 GENERAL FORMULATION

As we have seen in the previous examples, goal programming is just a particular formulation of the general linear programming problem. If cardinal weights are used in the objective function, the solution is obtained by applying the simplex method once; if ordinal weights are used instead, the simplex method must be applied several times.

CONSTRAINTS

A goal-programming model has at least some constraints, called *goal constraints*, that contain goal variables. The goal variables measure the deviation between management goal levels and actual outcomes. The general formulation of goal constraints is:

$$\sum_{j=1}^{n} a_{ij} X_j + D_i^- - D_i^+ = g_i \quad \text{for } i = 1, \ldots, G \quad (10.6)$$

where:

X_j is the jth activity (decision) variable

a_{ij} is the (constant) contribution to goal i per unit of activity j

g_i is a constant measuring the target of goal i, of which there are G

D_i^- is a goal variable that measures the amount by which the contribution of all activities to goal i falls short of the target

D_i^+ is the amount by which that contribution exceeds the target

All activities, X_j, and all goal variables, D_i^- and D_i^+, are greater than or equal to zero.

As long as both D_i^- and D_i^+ are present in a goal constraint, no infeasibility may result from that constraint. The goal variables always fill the gap between the goal level and what is actually achieved.

Other constraints may be present, of the usual linear programming variety; that is:

$$\sum_{j=1}^{n} a_{ij} X_j \leq, =, \text{ or} \geq b_i \quad \text{for } i = G+1, \ldots, m \quad (10.7)$$

OBJECTIVE FUNCTION

The general objective of goal programming is to minimize the sum of the weighted deviations from all goals. Thus the general form of the objective function is:

$$\min Z = \sum_{i=1}^{G} (w_i^- D_i^- + w_i^+ D_i^+) \quad (10.8)$$

where w_i^- and w_i^+ are the weights per unit of deviations D_i^- and D_i^+, respectively. The minimization is done by varying the activities and the goal variables simultaneously.

The weights fulfill two purposes. They express all deviations from goals in a common unit of measurement, and they reflect the relative importance of each goal. Deviations that concern the decision makers most get the larger weights relative to the others. Deviations that are of no concern or are looked at favorably may be omitted from the objective function altogether.

If deviations from some goals are unacceptable, the corresponding goal variable may be omitted from the appropriate goal constraint. For example, if goal g_i must not be exceeded, then D_i^+ may be omitted from the constraint, thus forcing D_i^+ to be zero. This procedure may, however, lead to an infeasibility, since the constraint is then equivalent to:

$$\sum_{j=1}^{n} a_{ij} X_j \leq g_i$$

The choice of weights for Objective Function (10.8) can be simplified by considering the relative, rather than the absolute, deviations with respect to goals. The new expression of the objective function is then:

$$\min Z = \sum_{i=1}^{G} \frac{u_i^- D_i^- + u_i^+ D_i^+}{g_i}$$

where each weight u_i now applies to a relative deviation from goal i.

ORDINAL WEIGHTING OF GOALS

This procedure consists of subjectively ranking all goals in order of priority. Then, all resources will be used to approach the top-priority goal as closely as possible. Any remaining resources are then used to approach the second goal as closely as possible. This is continued until the last goal is considered or until the best solution does not change.

In terms of the general objective function of Equation (10.8), ordinal weighting is equivalent to giving to the goal at the top of the list a weight that is infinitely large relative to the second, giving the second a weight infinitely large with respect to the third, and so on down to the bottom of the list.

The actual calculations proceed as follows: Assume that goal 1 has top priority, goal 2 has second priority, and so on. Assume further that the stakeholders are concerned about underachieving goal 1 and overreaching goal 2. Then, the first goal-programming problem solved is:

$$\min D_1^-$$

subject to Constraints (10.6) and (10.7).

Let D_1^{-*} be the best value of D_1^-, indicating how close one can get to goal 1. Then solve a second goal-programming problem:

$$\min D_2^+$$

subject to Constraints (10.6) and (10.7) and $D_1^- = D_1^{-*}$. This last constraint ensures that resources are allocated to achieving goal 2 only after goal 1 is satisfied to the fullest possible extent.

This procedure continues until all goals have been considered. Computer programs are available to go through these calculations automatically, once goals have been ranked. The general formulation of Objective Function (10.8) admits a mixture of ordinal and cardinal weights. In particular, some goals may have the same rank but different weights.

In using ordinal ranking of goals one must keep in mind its drastic implications, in fact giving infinite weight to a goal relative to another. Few forestry decisions involve such a drastic emphasis. More often, stakeholders seek a balance between goals. Cardinal weights reflect their willingness to make specific trade-offs. Even if exact cardinal weights cannot be specified, useful sets of weights can be arrived at by trial and error.

10.10 CONCLUSION: GOAL VERSUS LINEAR PROGRAMMING

Goal programming has two advantages relative to standard linear programming models for management problems with multiple objectives. First, all goals are represented in the same manner, by goal constraints and variables. All goal variables are in the objective function. This objective function minimizes the "cost" of deviating from the various goals. Second, as long as all goal variables are present in the model, goals can be set at any level without leading to infeasibilities. In practice, this is a considerable advantage relative to standard linear programming models. Goal programming can also help determine goal levels that are feasible and efficient.

On the other hand, determination of the relative importance of the various goals is difficult. Furthermore, the dual solution gives a useful measure of the trade-off between the goal in the objective function and those expressed by constraints in a standard linear programming model. To gain the most out of this feature, one should always try to work with an objective function expressed in units that are readily understood by most people.

For example, consider a management plan for a national forest developed with linear programming. The objective function is the value of the forest in terms of timber production only, expressed in dollars. But the forest must also

Multiple Objectives Management with Goal Programming

provide recreation opportunities, which may in fact be the main objective. This is expressed by a constraint to provide a specified number of visitor days. Then, the shadow price for the recreation constraint measures the marginal value of one unit of recreation, in dollars per visitor day. This is a very useful measure because it is expressed in a unit that most people can grasp. If the shadow price seems too low or too high, the recreation goal can be adjusted until a value of recreation acceptable to all parties involved is obtained.

PROBLEMS

10.1 Set up the river pollution problems shown in Figures 10.1 and 10.3 on two separate spreadsheets. Set up the corresponding Solver parameters, and verify the best solutions.

10.2 (a) Set up your own goal-programming spreadsheet model for the river pollution problem, as in Figure 10.3. Assume that the goals have been ranked, with the employment goal being judged most important, the pollution goal next most important, and the revenue goal least important.

(b) Change the objective function to minimize the underachievement of the employment goal (that is, to minimize L^-). What is the best production level for each type of pulp? How are the pollution and revenue goals met by this solution?

(c) Add constraints to force the underachievement of the employment goal to be equal to the value found in part (a). Then change the objective function to minimize the overachievement of the pollution goal (that is, to minimize P^+). Does the production level for either type of pulp change?

10.3 (a) Set up your own goal-programming spreadsheet model for the river pollution problem, as in Figure 10.3. Assume that the goals have been ranked, with the pollution goal being given top priority, the employment goal second priority, and the revenue goal last priority.

(b) Change the objective function to minimize the overachievement of the pollution goal (that is, to minimize P^+). What is the best production level for each type of pulp? How are the pollution and revenue goals met by this solution?

(c) Add constraints to the model that force P^- and P^+ to be equal to the value found in part (a). Then change the objective function to minimize underachievement of the employment goal (that is, to minimize L^-). What is the best production level for either type of pulp?

(d) Add other constraints that force L^- and L^+ to be equal to the value found in part (c). Then change the objective function to mimimize the underachievement of the revenue goal (that is, to minimize R^-). Does the production level for either type of pulp change?

10.4 Write the algebra of the growth equations, the constraints, and the objective function for the goal-programming spreadsheet model for the conversion of an even-aged forest to a diverse landscape (Figure 10.4). You may need to refer back to the spreadsheet model in Figure 6.2, on which this model was based.

10.5 (a) Set up your own goal-programming spreadsheet model for the conversion of an even-aged forest to a diverse landscape like the one shown in Figure 10.4. Using the same data, verify that your results are the same as in Figure 10.4.

(b) Modify the model to require conversion of the initial forest *exactly* to the desired forest structure after 30 years (that is, at the beginning of period 4) while producing at least 150,000 m³ of timber. Is there a feasible solution to this problem?

(c) Modify the model used in part (b) by removing the lower bound on timber production during the conversion and maximizing timber production. Compare the solution with this model to the solution in Figure 10.4. Explain the differences.

10.6 Write the algebra of the growth equations, the constraints, and the objective function for the goal-programming spreadsheet model for approaching a desirable stand structure in an uneven-aged forest (Figure 10.6). You may need to refer back to the spreadsheet model in Figure 8.7, on which this model was based.

10.7 (a) Set up your own goal-programming spreadsheet model for approaching a desirable stand structure in an uneven-aged forest like the one shown in Figure 10.6. Using the same data, verify that your results are the same as in Figure 10.6.

(b) Modify the model to find the land expectation value for this management policy. (*Hint*: Refer to Section 9.1 and use the same values per tree and interest rate as in Figure 9.1.) What is the opportunity cost of the policy described by the data in Figure 10.6 relative to the unconstrained land expectation value maximizing policy found in Figure 9.1?

(c) Add a constraint to the model used in part (b) to force the land expectation value to be no less than half of its unconstrained maximum. How does the solution change?

10.8 Write the algebra of the growth equations, the constraints, and the objective function for the goal-programming spreadsheet model for approaching a desirable stand structure (Figure 10.6), but for a cutting cycle of 10 years instead of 5 years.

10.9 Set up your own goal-programming spreadsheet model for approaching a desirable stand structure for an uneven-aged forest like the one shown in Figure 10.6, but for a cutting cycle of 10 years instead of 5 years. How does the length of the cutting cycle affect the best harvest and the best stocking?

10.10 Modify the goal-programming spreadsheet model for approaching a desirable stand structure (Figure 10.6) to minimize the largest deviation from the goal stand state. What is the effect on the best harvest and the best stock?

ANNOTATED REFERENCES

Boscolo, M., and J. Buongiorno. 1997. Managing a tropical rainforest for timber, carbon storage and tree diversity. *Commonwealth Forestry Review* 76(4):246–254. (Combines goal programming with the minimax criterion.)

Bottoms, K.E., and E.T. Bartlett. 1975. Resource allocation through goal programming. *Journal of Range Management* 28(6):442–447. (Application of goal programming to multiple-use management of a state forest with environmental and production goals.)

Buongiorno, J., J.L. Peyron, F. Houllier, and M. Bruciamacchie. 1995. Growth and management of mixed-species, uneven-aged forests in the French Jura: Implications for economic returns and tree diversity. *Forest Science* 41(3):397–429. (Source of the goal-programming model used in Section 10.8.)

Dane, C.W., N.C. Meador, and J.B. White. 1977. Goal programming in land use planning. *Journal of Forestry* 76(6):325–329. (Application of goal programming to national forest planning.)

Daz-Balteiro, L., and C. Romero. 1998. Modeling timber harvest–scheduling problems with multiple criteria: An application in Spain. *Forest Science* 44(1):47–57. (Describes a harvest–scheduling model with three criteria: NPV over the planning period, even flow, achieving area control regulation, and ending inventory.)

Dyer, A.A., J.G. Hof, J.W. Kelly, S.A. Crim, and G.S. Alward. 1979. Implications of goal programming in forest resource allocation. *Forest Science* 25(4):535–543. (Compares linear-programming and goal-programming models for forest planning.)

Field, D.B. 1973. Goal programming for forest management. *Forest Science* 19(1):125–135. (Application of goal programming to management of a small forest property.)

Gorte, R.W. 1999. Multiple use in the national forests: Rise and fall or evolution? *Journal of Forestry* 97(10). (Discusses the philosophical and political issues involved in the tension between advocates of multiple-use and ecosystem management.)

Hotvedt, J.E. 1983. Application of linear goal programming to forest harvest scheduling. *Southern Journal of Agricultural Economics* 15(1):103–108. (Application of goal programming to industrial forest management with multiple financial goals.)

Howard, A.F., and J.D. Nelson. 1993. Area-based harvest scheduling and allocation of forest land using methods for multiple-criteria decision making. *Canadian Journal of Forest Research* 23:151–158. (Describes how approaches other than goal programming to dealing with complex choice problems with multiple criteria and multiple stakeholders could be used to improve timber harvest–scheduling and land allocation decisions.)

Kangas, A., J. Kangas, and J. Pykäläinen. 2001. Outranking methods as tools in strategic natural resources planning. *Silva Fennica* 35(2):215–227. (Reviews the European experience with approaches, other than goal programming, to dealing with complex choice problems with multiple criteria and multiple stakeholders.)

Liu, G., and L.S. Davis. 1995. Interactive resolution of multiobjective forest planning problems with shadow price and parametric analysis. *Forest Science* 41(3):452–469. (Describes a alternative approach to a multiple-use planning problem originally addressed as a goal-programming problem.)

Mendoza, G.A. 1987. Goal-programming formulations and extensions: An overview and analysis. *Canadian Journal of Forest Research* 17:575–581. (Review of the use and potential applications of goal programming in forest resource management.)

Mendoza, G.A. 1988. Determination of efficient range of target levels in multiple-objective planning. *Natural Resource Modeling* 2(4):653–667. (Describes a nonoptimization approach to determining efficient dimensions for the decision space of a goal-programming model.)

Mitchell, B.R., and B.B. Bare. 1981. A separable goal programming approach to optimizing multivariate sampling designs for forest inventory. *Forest Science* 27(1):147–162. (Application of goal programming to optimization of forest inventory sampling effort.)

Prato, T. 1999. Multiple-attribute decision analysis for ecosystem management. *Ecological Economics* 30:207–222. (Describes how approaches, other than goal programming, to dealing with complex choice problems with multiple criteria and multiple stakeholders could be used to improve watershed management.)

Pomerol, J., and S. Barba-Romero. 2000. *Multicriterion Decision in Management: Principles and Practice* . Kluwer Academic, Boston. 395 pp. (Basic reference on discrete multicriterion decision making.)

Ragsdale, C.T. 1998. *Spreadsheet Modeling and Decision Analysis: A Practical Introduction to Management Science* . South-Western College Publishing, Cincinnati, OH. 742 pp. (Chapter 7 shows how to use spreadsheets to formulate and solve goal-programming models.)

Romesburg, H.C. 1974. Scheduling models for wilderness recreation. *Journal of Environmental Management* 2(2):159–177. (Application of goal programming to recreational use scheduling in wilderness areas.)

Rustagi, K.P., and B.B. Bare. 1987. Resolving multiple-goal conflicts with interactive goal programming. *Canadian Journal of Forest Research* 17:1401–1407. (Describes a two-phase interactive goal-programming approach.)

Schuler, A.T., H.H. Webster, and J.C. Meadows. 1977. Goal programming in forest management. *Journal of Forestry* 75(6):320–324. (Application of goal programming to multiple-use management of a national forest with recreation, hunting, timber, and grazing goals.)

CHAPTER **11**

Forest Resource Programming Models with Integer Variables

11.1 INTRODUCTION

The forestry applications of linear and goal programming presented in the previous chapters all had continuous variables. For example, in even-aged management models, the areas cut in different age classes could take any fractional value. The number of trees per unit area could also take a fractional value in uneven-aged management models, because it referred to an average number of trees over an area usually larger than one hectare, rather than to the number of trees on a particular hectare of land.

However, there are many forest management decisions that deal with items that cannot be divided. For example, a whole person may have to be assigned to a particular job. In bidding on a particular timber sale, one must decide whether to bid for the entire sale or not to bid at all. In building a network of forest roads, the sections of roads must usually link specific points in a continuous network. Building only a portion of a road section would be pointless. Similarly, bridges must be built completely or not at all, although different kinds of bridges may be chosen.

Situations like these can be addressed using models with integer variables. Methods exist to find, or at least approximate, the best value of such integer variables, given a particular objective function and a set of constraints. If only integer variables are involved, we have a *pure integer* programming problem. If both integer and continuous variables are used in a model, we have a *mixed integer* programming problem.

In the next section we shall first see briefly how integer programming differs from standard linear programming. We shall then study in detail some applications of pure and mixed integer programming in forestry.

11.2 SHORTCOMINGS OF THE SIMPLEX METHOD WITH INTEGER VARIABLES

Surprisingly perhaps, pure or mixed integer programming problems are more difficult to solve than linear programming problems with only continuous variables. For this reason, it is tempting just to solve an integer programming problem with the simplex method and then round off the solution. This approach does sometimes yield solutions sufficiently close to the optimal solution for all practical purposes. The rounding strategy is especially useful when there are many similar alternatives to choose from. However, a rounded solution may not be feasible, or if feasible it may be far from optimal. The following example will show why this may occur.

THE CONSULTANT'S PROBLEM

A forestry consulting firm has the opportunity of contracting for five different projects. Three of the projects are located in Georgia and two in Michigan. Each the Georgia projects would require 1 person-year of work and return a profit of $10,000. Each of the Michigan projects would require 10 person-years and return $50,000. The firm has a staff of 20 people. For which projects should they contract to maximize their total profit?

MODEL FORMULATION

This simple problem is clearly of the integer type because it is not possible to contract for part of a project. Let X_g be the number of projects in Georgia that the consultant takes on. The possible values of X_g are 0, 1, 2, or 3. Similarly, let X_m be the number of projects taken on in Michigan. Then, $X_m = 0$, 1, or 2.

Forest Resource Programming Models with Integer Variables

With these decision variables, the problem of the consulting firm can be expressed as follows: Find X_g and X_m, both positive integers, such that:

$$\max Z = X_g + 5X_m \quad (\$10{,}000)$$

subject to:

$$X_g \leq 3 \quad \text{(projects)}$$
$$X_m \leq 2 \quad \text{(projects)}$$
$$X_g + 10X_m \leq 20 \quad \text{(person-years)}$$

GRAPHICAL SOLUTION

This problem can be solved graphically because it has just two decision variables. The method is similar to the one used in Chapter 3 for linear programming with continuous variables. Figure 11.1 shows the boundaries of the feasible region for this problem, as defined by the horizontal axis, the vertical line $X_g = 3$, and the line $X_g + 10X_m = 20$. This last goes through the point D, with coordinates ($X_g = 0$, $X_m = 2$) and the point A, with coordinates ($X_g = 2$, $X_m = 1.7$). The constraint $X_m \leq 2$ is not graphed because it is redundant.

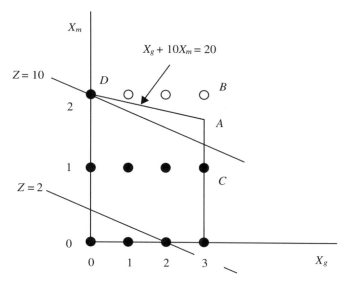

FIGURE 11.1 Graphic solution of the consultant's problem.

However, in contrast to the linear programming case, the feasible region is not the entire area within these boundaries. Only the grid of points highlighted by the black dots in Figure 11.1 represents feasible solutions. Only at these points are all the constraints satisfied, including the requirement that X_g and X_m be integers.

The best solution is then represented by the black dot in Figure 11.1 at which the value of the objective function is highest. To find it, we draw the graph of the objective function for an arbitrary value of Z, say $Z = 2 = X_g + 5X_m$. This line goes through the points with coordinates $(X_g = 2, X_m = 0)$ and $(X_g = 0, X_m = 0.4)$.

The value of the objective function increases regularly as the line representing the objective function gets farther from the origin. The value of the objective function is highest when the line goes through point D. This point corresponds to the best solution:

$$X_g^* = 0 \text{ projects in Georgia}$$
$$X_m^* = 2 \text{ projects in Michigan} \qquad (11.1)$$
$$Z^* = 10 \ (\$10,000)$$

Problems Arising from Rounded Solutions

If instead we had solved the consultant's problem by ordinary linear programming, the best solution would have corresponded to point A, and the "best" value of Z identified would have been impossible to achieve. Clearly, that solution is not feasible, because it is not an integer solution. It would suggest taking on 1.7 projects in Michigan, which is not possible.

Assume that we recognize this and round the best linear programming solution to the nearest integer. This would then lead to the solution corresponding to point B. This is an integer solution, but it is not feasible, because the solution at point B requires more than the 20 workers (we assume that the firm does not want to hire more staff).

Suppose that we see this and decide to round the linear programming solution to the "nearest"-integer feasible solution. This solution would correspond to point C in Figure 11.1. But the value of the objective function at point C is only $Z_c = 3 + 5(1) = 8$, that is, $80,000. This is $20,000 less than the true best solution at point D.

This example shows the importance of finding exact solutions of programming problems with integer variables. Unfortunately, integer programs can be very difficult to solve, especially when there are many integer variables. Integer programs that deal with *binary* variables, that is, variables that can take only the values 0 or 1, are generally easier to solve. They also have numerous applications in forest management. In the remainder of this chapter we will deal exclusively with models having binary variables.

11.3 CONNECTING LOCATIONS AT MINIMUM COST

To start, it is worth observing that some problems with binary variables have such simple solutions that they do not require sophisticated programs to solve them. In fact, they can sometimes be solved by hand. An example of such a simple yet practical problem is the *minimum-spanning tree*. It arises every time we want to connect a set of points (locations) so that the total "cost" or "length" of the connections is minimized.

LOGGING OKUME WITH LEAST ENVIRONMENTAL IMPACT

A valuable tree species cut in the tropical forest of West Africa is okume (*Aucoumea klaineana*). Okume wood is easy to peel into veneer sheets, and it is used to make high-quality marine plywood, among other things. Okume trees are typically found in clusters, amid vast areas of forest stocked with trees of little commercial value. Roads have to be built to reach the okume stands, but road building disturbs the surrounding virgin forest. Assume that a logging company has obtained a forest concession for the exploitation of okume in an area of virgin forest. However, they are under strict injunction to minimize the disturbance their logging will cause.

The forest survey has produced a map of the okume stands, sketched in Figure 11.2. Each circle shows the location of a stand of okume trees. A forest engineer has started planning a possible road network that could connect the stands to the Taiwani river. There the logs would be tied into rafts and floated down to a port where they would be loaded on cargo ships. Not all the roads that have been drawn on the map are necessary, but only those shown are possible. The length of each road segment is shown in kilometers. The object is to find the shortest road network that connects all of the okume stands to the Taiwani.

MANUAL SOLUTION

This is a typical minimum-spanning-tree problem. The solution requires only pencil and paper and some attention. The shortest road network that connects all the okume stands can be found by applying the following algorithm:

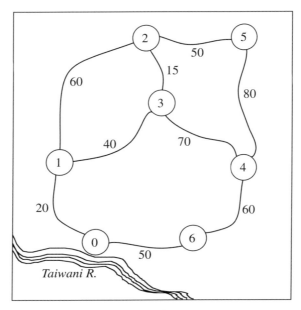

FIGURE 11.2 Possible roads for the exploitation of okume stands. Road lengths are in kilometers.

Select any stand arbitrarily, and connect it to the nearest stand.
Pick the unconnected stand closest to a connected stand, and connect the two. Repeat until all stands have been connected.

Any tie can be broken arbitrarily, and the result will still be an optimum solution. A tie simply shows that there is more than one best solution. Let us apply this algorithm to the network of possible roads shown in Figure 11.2.

Start arbitrarily at stand 5. The closest stand is 2. Thus, connect stands 2 and 5.
The unconnected stand closest to either stand 2 or stand 5 is stand 3, which is closest to 2. Thus, connect stands 2 and 3.
The unconnected stand closest to either stand 5, stand 2, or stand 3 is stand 1, which is closest to 3. Thus, connect stands 3 and 1.
The unconnected stand closest to either stand 5, stand 2, stand 3, or stand 1 is stand 0, which is closest to 1. Thus, connect stands 1 and 0.
The unconnected stand closest to either stand 5, stand 2, stand 3, stand 1, or stand 0 is stand 6, which is closest to 0. Thus, connect stands 0 and 6.

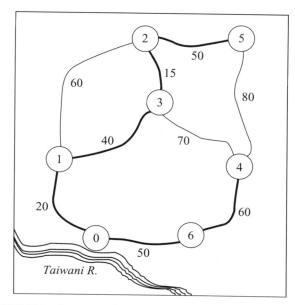

FIGURE 11.3 Shortest road network for the harvesting of okume stands.

The unconnected stand closest to stand 5, stand 2, stand 3, stand 1, stand 0, or stand 6 is stand 4, which is closest to 6. Thus, connect stands 4 and 6.
All locations have been connected. Stop.

The bold line in Figure 11.3 shows the shortest road network that connects all the okume stands. The total length of the network is 235 km. Because there were no ties, no other system of the same length could connect all the stands to the Taiwani river. You may verify that the choice of initial stand does not affect the solution by starting the algorithm with a stand other than stand 5.

Naturally, the same approach would be applicable, regardless of the measure of "distance" between locations. For example, the physical distance could be replaced by the estimated cost of building a particular road section. Then the objective would be to minimize the cost of connecting all stands, if the company had the option to mimimize cost rather than the environmental impact. In that case, if costs estimates were available, they would be better than simple distance. For a network of equal length, the cost will be much higher if a work of art, such as a bridge, is needed.

11.4 ASSIGNING FORESTERS TO JOBS

Another useful model with binary variables that has a simple solution is the so-called assignment model. The model is useful to assign different people, machines, or objects to different tasks or locations, or land to alternative uses, in order to optimize an objective that depends on the way the assignment is done.

PROBLEM DEFINITION

As an example, assume that you are a forest supervisor who has just hired three new graduate foresters. You have examined their application materials carefully, and you have interviewed them all. On this basis you believe that although they are all good, some will perform better at particular tasks than others.

Four tasks need additional staff urgently: timber sales management, public relations, fire fighting, and inventory. Because of the need for additional staff in all tasks, you do not want to assign more than one forester to a task. How should you assign each new forester to a task in order to maximize the productivity of all three?

DATA

To answer that question, you need some measure of productivity. Clearly, not much is available besides your subjective judgment. You might try to quantify that judgment by giving a score of 0–10 to each combination of person and task. Thus, if you expected that a particular individual would excel at a certain job, you would give that combination a score of 10. If you expected that an assignment would be a disaster, you would give it a score of 0. Any value between zero and 10 is possible, as are ties.

Table 11.1 shows the scores for all possible assignments of foresters to tasks. For example, Peter is expected to do very well in public relations (score of 9) but poorly in inventory work (score of 3).

TABLE 11.1 Expected Performance of Foresters in Different Tasks

Forester	Service			
	1 Timber sales	2 Public relations	3 Fire control	4 Inventory
1 Peter	5	9	8	3
2 Paul	7	6	4	6
3 Mary	6	9	5	8

Programming Formulation

Your problem is to assign the foresters to tasks in such a way that the total score, your measure of the overall performance of the three people together, is maximized. This problem can be expressed as an integer programming problem.

Define a set of decision variables X_{ij}, each equal to 1 if forester i is assigned to task j and 0 otherwise. The subscript i varies from 1 to 3, and j varies from 1 to 4, as shown in Table 11.1. Thus, there are 12 decision variables in this problem. All the possible assignments of Peter can be expressed by the following inequality:

$$X_{11} + X_{12} + X_{13} + X_{14} \le 1 \tag{11.2}$$

Because each variable in this constraint may take only the value 0 or 1 and all the variables must add up to no more than 1, only one variable may take the value 1. Thus, Constraint (11.2) says that there is only one Peter to assign to a task, although Peter might not be assigned to any task.

In the same manner, the equation that describes the possible assignments of foresters Paul and Mary are:

$$\begin{aligned} X_{21} + X_{22} + X_{23} + X_{24} &\le 1 \quad \text{for Paul} \\ X_{31} + X_{32} + X_{33} + X_{34} &\le 1 \quad \text{for Mary} \end{aligned} \tag{11.3}$$

We must now express the fact that each task may be assigned at most one forester, with the possibility that a task will not receive additional staff. For example, the constraint

$$X_{11} + X_{21} + X_{31} \le 1 \tag{11.4}$$

says that at most one forester may be assigned to timber sales. The less-than-or-equal-to inequality recognizes the possibility that nobody might be assigned to timber sales, in which case all three variables will be zero. The right-hand-side value of 1 ensures that timber sales will receive at most one additional staff.

The corresponding constraints for the three other tasks are:

$$\begin{aligned} X_{12} + X_{22} + X_{32} &\le 1 \quad \text{for public relations} \\ X_{13} + X_{23} + X_{33} &\le 1 \quad \text{for fire control} \\ X_{14} + X_{24} + X_{34} &\le 1 \quad \text{for inventory} \end{aligned} \tag{11.5}$$

Last, the objective function is written as follows. Let p_{ij} be the measure of the expected performance of forester i if he or she is assigned to task j. For example, Table 11.1 shows that $p_{23} = 4$. Then the contribution to overall performance of

TABLE 11.2 Linear Programming Tableau for the Assignement of Foresters to Tasks

	X_{11}	X_{12}	X_{13}	X_{14}	X_{21}	X_{22}	X_{23}	X_{24}	X_{31}	X_{32}	X_{33}	X_{34}	
Z	5	9	8	3	7	6	4	6	6	9	5	8	
Peter	1	1	1	1									<=1
Paul					1	1	1	1					<=1
Mary									1	1	1	1	<=1
Timber sales	1				1				1				<=1
Public relation		1				1				1			<=1
Fire control			1				1				1		<=1
Inventory				1				1				1	<=1

a particular assignment is $p_{ij} X_{ij}$, and the contribution of all possible assignments is the sum of $p_{ij} X_{ij}$ over all possible combinations of i and j. Thus, in our particular example, the expression of the objective function is:

$$Z = 5X_{11} + 9X_{12} + 8X_{13} + 3X_{14} + 7X_{21} + 6X_{22} + 4X_{23} + 6X_{24}$$
$$+ 6X_{31} + 9X_{32} + 5X_{33} + 8X_{34} \qquad (11.6)$$

Objective Function (11.6) and Constraints (11.2) to (11.5) constitute a standard linear program, except for the fact that each variable is limited to the integer value 0 or 1. This linear program is presented in Table 11.2. To emphasize its particular structure: two 1's appear in each column, and the right-hand side of every constraint is 1.

It turns out that the best solution of an assignment problem solved as if it were a standard linear program is always such that variables take the value of either 0 or 1, even though the variables are not explicitly constrained to be binary. Thus, the standard simplex method of linear programming can be used to solve assignment models of this kind.

SPREADSHEET SOLUTION

Figure 11.4 shows a spreadsheet set up for this assignment problem. Cells B4:E6 contain the scores of each forester for each task; these are input data, indicated in bold characters. Cells B10:E12 contain the decision variables, the X_{ij} in the notation used above. Cells F10:F12 contain the formulas of the left-hand side of the constraints on the foresters [Equations (11.2) and (11.3)]. Cells B13:E13 contain the formulas of the left-hand side of the constraints on the tasks [Equations (11.4) and (11.5)]. Cell A16 contains the formula of the objective function: the total score of the assignments.

The corresponding Solver parameters are in Figure 11.5. The objective is to maximize the target cell, A16, by changing the decision variables in cells

Forest Resource Programming Models with Integer Variables

	A	B	C	D	E	F	G	H
1	ASSIGNING FORESTERS							
2		Score of forester in:						
3	Forester	1 Timber	2 Public	3 Fire	4 Inventory			
4	1 Peter	5	9	8	3			
5	2 Paul	7	6	4	6			
6	3 Mary	6	9	5	8			
7								
8		Assignment of foresters to:						
9	Forester	1 Timber	2 Public	3 Fire	4 Inventory	Total		
10	1 Peter	0.00	0.00	1.00	0.00	1	<=	1
11	2 Paul	1.00	0.00	0.00	0.00	1	<=	1
12	3 Mary	0.00	1.00	0.00	0.00	1	<=	1
13	Total	1.00	1.00	1.00	0.00			
14		<=	<=	<=	<=			
15		1	1	1	1			
16	24	Z(max)						
17								
18			Key cell formulas					
19	Cell formula					Copy to		
20	B13	=SUM(B10:B12)				B13:E13		
21	F10	=SUM(B10:E10)				F10:F12		
22	A16	=SUMPRODUCT(B4:E6,B10:E12)						

FIGURE 11.4 Spreadsheet model to assign foresters to tasks.

FIGURE 11.5 Solver parameters to assign foresters to tasks.

B10:E12. The first set of constraints indicates that the decision variables must be nonnegative. There is no constraint forcing a decision variable to be 0 or 1; the problem is solved as an ordinary linear program. The second set says that each task may be assigned at most one forester [Constraints (11.4) and (11.5)]. The third set of constraints recognizes that there is only one of each forester [Constraints (11.2) and (11.3)].

Figure 11.4 shows the best solution. The highest measure of performance (24 points) is achieved by assigning Peter to fire protection, Paul to timber sales, and Mary to public relations. This best though not unique, solution has some interesting features.

First, although Peter is better at public relations than at any other task, as part of a team with Paul and Mary it is best to assign Peter to fire protection. Second, no one is assigned to the inventory task. The best solution is to assign each forester entirely to one task, although there is no explicit constraint in the model to force this to happen. Indeed, assigning foresters partly to different tasks could satisfy all the constraints. Yet the solution is integer. This is true of all assignment models, regardless of the objective function and number of constraints. However, fractional solutions may result if additional constraints are imposed on the system. In that case, special integer programming algorithms are necessary to solve the problem. The next example will illustrate this more general situation.

11.5 DESIGNING AN EFFICIENT ROAD NETWORK

Although the minimum-spanning-tree and the assignment models have many applications, they deal only with special cases. The problems must have very special structures to yield integer solutions with simple solution methods. Solving more general problems requires specialized integer programming algorithms and related computer programs. To illustrate their application, we will study a problem similar to one solved by Kirby (1975). It deals with the development of a road network suitable to serve a set of multiple-use forestry projects. Roads are a big issue in forestry because the need for access must be balanced against the negative environmental effects of roads and road building. It is thus often necessary to design road networks as short as possible while meeting various management objectives. Optimization methods can be of great help in this process. Because each road section must either be built to connect two locations or not be built at all, integer programming with (0, 1) variables is a natural way to tackle such problems.

Forest Resource Programming Models with Integer Variables

PROBLEM DEFINITION AND DATA

The general setting for our example is illustrated in Figure 11.6. This figure shows a road network serving four potential multiple-use forestry projects named Eagle, Highland, Tall Pine, and Golfech. The roads have not yet been built; the map represents only the possible ways the projects can be connected to the existing county road, represented by a bold line.

Each project must be done completely or not at all. But not all four projects need to be done; thus the simple spanning-tree method cannot be used. On the other hand, the projects that are done must be connected to the existing county road because they will require some form of road access. The cost of building each road section depends on its length, its topography, and any necessary work of art. For example, road section 1 is especially expensive because it includes a bridge across Stony Brook. The civil engineers attached to the project have estimated the costs of each road section, shown in Table 11.3. These costs are the cumulative discounted cost of building and maintaining the roads over the entire life of the project.

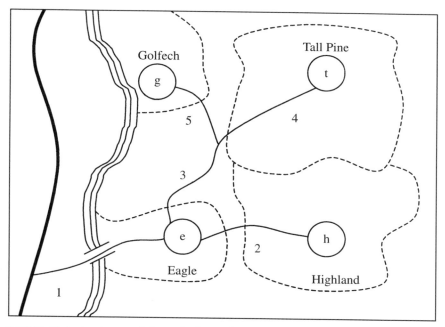

FIGURE 11.6 Road network for multiple-use forest development. Dotted lines indicate the project boundaries.

TABLE 11.3 Cost of Building Each Road Section

	Section				
	1	2	3	4	5
Cost (10^6 $)	0.8	0.4	0.3	0.2	0.4

TABLE 11.4 Output and Cost, by Project

	Project			
Result	Eagle, e	Highland, h	Tall Pine, t	Golfech, g
Recreation (10^3 rvd)	1	1	2	3
Timber (10^3 m^3)	6	8	13	10
Cost (10^6 $)	0.7	0.1	0.5	0.8

rvd: recreation visitor days

On completion, each project is expected to produce the amounts of timber and recreation shown in Table 11.4. The expected cost of each project is also shown in this table. Again, this is the discounted value of expected costs over the entire life of a project. Thus, project costs in Table 11.4 are comparable to road costs in Table 11.3.

The following objectives have been set for the set of projects and for the road network linking them:

All camps taken together must be able to accommodate 2,000 recreation visitor days (rvd) per year.

The timber production from all projects should be at least 17,000 m³ per year.

These goals must be met efficiently. That is, we seek the set of projects and the road network that meet the timber and recreation goals at least cost.

Because the cost of roads is a large part of the cost of the entire project, this last objective will act to keep the total length of the road network and the associated environmental impact low.

Decision Variables

As for an ordinary linear programming problem, formulating a model to solve this problem requires specification of the decision variables, objective function, and constraints. In this case, all decision variables take the value 0 or 1, since a project must either be done or not and a road section must be either built or not. Specifically, let Y_1, Y_2, Y_3, Y_4, and Y_5 designate the decision to build particular

Forest Resource Programming Models with Integer Variables

road sections or not. For example $Y_1 = 1$ means that road section 1 is built; otherwise $Y_1 = 0$.

Similarly, let X_e, X_h, X_t, and X_g refer to the decision to do particular projects or not. For example, project h (Highland) is done if $X_h = 1$; otherwise $X_h = 0$.

In this formulation the problem is a pure integer programming problem, and all variables are binary: they can take only the value 0 or 1. This is one of the easier kinds of pure integer programming problems to solve.

OBJECTIVE FUNCTION

The objective function is very similar to those we have used in standard linear programs. The general expression of the cost of building the roads, in terms of the decision variables, is:

$$0.8Y_1 + 0.4Y_2 + 0.3Y_3 + 0.2Y_4 + 0.4Y_5 \quad (\$10^6)$$

while the cost of doing the projects is:

$$0.7X_e + 0.1X_h + 0.5X_t + 0.8X_g \quad (\$10^6)$$

So the expression of the objective function, the total cost, is:

$$\min Z = 0.8Y_1 + 0.4Y_2 + 0.3Y_3 + 0.2Y_4 + 0.4Y_5$$
$$+ 0.7X_e + 0.1X_h + 0.5X_t + 0.8X_g \quad (\$10^6) \quad (11.7)$$

The object of the problem is to find the value of the decision variables that make this total cost smallest while meeting all the constraints.

CONSTRAINTS

Some of the constraints are similar to the familiar constraints of standard linear programming. For example, the timber production goal is readily expressed as:

$$6X_e + 8X_h + 13X_t + 10X_g \geq 17 \quad (10^3 \text{ m}^3/\text{y}) \quad (11.8)$$

Our choice of units throughout the formulation of this problem is meant to avoid very large and very small numbers. This is useful in any programming problem, but especially so in integer programming.

Note that the contribution of a particular project to this goal is either its total potential output, shown in Table 11.4, or nothing. For example, $X_t = 1$ implies that project Tall Pine is done and produces 13,000 m³ of timber a year. $X_t = 0$ implies no production from Tall Pine.

Similarly, the recreation goal is expressed by the following constraint:

$$1X_e + 1X_h + 2X_t + 3X_g \geq 2 \quad (10^3 \text{ rvd/y}) \tag{11.9}$$

The road-building options require a new form of constraint, for which integer (0, 1) variables are eminently suited. Consider first the road sections that end at a project, say section 1. This section must be built if and only if project Eagle is done. This is ensured by the following constraint:

$$X_e \leq Y_1$$

This constraint forces Y_1 to take the value 1 (section 1 is built) if X_e is equal to 1 (project Eagle is done). On the other hand, if $X_e = 0$ (project Eagle is not done), then cost minimization will result in $Y_1 = 0$ (section 1 is not built), unless the road section is needed as part of the network accessing other projects that are done. The constraints that ensure that each project is served by a road are:

$$\begin{aligned} X_e &\leq Y_1 \\ X_h &\leq Y_2 \\ X_t &\leq Y_4 \\ X_g &\leq Y_5 \end{aligned} \tag{11.10}$$

Some road sections may have to be built not because they end up at a project themselves, but because they collect traffic from other sections. Consider road section 1. We have just recognized that it must be built if project Eagle is done. But section 1 must also be built if either section 3 or 2 is built, even if project Eagle is not done. The following constraint expresses this possibility:

$$Y_2 + Y_3 \leq 2Y_1$$

This ensures that if either Y_2 or Y_3 or both are equal to 1 (that is, either section 2 or 3 or both are built), then Y_1 must equal 1 (section 1 is built). It is necessary to multiply Y_1 by 2 to allow for the fact that both Y_2 and Y_3 may equal 1. Any number larger than 2 would also work, but numerical solutions are easiest with coefficients that are close to 1 in absolute value. Minimizing the objective function will result in $Y_1 = 0$ (road section 1 is not built) if $Y_2 = Y_3 = 0$ (neither section 2 nor 3 is built), unless project e (Eagle) is done. A similar constraint is needed to model the branching of road section 3 into sections 4 and 5. In summary, the constraints that ensure that a collector road is built if branches are

Forest Resource Programming Models with Integer Variables

built are:

$$Y_2 + Y_3 \leq 2Y_1$$
$$Y_4 + Y_5 \leq 2Y_3$$
(11.11)

SPREADSHEET FORMULATION AND SOLUTION

Figure 11.7 shows the spreadsheet version of the integer programming model with Objective Function (11.7) and Constraints (11.8) to (11.11). The bold numbers are input data. Other numbers are values of the decision variables or the results of formulas that depend on the decision variables. Cells B4:J4 contain the values of the decision variables. Cell K5 contains the formula of Objective Function (11.7). Cells K6:K7 contain the formulas for the left-hand side of the constraints on timber production [Equation (11.8)] and recreation [Equation (11.9)]. Cells K8:K11 contain the formula of the left-hand side of Road Constraints (11.10), and cells K12:K13 contain the formula of the left-hand side of Road Constraints (11.11).

The corresponding Solver parameters in Figure 11.8 indicate that the target cell is K5, to be minimized by changing the decision variables in cells B4:J4. The first set of constraints indicates that the decision variables are binary, that is, 0 or 1. The second set of constraints sets the lower bounds on timber production and recreation. The last set expresses the constraints on the road sections [Equations (11.10) and (11.11)].

	A	B	C	D	E	F	G	H	I	J	K	L	M	N
1	EFFICIENT ROAD NETWORK, NONDIVISIBLE PROJECTS													
2		Road section:					Project:							
3		Y_1	Y_2	Y_3	Y_4	Y_5	X_e	X_h	X_t	X_g				
4		1	1	1	1	0	0	1	1	0	Total			
5	Cost	0.8	0.4	0.3	0.2	0.4	0.7	0.1	0.5	0.8	2.3	min		(10^6)
6	Timber						6	8	13	10	21	>=	17	(10^3 m³/y)
7	Recreation						1	1	2	3	3	>=	2	(10^3 rvd/y)
8	1 to e	-1					1				-1	<=	0	
9	2 to h		-1					1			0	<=	0	
10	4 to t				-1				1		0	<=	0	
11	5 to g					-1				1	0	<=	0	
12	1 to 2 or 3	-2	1	1							0	<=	0	
13	3 to 4 or 5			-2	1	1					-1	<=	0	
14														
15						Key cell formulas								
16	Cell formula										Copy to			
17	K5	=SUMPRODUCT(B5:J5,B$4:J$4)									K5:K13			

FIGURE 11.7 Spreadsheet model to design a road network.

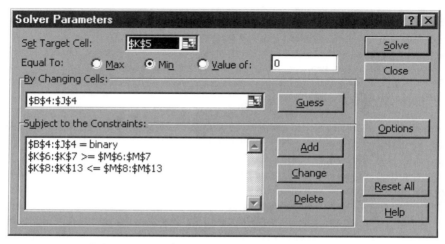

FIGURE 11.8 Solver parameters to design a road network.

The best solution is in Figure 11.7. It shows that all the objectives could be met at least cost by doing only projects h (Highland) and t (Tall Pine). These projects should be connected to the existing county road by building road sections 1, 2, 3, and 4. Timber production would then exceed the required amount by 4,000 m³/y, and the recreation goal would be exceeded by 1,000 rvd/y. The total cost would be $2.3 million.

Avoiding Adjacent Land Disturbances

As shown in Figure 11.6, the Highland and Tall Pine project areas share a long common boundary. Because all the projects involve building a road and timber harvesting, it may be desirable, for esthetic reasons, to avoid doing projects on adjacent land areas. This requirement is expressed with the following *adjacency constraint*:

$$X_t + X_h \leq 1$$

Because X_t and X_h may take only the value 0 or 1, the adjacency constraint allows only one of them to take the value 1, thus ensuring that either project t (Tall Pine) or project h (Highland) is done, but not both.

This setup still admits a solution that meets the timber production and recreation goals, at somewhat higher cost (see Problem 11.5). Adjacency constraints of this kind could be added for other projects as appropriate. In large problems, however, one must keep in mind that the more constraints there are, the more difficult it is to solve the model.

11.6 MODELS WITH INTEGER AND CONTINUOUS VARIABLES

In previous chapters we have worked with linear programming models with only continuous variables, and so far in this chapter our models have had only binary integer (0, 1) variables. However, there are many applications where both types of variables are needed in the same model. We shall illustrate this in the context of our road network design problem.

Partially Completed Projects

In the model of the previous section we assumed that each of the projects shown in Figure 11.6 had to be done completely or not at all. However, the projects may be divisible and need not be completed as planned. In land use planning, it is common to measure the size of a project by the amount of the available land that it uses.

With this interpretation, the decision variables X_e, X_h, X_t, and X_g are not integer (0, 1) variables, but continuous variables that may take any value between 0 and 1. For example, $X_e = 0.8$ means that project Eagle is done using 80% of the land that could be allocated to the project.

Let us assume further that the output and cost of each project are directly proportional to the level of completion of the project. For example, using the data in Table 11.4, if $X_e = 0.8$, then the Eagle project is expected to produce:

$$0.8 \times 1 = 0.8 \times 10^3 \text{ rvd/y of recreation}$$
$$0.8 \times 6 = 4.8 \times 10^3 \text{ m}^3\text{/y of timber}$$

and to cost:

$$0.8 \times 0.7 = \$0.56 \text{ million}$$

Assume that the objectives remain the same. Then the spreadsheet formulation of the problem, shown in Figure 11.9, is exactly as in Figure 11.7. The difference is in the Solver parameters (Figure 11.10), where the first set of constraints specifies that only the decision variables referring to roads, Y_1 to Y_5, are binary. The second and third set of constraints indicate that variables referring to projects, X_e to X_g, can take any value between 0 and 1.

The best solution of the model with divisible projects is shown in Figure 11.9. Project h (Highland) is completely done, but only 70% of project t (Tall Pine) is completed. Doing only part of Tall Pine decreases total cost by $200,000 but still meets the timber production goal exactly, while the recreation goal is exceeded by 400 rvd/y.

	A	B	C	D	E	F	G	H	I	J	K	L	M	N
1	EFFICIENT ROAD NETWORK, DIVISIBLE PROJECTS													
2		Road section:					Project:							
3		Y_1	Y_2	Y_3	Y_4	Y_5	X_e	X_h	X_t	X_g				
4		1	1	1	1	0	0.0	1.0	0.7	0.0	Total			
5	Cost	0.8	0.4	0.3	0.2	0.4	0.7	0.1	0.5	0.8	2.1	min		(10^6)
6	Timber						6	8	13	10	17	>=	17	(10^3 m³/y)
7	Recreation						1	1	2	3	2.4	>=	2	(10^3 rvd/y)
8	1 to e	-1					1				-1	<=	0	
9	2 to h		-1					1			0	<=	0	
10	4 to t				-1				1		-0	<=	0	
11	5 to g					-1				1	-0	<=	0	
12	1 to 2 or 3	-2	1	1							0	<=	0	
13	3 to 4 or 5			-2	1	1					-1	<=	0	
14														
15						Key cell formulas								
16	Cell formula										Copy to			
17	K5	=SUMPRODUCT(B5:J5,B$4:J$4)									K5:K13			

FIGURE 11.9 Spreadsheet model to design a road network, with divisible projects.

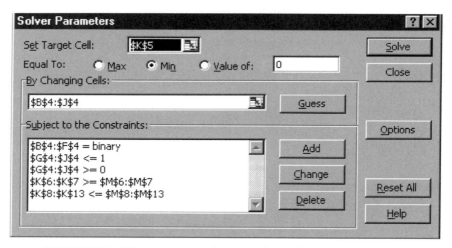

FIGURE 11.10 Solver parameters to design a road network, with divisible projects.

Start-up Costs

A useful application of mixed integer programming is to model discontinuous functions. As an example, consider the cost of project Golfech. In the previous section, we assumed that costs were directly proportional to the level of completion

Forest Resource Programming Models with Integer Variables

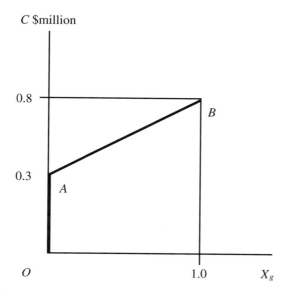

FIGURE 11.11 Cost function for a project with large start-up costs.

of the project, as indicated by the value of X_g. This would not be a satisfactory representation of costs if the costs of starting the project were very large. In that case, the cost function could look like line *OAB* in Figure 11.11. If the project is started, $0.3 million is needed for basic infrastructure. Thereafter, costs increase in proportion to the level of completion.

Using the decision variables of the previous section, the cost function of project Golfech is:

$$C = \begin{cases} 0.3 + 0.5X_g & \text{if the project is started} \\ 0 & \text{otherwise} \end{cases}$$

Note that we cannot simply add $0.3 + 0.5X_g$ to the objective function, because the $0.3 million is spent only if the project is started. This problem can be solved by introducing a binary integer (0, 1) variable, S_g. S_g takes the value 1 if project Golfech is started, 0 otherwise. The new objective function is:

$$\min Z = 0.8Y_1 + \cdots + 0.5X_t + 0.3S_g + 0.5X_g \qquad (\$10^6)$$

to which we add this additional constraint:

$$X_g \leq S_g \qquad (11.13)$$

Now, since S_g can take only the value 0 or 1, it must equal 1 when X_g acquires a positive value (that is, when project Golfech is undertaken at any scale). This in turn causes the objective function to increase by \$0.3 million, the project start-up cost. On the other hand, if X_g is 0 (project Golfech is not undertaken at any scale), minimization of the objective function will force S_g to be equal to 0, as desired.

TIMBER OR RECREATION

The previous examples showed some of the problems that may be modeled with binary integer (0, 1) variables. There are many more possibilities. For example, assume that the management objectives are to meet either the timber production goal of 17,000 m³/y or the recreation goal of 2,000 rvd/y, but not necessarily both, if that could decrease total cost.

This situation can be represented by changing the timber production and the recreation constraints as follows:

$$6X_e + 8X_h + 13X_t + 10X_g + mD \geq 17 \qquad (10^3 \text{ m}^3/\text{y}) \qquad (11.8)^*$$

$$1X_e + 1X_h + 2X_t + 3X_g + m(1-D) \geq 2 \qquad (10^3 \text{ rvd/y}) \qquad (11.9)^*$$

where m is an arbitrary number, larger than either the timber or the recreation goal, and D is an additional binary (0, 1) decision variable. The program seeks the values of D, of the road variables, Y, and the project variables, X, that minimize total cost. If the best solution is $D = 1$, the recreation objective is met, but not necessarily the timber objective. If $D = 0$, the timber objective is met, but not necessarily the recreation objective.

The corresponding spreadsheet model is in Figure 11.12. It is the same as Figure 11.9, except for the addition of the option of satisfying either the timber goal or the recreation goal. The binary variable D is in cell M2. The constant $m = 100$ is in cell M3. Cells K6 and K7 contain the left-hand side of the revised timber and recreation constraints [Equations (11.8)* and (11.9)*, respectively].

The Solver parameters are in Figure 11.13. The only changes with respect to Figure 11.8 are the addition of M2 to the list of changing cells and the constraint that M2 is binary. The best solution is to do project t (Tall Pine) entirely, thus meeting the recreation objective, but not the timber objective.

Forest Resource Programming Models with Integer Variables 225

	A	B	C	D	E	F	G	H	I	J	K	L	M	N
1	EFFICIENT ROAD NETWORK, TIMBER OR RECREATION													
2		Road section:					Project:				$D=$	1		
3		Y_1	Y_2	Y_3	Y_4	Y_5	X_e	X_h	X_t	X_g		$m=$	100	
4		1	0	1	1	0	0.0	0.0	1.0	0.0	Total			
5	Cost	0.8	0.4	0.3	0.2	0.4	0.7	0.1	0.5	0.8	1.8	min		(10^6)
6	Timber						6	8	13	10	113	>=	17	(10^3 m^3/y)
7	Recreation						1	1	2	3	2	>=	2	(10^3 rvd/y)
8	1 to e	-1					1				-1	<=	0	
9	2 to h		-1					1			0	<=	0	
10	4 to t				-1				1		0	<=	0	
11	5 to g					-1				1	0	<=	0	
12	1 to 2 or 3	-2	1	1							-1	<=	0	
13	3 to 4 or 5			-2	1	1					-1	<=	0	
14														
15					Key cell formulas								
16	Cell formula											Copy to		
17	K5	=SUMPRODUCT(B5:J5,B$4:J$4)									K8:K13			
18	K6	=SUMPRODUCT(B6:J6,B$4:J$4)+M2*M3												
19	K7	=SUMPRODUCT(B7:J7,B$4:J$4+(1-M3)*M2												

FIGURE 11.12 Spreadsheet model to design a road network for timber or recreation.

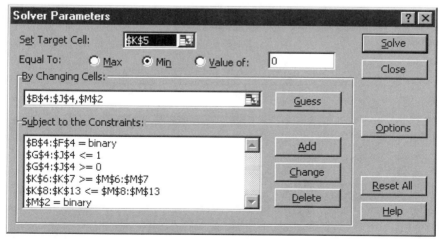

FIGURE 11.13 Solver parameters to design a road network for timber or recreation.

11.7 CONCLUSION

The methods described in this chapter are useful to model and solve practical forest management problems that involve integer variables. Nevertheless, some of the techniques should be used cautiously. General programming problems with integer variables are difficult to solve, even with powerful computers. Problems with only (0,1) variables are the easiest to solve. In some cases, hand computations or ordinary linear programming will suffice, as illustrated by the minimum-spanning-tree and assignment models. Large mixed integer programming problems can be hard to solve, even with a modest number of integer (0,1) variables. Therefore, in large models, the possibility of working with ordinary linear programming and rounding the final solution should not be neglected, as long as the real possibility of obtaining suboptimal or infeasible solutions is kept in mind. The simplex method and its variants are still the most powerful mathematical programming algorithms for solving problems with multiple constraints.

PROBLEMS

11.1 The paper division of a large forest products company is considering building new pulp mills, liner board mills, and newsprint mills. Suitable locations for each kind of mill have been identified in Oregon and Georgia. The cost of construction and the expected net present value of operation for each new mill are shown in the table. The paper division has been authorized by its parent company to spend up to $900 million on new mills, to be selected with the objective of maximizing the total net present value of their future operation.

Mill Construction Cost and Net Present Value

Mill	Construction cost ($10^6)	Net present value ($10^6)
Oregon pulp mill	500	25
Georgia pulp mill	450	22
Oregon liner board mill	260	12
Georgia liner board mill	270	10
Oregon newsprint mill	150	14
Georgia newsprint mill	170	14

(a) Write the algebra for the objective function and constraints of an integer programming model to determine which mills the paper division should build. (*Hint:* Allow a mill to be built only completely or not at all by expressing the decision to build or not to build as a binary variable.)

(b) Set up a spreadsheet model for solving the model you formulated in part (a). Which mills should the paper division build?

(c) Modify the spreadsheet model you set up in part (b) to build only one mill of each type. How does the solution change?

11.2 A private developer wants to build a park for recreational vehicles with nine parking sites. These sites are represented in the figure by numbered circles. Each site must be connected to a utilities network, and all possible network connections between the sites are shown in the figure. The length of each connection is indicated in units of 100 m. Construction costs for the utility network will be directly proportional to its total length.

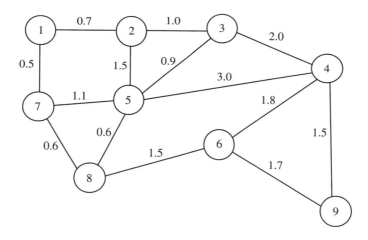

(a) Use the minimum-spanning-tree method to identify the least expensive utility network that would connect all nine sites.

(b) Suppose that parking sites 2 and 3 are separated from the others by a stream. As a result, the costs to connect sites 1 and 2, 2 and 5, 5 and 3, or 3 and 4 would double. Use the minimum-spanning-tree method to identify the least expensive utility network that would connect all nine sites.

11.3 The firefighter in charge of preparations for several prescribed burns has to build a fireline (a cleared strip to control fire spread) around each area to be burned. He has three machines to do this: a small dozer, a large dozer, and a brush cutter. Because of differences in vegetation and terrain, the rate at which each machine can construct a fireline is different for each area, as shown in the table. The firefighter wants to assign equipment to prescribed burn areas to maximize the total hourly fireline construction rate.

Fireline Construction Rates (m/hr)

Equipment	Prescribed burn area			
	1	2	3	4
Small dozer	100	75	50	25
Large dozer	150	100	80	0
Brush cutter	55	70	45	40

(a) Write the algebra for the objective function and constraints of a linear programming model to determine which machine should be assigned to each prescribed burn area. Assume that only one machine can be assigned to a given area.
(b) Set up a spreadsheet model for solving the model you formulated in part (a). What assignments should the firefighter make?

11.4 A logging contractor with three logging crews needs to assign each crew to one of five timber sales. The terrain and stand characteristics of these sales vary significantly. One of the crews is equipped for cable logging on steep slopes, and the other two are equipped for conventional logging. One crew is less experienced than the other and therefore is less productive. The contractor's estimates of the daily harvesting rates for crews at each timber sale are shown in the table. The contractor wants to assign crews to timber sales to maximize the total daily harvesting rate.

Daily Harvesting Rates (m³/day)

Crew	Timber sale				
	1	2	3	4	5
Cable	75	40	65	35	20
Experienced conventional	20	60	15	55	30
Less experienced conventional	15	55	10	55	25

(a) Write the algebra for the objective function and constraints of a linear programming model to determine which crew should be assigned to which timber sale. Assume that only one crew can be assigned to any timber sale.
(b) Set up a spreadsheet model for solving the model you formulated in part (a). What assignments should the logging contractor make?

11.5 (a) Set up your own spreadsheet model to design a road network like the one shown in Figure 11.7. Using the same data and assumptions, verify that your results are the same as in Figure 11.7.

(b) The Highland and Tall Pine projects are adjacent (Figure 11.6 shows their common boundary). Modify the model so that only one of these two projects can be done. How does the solution change?

11.6 Consider the spreadsheet model to design a road network shown in Figure 11.7. Assume that it is possible to build road section 1 at either a high or a low standard. A high-standard road would have a higher carrying capacity but would be more costly to build, as shown in this table.

Carrying Capacity and Cost for Road Section 1

	Standard	
	Low	High
Carrying capacity (10^3 tons/y)	25	50
Cost ($\$10^6$)	0.4	0.8

Modify the model to allow for this possibility by doing the following:

(a) Replace the variable Y_1 by two binary variables, one corresponding to each possible standard.

(b) Add a constraint to indicate that the section must be built at either high or low standard.

(c) Add another constraint to ensure that the carrying capacity of the section is high enough to support the traffic generated by the projects that are done, as indicated in this next table. Find the best road network, the corresponding projects, and the best standard for road section 1.

Traffic Generated by Projects

	Project			
	e	h	t	g
Traffic (10^3 tons/y)	7	8	15	15

11.7 (a) Set up your own spreadsheet model to design a road network with divisible projects like the one shown in Figure 11.9. Using the same data, verify that your results are the same as in Figure 11.9.

(b) Modify the model to allow for a start-up cost of $0.2 million for project h (Highland) and a variable cost between $0 and $0.1 million that is directly proportional to the level of completion of the project. How does the solution change?

Problems 11.8 to 11.10 are adapted from Kirby (1975).

11.8 The director of a state forest is considering six multiple-use projects that would provide both timber and hunting opportunities. The projects are represented in the figure by lettered circles. Each project must be connected by a road to the existing road shown as a solid line. The dashed lines are the road sections that might be built. Each road section is identified by a number.

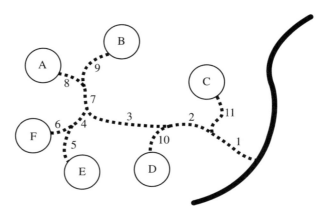

The cost of building and maintaining each road section is:

	Road section										
	1	2	3	4	5	6	7	8	9	10	11
Cost ($10^3)	75	50	65	40	45	70	50	40	20	50	25

The benefits of doing each project are:

	Project					
	A	B	C	D	E	F
Hunting days (100/y)	2	3	1	1	2	3
Timber (10^3 m^3/y)	6	9	13	10	8	3
Revenues (10^3)	70	130	140	130	110	100

The Director's objectives are:

1. To minimize road construction costs
2. To provide at least 400 hunting days/year
3. To harvest at least 30,000 m^3/y of timber
4. To get enough revenues from the projects to cover the total costs of road construction.

(a) Write the algebra for the objective function and constraints of an integer programming model to determine which projects to undertake and which roads to build. Assume that each project must be done completely or not at all.

(b) Set up a spreadsheet model for solving the model you formulated in part (a). What would be the timber production and the number of hunting days provided according to this solution? By how much would the total net present value of the projects undertaken exceed the total road cost?

11.9 Consider the state forest planning problem described in Problem 11.8. Assume that local environmentalists are lobbying the director to scale back the size of the projects and that local sawmills and logging contractors oppose this. Modify the spreadsheet model to allow projects to be divisible. Assume that the outputs and revenues of projects are directly proportional to the scale at which they are undertaken. For example, if one-third of project A was undertaken, the amount of timber produced would be $6/3 = 2$ (10^3 m³/y). How does the new solution compare to the solution of Problem 11.8?

11.10 For the forest development project described in Problem 11.8, road section 1 will have to bear traffic to and from all of the areas. Thus, it may be necessary to build this segment to a higher standard than others.

The carrying capacity and construction costs for road section 1 for three different standards are:

	Standard		
	Low	Medium	High
Carrying capacity (10^3 tons/y)	40	50	70
Cost ($\$10^3$)	25	50	75

The traffic that would result from each project is:

	Project					
	A	B	C	D	E	F
Traffic (10^3 tons/y)	8	4	14	13	10	7

Modify the model developed in Problem 11.8 (where both road sections and multiple-use projects were treated as all-or-nothing decisions) to find the best standard for road section 1. To accomplish this, do the following:

(a) Replace the variable corresponding to road section 1 with three (0, 1) variables whose sum is constrained to be less than or equal to 1.

(b) Add a constraint to ensure that the carrying capacity of section 1 is high enough to support the traffic generated by the projects that are done.

How does the solution compare to the solution from Problem 11.8? What happens to road construction costs? Why?

ANNOTATED REFERENCES

Bevers, M., and J. Hof. 1999. Spatially optimizing wildlife habitat edge effects in forest management linear and mixed-integer programs. *Forest Science* 45(2):249–258. (Describes a model to consider the edge effects on wildlife of forest management activities.)

Buongiorno, J., N.H.H. Svanqvist, and P. Wiroatmodjo. 1981. Forestry sector development planning: A model for Indonesia. *Agricultural Systems* 7:113–135. (Uses an integer programming model with fixed costs to find the best location for a port for timber exports.)

Carter, D.R., M. Vogiatzis, C.B. Moss, and L.G. Arvanitis. 1997. Ecosystem management or infeasible guidelines? Implications of adjacency restrictions for wildlife habitat and timber production. *Canadian Journal of Forest Research* 27:1302–1310. (Uses an integer programming model to analyze adjacency constraints.)

Davis, L.S., K.N. Johnson, P.S. Bettinger, and T.E. Howard. 2001. *Forest Management: To Sustain Ecological, Economic, and Social Values*. McGraw-Hill, New York. 804 pp. (Chapter 14 discusses the use of integer programming models to address road network decisions and adjacency constraints.)

Jones, J.G., J.F.C. Hyde, III, and M.L Meacham. 1986. *Four Analytical Approaches for Integrating Land Management and Transportation Planning on Forest Lands*. U.S. Forest Service Research Paper INT-361. Intermountain Research Station, Ogden, UT. 16 pp. (Discusses the use of mixed-integer programming models for planning harvesting, road construction, and other forest activities on a national forest.)

Kirby, M. 1975. Land use planning, transportation planning, and integer programming. Pages 271–284 in *Systems Analysis and Forest Resource Management*. Society of American Foresters, Bethesda, MD. 457 pp. (Basis of road network model in Section 11.5.)

Mäki, S., R. Kalliola, and K. Vuorien. 2001. Road construction in the Peruvian Amazon: Process, causes and consequences. *Environmental Conservation* 28(3):199–214. (Discusses the institutional, economic, and ecological aspects of new road networks.)

McDill, M.E., and J. Braze. 2000. Comparing adjacency constraint formulations for randomly generated forest planning problems with four age-class distributions. *Forest Science* 46(3):423–436. (Discusses alternative specifications for adjacency constraints.)

Murray, A.T., and S. Snyder. 2000. Spatial modeling in forest management and natural resource planning. *Forest Science* 46(2):153–156. (Introduction to a special issue of *Forest Science* containing a number of articles that use integer programming models to design natural reserves or analyze road networks and adjacency constraints.)

Nelson, J., and J.D. Brodie. 1990. Comparison of a random search algorithm and mixed integer programming for solving area-based forest plans. *Canadian Journal of Forest Research* 20:934–942. (Presents a nonoptimization approach for solving problems in which land management and road network decisions are represented with integer variables.)

Ragsdale, C.T. 1998. *Spreadsheet Modeling and Decision Analysis: A Practical Introduction to Management Science*. South-Western College Publishing, Cincinnati, OH. 742 pp. (Chapter 6 shows how to use spreadsheets to formulate and solve integer programming models.)

Snyder, S., and C. ReVelle. 1996. The grid-packing problem: Selecting a harvesting pattern in an area with forbidden regions. *Forest Science* 42(1):27–34. (The forest is conceptualized as a regular grid, many cells of which cannot be harvested because of environmental constraints. Management actions taken in cells are represented by integer variables.)

Walters, K.R., and E.S. Cox. 2001. An empirical evaluation of spatial restrictions in industrial harvest scheduling: The SFI Planning problem. *Southern Journal of Applied Forestry* 25(2):60–68. (Describes a mixed integer programming model linked to a geographic information system for determining harvesting plans compliant with industry standards for sustainable forest management, including restrictions such as adjacency constraints.)

Weintraub, A., G. Jones, M. Meacham, A. Magendzo, A. Magendzo, and D. Malchuk. 1995. Heuristic procedures for solving mixed integer harvest scheduling–transportation planning models. *Canadian Journal of Forest Research* 25:1618–1626. (Presents a nonoptimization approach for solving problems in which land management and road network decisions are represented with integer variables.)

CHAPTER 12

Project Management with the Critical Path Method (CPM) and the Project Evaluation and Review Technique (PERT)

12.1 INTRODUCTION

Often, forestry systems can be symbolized by a set of nodes connected by branches. Nodes may represent geographic points, such as stands of trees, campgrounds, and water reservoirs, or activities, such as the drafting of an environment statement. Branches may represent actual physical connections, such as roads and rivers, or constraints, such as the need to complete one task before starting another.

We have already seen an application of networks in Chapter 11. There we observed that a simple method, the minimum-spanning-tree algorithm, could be used to determine the shortest road network that connected forest stands. This method allowed us to solve simply a complex integer programming problem.

In this chapter, we shall study one of the most important applications of networks: the management of time and resources in projects with interdependent activities. Forestry projects, such as the preparation of a timber sale, the construction of a recreation area, and the development of a management plan, involve many activities. Some of these activities may run in parallel, others must be sequenced properly. Forest managers are responsible for scheduling activities so as to avoid bottlenecks and meet project deadlines. Sometimes, they must predict the most likely date for the completion of a project, taking into account all the things to be done and how they interact.

Two techniques designed to help project managers are the critical path method (CPM) and the project evaluation and review technique (PERT). CPM, developed by the du Pont and Remington Rand Univac companies in the late 1950s, focused on the trade-off between the duration of activities and their cost, and could be used to allocate resources to tasks to complete a project by a given deadline. PERT, developed at about the same time by consultants of the U.S. Navy to help them manage the Polaris weapons system program, was designed to evaluate the probability of meeting project deadlines, given probabilistic estimates of the duration of each activity in the project.

The distinction between the two techniques has become blurred over time, and neither the probabilistic elements of PERT nor the time/cost trade-offs of CPM are as important to most project managers as their developers anticipated. Accordingly, we shall emphasize the common elements of PERT and CPM that are most useful in project management in practice, touching only briefly on the probabilistic applications of PERT.

12.2 A SLASH-BURN PROJECT

The first step in using CPM/PERT is to identify all of the activities involved in a project and any precedence relationships among them. A precedence relationship is a requirement that a particular activity be completed before work begins on some other activity. In our discussion of CPM/PERT, we shall use an example described by Davis (1968). It deals with the burning of slash left at a logging site. Burning the slash is one way to prepare a site for planting or aerial seeding.

This project can be divided into six distinct activities. These are shown in Table 12.1, along with their expected duration and the activities that must be

TABLE 12.1 Activities and Their Precedents for a Slash-Burn Project

Activity	Duration (days)	Preceding activities
A. Prepare external firebreaks	5	
B. Fell internal hardwoods	8	
C. Fell snags in vicinity	4	
D. Check pumps and equipment	2	
E. Apply chemical fire retardant	2	A. Prepare external firebreaks
F. Install firing devices	3	A. Prepare external firebreaks B. Fell internal hardwoods

completed before they can begin. For example, the manager of this project expects that it will take two days to apply the chemical fire retardant to keep the fire from spreading to the rest of the forest. But that activity cannot start before the external firebreaks have been built. To simplify notation, we use a letter to label each activity. For example, D stands for "Check pumps and equipment."

12.3 BUILDING A CPM/PERT NETWORK

Once all the activities of a project, their expected duration, and their precedence relations have been identified, a CPM/PERT network can be drawn to represent this project. Each node (box) of the network represents an activity. For example, in Figure 12.1 the node labeled A stands for "Prepare the external firebreaks." Each arc (arrow) of the network represents a precedence relation. For example, in Figure 12.1 the arrow going from node A to node F means that activity A must be completed before activity F may start.

Every CPM/PERT network must have a single starting activity and a single ending activity. If the project starts or ends with several simultaneous activities, then we must add an artificial beginning or ending activity of zero duration. For example, in Figure 12.2, activities A, B, C, and D may all start at the same time. To make the starting activity unique we add the activity "Start," of zero duration. Similarly, C, D, E, and F have no succeeding activity. Adding the activity "End," of zero duration, results in a unique ending activity.

All paths through the network lead from the project's beginning to its end. There is no circular path from one activity back to the same activity. CPM/PERT

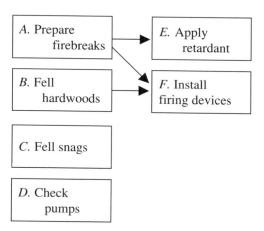

FIGURE 12.1 Activities and precedence among activities for the slash-burn project.

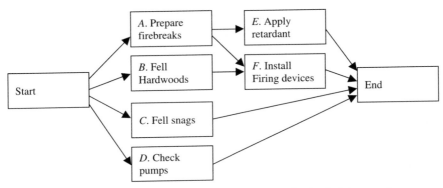

FIGURE 12.2 Adding beginning and ending activities for the slash-burn project.

networks are usually drawn with the starting event at the top or left-hand side of the network, and its end on the bottom or right-hand side.

There are other ways to draw CPM/PERT networks. For example, some people represent activities by arcs (arrows), using nodes to represent the beginning and the end of each activity. There is no substantive difference between the two representations, but the activities-as-nodes approach used here is somewhat simpler and more popular than the activities-as-arcs approach.

Constructing a CPM/PERT network is a useful exercise in and of itself for project managers. It forces them to identify the activities that a project will involve, how long they are likely to take, and what activities require others to be completed first. Project managers can always use this information whether or not they use CPM/PERT. But once this information is in hand, drawing a network is a simple task. The network helps everyone involved in a project visualize the "big picture."

However, CPM/PERT goes well beyond constructing networks. It can also provide the following useful data:

The *earliest start time* and the *earliest finish time* of each activity—These are the earliest times at which each activity could start and end if all preceding activities were completed as quickly as possible.

The *earliest finish time* of the project—This is the minimum time required to complete a project.

The *latest start time* and the *latest finish time* of each activity—These are the latest times at which each activity could start and end without increasing the earliest finish time of the project.

The *slack time* of each activity—This is the time that any activity's start time can be delayed without increasing the earliest finish time of the project.

Project Management with CPM and PERT 239

The *critical activities*—These are the activities that have no slack time. Their start time cannot be delayed without postponing project completion beyond the earliest finish time.

There are different methods of performing the CPM/PERT computations to get this information. Here we will use linear programming, a method we already learned in previous chapters.

12.4 EARLIEST START AND FINISH TIMES

Consider the slash-burn example described by the data in Table 12.1 and the network in Figure 12.2. We use linear programming to find the earliest start and finish times of all the project activities, and thus the earliest finish time for the project.

LINEAR PROGRAMMING FORMULATION

Let $ES_{start}, ES_A, ES_B, \ldots, ES_{end}$ be variables referring to the earliest start time of each activity in Figure 12.2 and $EF_{start}, EF_A, EF_B, \ldots, EF_{end}$ be variables designating their earliest finish times.

Activity Constraints

For each activity, the earliest finish time is equal to the earliest start time plus the expected duration of that activity:

$$EF_{start} = ES_{start} + 0$$
$$EF_A = ES_A + 5$$
$$EF_B = ES_B + 8$$
$$EF_C = ES_C + 4$$
$$EF_D = ES_D + 2 \quad (12.1)$$
$$EF_E = ES_E + 2$$
$$EF_F = ES_F + 3$$
$$EF_{end} = ES_{end} + 0$$

Arc Constraints

The earliest start time of each activity must at least equal the earliest finish time of each of the immediately preceding activities. This leads to one constraint for each arc in Figure 12.2:

$$\begin{aligned} ES_A &\geq EF_{start} \\ ES_B &\geq EF_{start} \\ ES_C &\geq EF_{start} \\ ES_D &\geq EF_{start} \\ ES_E &\geq EF_A \\ ES_F &\geq EF_A \\ ES_F &\geq EF_B \\ ES_{end} &\geq EF_E \\ ES_{end} &\geq EF_F \\ ES_{end} &\geq EF_C \\ ES_{end} &\geq EF_D \end{aligned} \quad (12.2)$$

Note that there are two constraints on ES_F, because both activity A and activity B must be finished before activity F can start.

Nonnegativity Constraints

The earliest start of all the activities must at least equal the earliest start of the project, which we set arbitrarily to 0. This leads to these constraints:

$$ES_{start}, ES_A, ES_B, ES_C, ES_D, ES_E, ES_F, ES_{end} \geq 0 \quad (12.3)$$

Objective Function

The objective is to find the earliest start of all the activities. This is equivalent to finding the minimum of the sum of the earliest start times of all the activities:

$$\min Z = ES_{start} + ES_A + ES_B + ES_C + ES_D + ES_E + ES_F + ES_{end} \quad (12.4)$$

Project Management with CPM and PERT

	A	B	C	D	E	F	G	H	I	J
1	SLASH BURNING EARLIEST START AND FINISH									
2						Arc				
3	Activity	Expected duration	Earliest start ES	Earliest finish EF		From activity	EF		To activity	ES
4	Start	0	0	0		Start	0		A	0
5	A	5	0	5		Start	0		B	0
6	B	8	0	8		Start	0		C	0
7	C	4	0	4		Start	0		D	0
8	D	2	0	2		A	5		E	5
9	E	2	5	7		A	5		F	8
10	F	3	8	11		B	8		F	8
11	End	0	11	11		C	4		End	11
12		Total	24			D	2		End	11
13		(min)				E	7		End	11
14						F	11		End	11
15										
16					Key cell formulas					
17	Cell	Formula							Copied to	
18	D4	=C4+B4							D4:D11	
19	C12	=SUM(C4:C11)								
20	G4	=VLOOKUP(F4,$A$4:$D$11,4,FALSE)							G4:G14	
21	J4	=VLOOKUP(I4,$A$4:$C$11,3,FALSE)							J4:J14	

FIGURE 12.3 Spreadsheet to compute the earliest start and finish times for the slash-burn project.

This minimization is accomplished by varying the earliest start of the activities, subject to the constraints.

Spreadsheet Solution

Figure 12.3 shows a spreadsheet to use linear programming to compute the earliest start and earliest finish times of each activity in the slash-burn example. The entries in bold are data, while the others result from the optimization. Cells A4:A11 contain the name of each activity, including the start and end activities. Cells B4:B11 contain the data on the expected duration of each activity, in days, taken from Table 12.1. A duration of 0 has also been entered for the start and end activities.

Cells C4:C11 contain the earliest start of each activity, the variables of the linear program. Cells D4:D11 contain the formulas for the earliest finish time of each activity, corresponding to Equations (12.1).

Cells F4:F14 contain the name of the activities at the beginning of each arc of the network in Figure 12.2, while the formulas in cells G4:G14 record the earliest finish times of these activities. For example, the VLOOKUP function in cell

G4 takes the content of cell F4 (the word "Start"), looks for its match in the first column of the range A4:D11 (the match is in cell A4), and returns the value in the fourth column of the range (the value in cell D4). The parameter "FALSE" induces the VLOOKUP function to match activity names regardless of their order.

Cells I4:I14 contain the names of the activities at the end of each arc in the network in Figure 12.2, while the formulas in cells J4:J14 record the earliest starts of these activities. For example, the VLOOKUP function in the cell J4 takes the content of cell I4 (the letter "A"), looks for its match in the first column of the range A4:C11 (the match is in cell A5), and returns the value in the third column of the range (the value in C5).

Cell C12 contains the expression of the objective function, the sum of the earliest start times for all activities, corresponding to Equation (12.4).

Figure 12.4 shows the Solver parameters to find the earliest start and finish times of each activity. The Solver minimizes the target cell, C12, by changing cells C4:C11, the earliest starts of the activities. The minimization is done subject to:

The nonnegativity constraints: C4:C11 >= 0, corresponding to Inequalities (12.3)

The constraints that the earliest start time of an activity must at least equal the earliest finish time of any immediately preceding activity: J4:J14 >= G4:G14, corresponding to Inequalities (12.2).

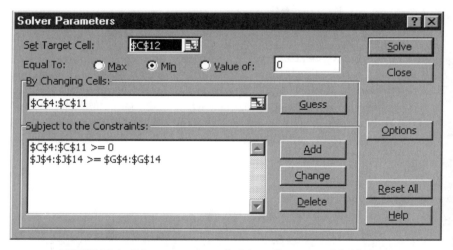

FIGURE 12.4 Solver parameters to find the earliest start and finish times.

The best solution is given in Figure 12.3, which shows that, for example, the installation of firing devices, activity F, may start, at the earliest, 8 days after the beginning of the project on day 0 (cell C10). The earliest finish time of activity F is 11 days (cell D10), which is also the earliest finish time of the slash-burn project (cell D11).

12.5 LATEST START AND FINISH TIMES

Continuing with the same example, we seek the latest start and finish times of each activity so that the project may be finished at the earliest time found in Figure 12.2. We again formulate the problem as a linear program and then solve it with a spreadsheet.

LINEAR PROGRAMMING FORMULATION

Let $LS_{start}, LS_A, LS_B, \ldots, LS_{end}$ be variables referring to the latest start time of each activity and $LF_{start}, LF_A, LF_B, \ldots, LF_{end}$ be variables designating the latest finish time of each activity.

Activity Constraints

For each activity, the latest finish time is equal to its latest start time plus the expected duration of that activity:

$$\begin{aligned} LF_{start} &= LS_{start} + 0 \\ LF_A &= LS_A + 5 \\ LF_B &= LS_B + 8 \\ LF_C &= LS_C + 4 \\ LF_D &= LS_D + 2 \\ LF_E &= LS_E + 2 \\ LF_F &= LS_F + 3 \\ LF_{end} &= LS_{end} + 0 \end{aligned} \quad (12.5)$$

In addition, to complete the project by its earliest finish time, the latest finish time of the end activity must be at most equal to its earliest finish:

$$LF_{end} \leq EF^*_{end} \tag{12.6}$$

Arc Constraints

The latest start time of each activity is at least equal to the latest finish time of the immediately preceding activities. This leads to one constraint for each arc in Figure 12.2:

$$\begin{aligned}
LS_A &\geq LF_{start} \\
LS_B &\geq LF_{start} \\
LS_C &\geq LF_{start} \\
LS_D &\geq LF_{start} \\
LS_E &\geq LF_A \\
LS_F &\geq LF_A \\
LS_F &\geq LF_B \\
LS_{end} &\geq LF_E \\
LS_{end} &\geq LF_F \\
LS_{end} &\geq LF_C \\
LS_{end} &\geq LF_D
\end{aligned} \tag{12.7}$$

Note that there are two constraints on LS_F, because both activity A and activity B must be finished before activity F can start.

Objective Function

The objective is to find the latest start time of each activity. This is equivalent to finding the maximum of the sum of the latest starts of all the activities:

$$\max Z = LS_{start} + LS_A + LS_B + LS_C + LS_D + LS_E + LS_F + LS_{end} \tag{12.8}$$

This maximization is accomplished by varying the latest start of the activities, subject to the constraints.

Project Management with CPM and PERT

	A	B	C	D	E	F	G	H	I	J
1	SLASH BURNING LATEST START AND FINISH									
2						Arc				
3	Activity	Expected duration	Latest start LS	Latest finish LF		From activity	LF		To activity	LS
4	Start	0	0	0		Start	0		A	3
5	A	5	3	8		Start	0		B	0
6	B	8	0	8		Start	0		C	7
7	C	4	7	11		Start	0		D	9
8	D	2	9	11		A	8		E	9
9	E	2	9	11		A	8		F	8
10	F	3	8	11		B	8		F	8
11	End	0	11	11		C	11		End	11
12		Total	28			D	11		End	11
13		(max)				E	11		End	11
14		Project earliest finish:		11		F	11		End	11
15										
16				Key cell formulas						
17	Cell	Formula							Copied to	
18	D4	=C4+B4							D4:D11	
19	C12	=SUM(C4:C11)								
20	G4	=VLOOKUP(F4,$A$4:$D$11,4,FALSE)							G4:G14	
21	J4	=VLOOKUP(I4,$A$4:$C$11,3,FALSE)							J4:J14	

FIGURE 12.5 Spreadsheet to compute the latest start and finish times for the slash-burn project.

SPREADSHEET SOLUTION

Figure 12.5 shows a spreadsheet to compute the latest start and finish times of each activity in the slash-burn example, using linear programming. Most of the spreadsheet is the same as that in Figure 12.3. The formulas are the same, but they now apply to the latest start and finish times instead of the earliest start and finish times.

Cells C4:C11 now contain the latest start time of each activity, the variables of the linear program. Cells D4:D11 contain the formulas for the latest finish times, corresponding to Equations (12.5).

Cell D14 contains the earliest finish time of the project, copied from cell D11 in Figure 12.3.

The formulas in cells G4:G14 record the latest finish of the activity at the beginning of each arc in Figure 12.2, and those in cells J4:J14 record the latest start of the activity at the end of each arc.

Cell C12 contains the sum of the latest start times for all activities, corresponding to Objective Function (12.8).

Figure 12.6 shows the Solver parameters to find the earliest start and finish times of each activity. The Solver maximizes the target cell, C12, by changing

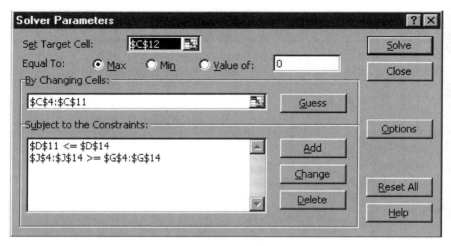

FIGURE 12.6 Solver parameters to compute the latest start and finish times.

cells C4:C11, the latest start times of the activities. The maximization is done subject to the following constraints:

The latest finish time of the end activity must be less than or equal to its earliest finish time: D11 <= D14, corresponding to Equation (12.6).

The latest start time of an activity must be at least equal to the latest finish time of the immediately preceding activities: J4:J14 >= G4:G14, corresponding to Equations (12.7).

The best solution is given in Figure 12.5, which shows that to complete the slash-burn project by its earliest finish time of 11 days (cell D14), the latest time at which activity F may start is 8 days after the beginning of the project (cell C10).

12.6 ACTIVITY SLACK AND CRITICAL PATH

The *slack time* of an activity is the difference between the latest and earliest starts of that activity. It is also equal to the difference between the latest and earliest finish times.

For example, the slack time for the felling of snags in the vicinity of the slash-burn, activity C, is, according to the results in Figures 12.3 and 12.5:

$$S_C = LS_C - ES_C = 7 - 0 = 7 \text{ days}$$

$$= LF_C - EF_C = 11 - 4 = 7 \text{ days}$$

	A	B	C	D	E	F	G	H
1	SLASH BURN PROJECT ACTIVITY SLACK							
2		Expected duration	Earliest			Latest		
3	Activity	(days)	Start	Finish		Start	Finish	Slack
4	A. Prepare external firebreaks	5	0	5		3	8	3
5	B. Fell internal hardwoods	8	0	8		0	8	0
6	C. Fell snags in vicinity	4	0	4		7	11	7
7	D. Check pumps and equipment	2	0	2		9	11	9
8	E. Apply chemical fire retardant	2	5	7		9	11	4
9	F. Install firing devices	3	8	11		8	11	0
10								
11			Key cell formulas					
12	Cell		Formula					Copied to
13	H4		=F4-C4					H4:H9

FIGURE 12.7 Spreadsheet to compute activity slacks for the slash-burn project.

This means that activity C could be delayed by 7 days after its earliest start time or finished 7 days after its earliest finish time without causing the project to extend beyond its earliest finish time.

Similarly, for the installation of firing devices, activity F, the slack time is:

$$S_F = LS_F - ES_F = 8 - 8 = 0 \text{ days}$$
$$= LF_F - LF_F = 11 - 11 = 0 \text{ days}$$

which means that this activity must start at its earliest start time and be completed at its earliest finish time to complete the project by the earliest finish. Because it has 0 slack time, this is a critical activity.

Figure 12.7 shows a spreadsheet that summarizes all the time data and computes the slack times for the activities of the slash-burn project. Cells B4:D9 contain the duration, earliest start time, and earliest finish time of the activities, copied from cells B5:D10 in Figure 12.3. Cells F4:G9 contain the latest start and finish times of the activities, copied from cells C5:D10 in Figure 12.5.

The slack times computed with the formulas in cells H4:H9 show two critical activities: B, the felling of internal hardwoods, and F, the installation of firing devices. These critical activities, together with the start and end activities constitute the *critical path* through the network in Figure 12.2.

The sum of the durations of the critical activities determines the earliest finish time of the project. Other things remaining the same, a lengthening of the duration of a critical activity will lengthen the earliest finish time of the project. Conversely, a shortening will reduce it, as long as the critical path remains the same.

Clearly, managers must monitor carefully the progress of all the critical activities, since they have the greatest potential for becoming bottlenecks and preventing a project from being completed on schedule.

Nearly critical activities (those with little slack) must be closely watched as well, since the duration of activities can rarely be forecast precisely. When the completion of critical or near-critical activities begins to slip beyond their latest finish time, it is often advisable to reallocate resources away from ongoing non-critical activities to the problem activities.

The critical path for a project may change when the expected duration of any activity is revised. For example, in the case of the slash-burn project, if preparing external firebreaks, activity A, were to require 9 days instead of 5, the new critical path would consist of activities A and F instead of B and F (see Problem 12.3). As activities are completed or estimates of activity durations are revised, project managers should update their CPM/PERT network. In that way, the network becomes a useful tool in the day-to-day administration of a project rather than just a monument to their initial optimism.

12.7 GANTT CHART

A Gantt chart is a useful way to display the CPM/PERT data for a project. Figure 12.8 is a Gantt chart for the slash-burn project. It shows the earliest start time, the expected duration, and the slack time for each activity.

The Gannt chart in Figure 12.8 can be built with Excel as follows:

1. Click on the chart wizard icon, and select the stacked bar chart as the chart type.
2. Select as data range the activity names, their duration, and the earliest start in cells A4:C9 in the spreadsheet in Figure 12.7.
3. Place the chart as a new sheet.
4. Reverse the order of the duration and earliest start series. To do this, click on the duration data on the chart (series 1). Then click on Format, selected data series, series order, move down.
5. Add the slack data series (series 3) by pulling down the chart menu, the add data command, and selecting cells H4:H9 in the spreadsheet in Figure 12.7.
6. Complete the chart by renaming series1, series2, and series3 as duration, earliest start time, and slack time, respectively, and formatting the chart as desired.

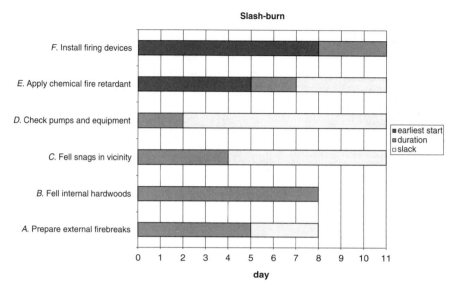

FIGURE 12.8 Gantt chart for the slash-burn project.

Although some Gannt charts do not do this, it is important to display the slack time. This allows a clear distinction of the critical activities and of some of the precedence relations. For example, in Figure 12.8, it is clear that activities B and F are critical. Furthermore, it is clear that activities A and B precede activity F. However, that activity A must precede activity E is not obvious from the Gantt chart, so the Gantt chart cannot fully substitute for the network in Figure 12.2. Gantt charts can, however, be supplemented with precedence arcs.

12.8 DEALING WITH UNCERTAINTY

The network computations we have been considering assume that the duration of each activity is known with certainty. The program evaluation and review technique (PERT) recognizes that the duration of each activity is in fact a random variable. PERT uses the same network representation and basic computations to find the critical activities of a project, but it adds a realistic measure of uncertainty to the results. Thus, instead of predicting that the earliest completion time for a project is 11 days, the PERT approach gives the probability of finishing the project within 11 days (or within some other time period).

	A	B	C	D	E	F
1	ACTIVITY DURATION STATISTICS					
2		Duration (days)				
3	Activity	Optimistic	Most likely	Pessimistic	Expected	Standard deviation
4	A. Prepare external firebreaks	2.5	6	10	6.1	1.3
5	B. Fell internal hardwoods	4	7	16	8.0	2.0
6	C. Fell snags in vicinity	2	5	8	5.0	1.0
7	D. Check pumps and equipment	1	3	4	2.8	0.5
8	E. Apply chemical fire retardant	1	3	4	2.8	0.5
9	F. Install of firing devices	1.5	2	6	2.6	0.8
10						
11			Key cell formulas			
12	Cell		Formula			Copied to
13	E4		=(B4+4*C4+D4)/6			E5:E9
14	F4		=(D4-B4)/6			F5:F9

FIGURE 12.9 Statistics for the duration of activities in the slash-burn project.

UNCERTAIN ACTIVITY DURATION

Duration uncertainty is modeled in PERT by using three estimates of duration for each activity: an optimistic duration, O, a pessimistic duration, P, and a most likely duration, L. Then, assuming a beta distribution for the duration, the expected duration of an activity is obtained as the following weighted average:

$$d = \frac{O + 4L + P}{6} \qquad (12.9)$$

And the standard deviation of the duration of that activity is:

$$s = \frac{P - O}{6} \qquad (12.10)$$

Figure 12.9 shows a spreadsheet to compute the expected value and the standard deviation of the duration of the activities for our slash-burn example.

PERT COMPUTATIONS WITH UNCERTAINTY

The computations of the earliest start and finish times of the activities, their latest start and finish times, and their slack proceed as before, based on the expected duration of each activity computed with Equation (12.9).

Project Management with CPM and PERT

For the slash-burn example, with the expected durations in Figure 12.9, the critical activities are B, the felling of internal hardwoods, and F, the installation of firing devices. The expected earliest time of completion of the project is 10.6 days (see Problem 12.4).

This result is close to the results found with the deterministic estimates of activity durations in Table 12.1. The real difference is that we can now make probabilistic statements regarding the project duration. For example, let us determine the probability that the project will be done in 8 days. As just noted, the duration of the project is determined by the critical activities, B and F.

Assuming that activities B and F are independent, the standard deviation of the duration of the project is given by:

$$S = \sqrt{s_B^2 + s_F^2} = 2.1 \text{ days}$$

where s_B and s_F are the standard deviations of the duration of the critical activities B and F.

The probability of a project duration of less than 8 days is the probability of a duration less than T standard deviations from the expected duration, where T is:

$$T = \frac{8 - 10.6}{2.1} = -1.2$$

This probability can be read from the normal distribution in Table 12.2, which shows that the probability of completing the slash-burn project in less than 8 days is between .10 and .20.

TABLE 12.2 Values of the Standard Normal Distribution

P	T
.05	−1.64
.10	−1.28
.20	−0.84
.30	−0.52
.40	−0.25
.50	0.00
.60	0.25
.70	0.52
.80	0.84
.90	1.28
.95	1.64

P is the probability that an observation from a normal distribution is T standard deviations or less from its expected value.

Another way of using the PERT statistics is to compute a confidence interval on the expected duration of 10.6 days. For example, we can compute upper and lower bounds so that there is a .60 probability that the project duration will be within those bounds. For this, we note from Table 12.2 that there is a .60 probability that an observation from a normal distribution is 0.84 standard deviations from the mean. Therefore, the 60% confidence interval for the duration of the slash-burn project is:

$$10.6 \pm 0.84 \times 2.1 \text{ days}$$

which implies a .60 probability that the project duration will be between approximately 9 and 12 days.

The PERT probabilistic calculations are only approximations, because they assume that the critical activities are independent. This may not be true if activities share resources. Furthermore, the use of the normal distribution is justified only if the number of critical activities is large. Still, even as rough approximations, the probability of project completion in a given time and the confidence interval on project duration are useful information to decision makers.

12.9 SUMMARY AND CONCLUSION

Network diagrams provide a useful way of conceptualizing many forest resource management problems. In CPM/PERT, network nodes represent time-consuming activities, while branches show how activities depend on each other, which ones must be finished before others can start. Drawing up the network of activities for a project is useful in and of itself to clarify the components of a project and their interrelations.

CPM/PERT computations have three phases: (1) finding the earliest start and finish times of all the activities; (2) finding the latest start and finish times of the activities; (3) determining the activities slack as the difference between the latest and earliest finish (or start) times. Linear programming can be used to do the computations.

Activities with zero slack are called critical activities. They form a unique path through the network and determine the earliest finish time of the project. Managers must closely monitor critical activities to avoid delays.

PERT computations are similar but use probabilistic estimates of activity duration. Based on the uncertainty attached to the duration of each activity, the method allows computation of the probability of completing a project by a certain date and of confidence intervals on the project duration.

Project Management with CPM and PERT   253

Apart from the quantitative data on start, finish, and slack times obtained from CPM/PERT, the approach has wider benefits. It requires a clear specification of project components and their relationships, and it provides a realistic way to examine them in light of deadlines. It can help avoid major project cost overruns. It offers a mechanism for communicating project plans to those inside and outside the project. It can let those inside the project see exactly where they fit into the big picture and foster a sense of teamwork. Even small projects can benefit from the CPM/PERT approach.

PROBLEMS

12.1 (a) Set up your own spreadsheet model to compute the earliest start and finish times for the slash-burn project, like the one shown in Figure 12.3. Using the same data, verify that your results are the same.

(b) Change the duration of activity B from 8 days to 10 days. How does this affect the earliest start and finish times for the other activities and for the project as a whole?

12.2 (a) Set up your own spreadsheet model to compute the latest start and finish times for the slash-burn project, like the one shown in Figure 12.5. Using the same data, verify that your results are the same.

(b) Change the earliest finish time for the project from 11 days to 15 days. How does this affect the latest start and finish times for the activities?

(c) Change the earliest finish time for the project from 11 days to 10 days. Explain what happens.

12.3 (a) Set up your own spreadsheet model to compute activity slack times for the slash-burn project, like the one shown in Figure 12.7. Using the same data, verify that your results are the same.

(b) Change the duration of activity A from 5 days to 9 days in the spreadsheets, corresponding to Figures 12.3, 12.5, and 12.7, in that order.

Does this change the earliest finish time for the project? Does this change the critical path? How are these two results linked?

12.4 Replace the durations of the activities in the spreadsheets corresponding to Figures 12.3, 12.5, and 12.7, in that order, with the expected durations for each activity shown in Figure 12.9. How does this change the earliest finish time and the critical path?

12.5 (Adapted from a problem originally formulated by Jeff Martin) The supervisor of a national forest is experiencing unexpected delays in the preparation of timber sales. Accordingly, he has directed his staff to use CPM to coordinate the next timber sale. The staff has identified the various activities involved, their precedence relationships, and the time necessary to do them, as shown in the table.

Activity	Expected duration (months)	Preceding activities
Timber inventory	4	None
Nontimber inventory	3	None
Initial sale design	2	Timber inventory
		Nontimber inventory
Transportation planning	21	Timber inventory
		Nontimber inventory
Initial environmental assessment	4	Timber inventory
		Nontimber inventory
Final environmental assessment	5	Initial sale design
		Transportation planning
		Initial environmental assessment
Final sale design	5	Final environmental assessment
Mark sale boundaries	2	Final sale design
Lay out roads	14	Final sale design
Mark timber	5	Mark sale boundaries
Cruise timber	4	Mark timber
Prepare advertisement	2	Lay out roads
		Cruise timber
Advertise sale	7	Prepare advertisement
Award sale	2	Advertise sale

(a) Using the information in the table, draw a network diagram showing the precedence relationships between activities. Remember that such a network must have a single starting activity and a single ending activity.
(b) Formulate a linear programming model to compute the earliest start and finish times for this project. What is the minimum time that will be needed to conduct the sale?
(c) Formulate a linear programming model to compute the latest start and finish times for this project.
(d) Use the results in (b) and (c) to compute the slack times for each activity.
(e) What is the critical path for this project?
(f) Summarize your results in a Gantt chart. For each activity, this chart should show the earliest start time, duration, latest start time, and slack time.

12.6 (Adapted from a problem originally formulated by George L. Martin) You have just been hired as director of a newly created national park. The park will cover almost 80,000 acres of old-growth redwood forest. Prior to its purchase, most of this land was owned by lumber companies. Therefore, it has very little infrastructure to support recreational use. Your immediate task is to plan, schedule, and direct the numerous activities that must be completed before the park can be opened to the public. You plan to use PERT to help manage this complex project. Accordingly, you have identified the various activities involved, their precedence relationships, and the time necessary to do them, as shown in the table. Note that there are three estimates of the activity for each activity: optimistic (O), most likely (L), and pessimistic (P).

Redwood Park Project

Activity	Duration (months)			Preceding activities
	O	L	P	
Establish ranger stations	1	1.5	2	None
Establish fire protection	0.5	1	2	Establish ranger stations
Purchase inholdings	4	6	8	Establish ranger stations
Phase out lumber operations	9	11	13	Establish ranger stations
Secure mineral rights	4	5	6	Establish ranger stations
Obtain scenic easements	8	10	12	Establish ranger stations
Construct signs	3	4	6	Establish fire protection
Construct temporary campgrounds	1	2	3	Establish fire protection
Designate temporary primitive campsites	2	4	6	Establish fire protection
Construct temporary headquarters	1	2	3	Establish fire protection
Repair existing trails	9	11	14	Establish fire protection Purchase inholdings
Repair boundary fences	10	12	16	Establish ranger stations
Post boundary	10	12	16	Establish ranger stations
Preliminary hydrological study	2	3	5	Establish ranger stations
Archaeological study	4	6	8	Phase out lumber operations
Range study	3	4	6	Phase out lumber operations
Preliminary environmental study	4	6	8	Construct signs Construct temporary campgrounds Designate temporary primitive campsites Construct temporary headquarters Repair existing trails Repair boundary fences
Preliminary master plan	3	5	7	Preliminary hydrological study Preliminary environmental study
Public hearings on master plan	1	2	3	Preliminary master plan
Final hydrological study	1	2	3	Public hearings on master plan
Final environmental study	3	4	5	Public hearings on master plan Secure mineral rights Obtain scenic easements

(continues)

Redwood Park Project (*continued*)

Activity	Duration (months)			Preceding activities
	O	L	P	
Final master plan	5	6	8	Archaeological study
				Range study
				Final hydrological study
				Final environmental study
Construct new trails	12	14	18	Public hearings on master plan
Designate permanent primitive campsites	5	6	7	Public hearings on master plan
Construct permanent campgrounds	10	12	13	Final master plan
Construct permanent headquarters	10	12	14	Final master plan
Construct vehicle access system	12	14	16	Final master plan

(a) Using the information in the table, draw a network diagram showing the precedence relationships among activities. Remember that such a network must have a single starting activity and a single ending activity.

(b) Set up a spreadsheet like Figure 12.9 to compute the expected value and the standard deviation of the duration of each activity. Use these expected durations in parts (c), (d), and (e).

(c) Formulate a linear programming model to compute the expected earliest start and finish times for this project. What is the expected minimum time until the park opening?

(d) Formulate a linear programming model to compute the expected latest start and finish times for this project.

(e) Using the results in parts (c) and (d), compute the slack time for each activity.

(f) What is the critical path for this project?

(g) Summarize your results in a Gantt chart. For each activity, this chart should show the expected earliest start time, duration, latest start time, and slack.

(h) Compute the standard deviation of the finish time for this project.

(i) Using a table of the standard normal distribution, determine the probability of getting the park ready for the public in either 52 months or 60 months.

ANNOTATED REFERENCES

Dane, C.W., C.F. Gray, and B.M. Woodworth. 1979. Factors affecting the successful application of PERT/CPM systems in a government organization. *Interfaces* 9(5):94–98. (Discussion of the reasons for the success or failure of attempts to use PERT/CPM on 18 national forests.)

Davis, J.B. 1968. Why not PERT your next resource management problem? *Journal of Forestry* 66(5):405–408. (Basis of the PERT model presented in this chapter.)

Devaux, S.A. 1999. *Total Project Control: A Manager's Guide to Integrated Project Planning, Measuring, and Tracking.* Wiley, New York. 318 pp. (Basic reference on PERT/CPM models.)

Dunn, R.A., and K.D. Ramsing. 1981. *Management Science: A Practical Approach to Decision Making.* Macmillan, New York. 527 pp. (Chapter 14 has a PERT/CPM application to the management of a ski resort.)

Dykstra, D.P. 1984. *Mathematical Programming for Natural Resource Management.* McGraw-Hill, New York. 318 pp. (Chapter 7 discusses PERT/CPM models.)

Husch, B. 1970. Network analysis in FAO international assistance forestry projects. *Unasylva* 24(4):18–28. (Discusses the use of PERT/CPM for forestry development projects.)

Meredith, J.R., and S.J. Mantel, Jr. 2000. *Project Management: A Managerial Approach.* Wiley, New York. 616 pp. (Basic reference on PERT/CPM models).

Ragsdale, C.T. 1998. *Spreadsheet Modeling and Decision Analysis: A Practical Introduction to Management Science.* South-Western College Publishing, Cincinnati, OH. 742 pp. (Chapter 15 shows how to use spreadsheets to formulate and solve PERT/CPM models.)

Winston, W.L. 1995. *Introduction to Mathematical Programming.* Duxbury Press, Belmont, CA. 818 pp. (Chapter 8 discusses PERT/CPM models.)

CHAPTER 13

Multistage Decision Making with Dynamic Programming

13.1 INTRODUCTION

In the preceding chapters, we have solved many forest management problems using linear programming. Linear programming is a very powerful method, and good software is readily available. However, linear programming has some limitations. First, it requires that the model variables be continuous. In Chapter 11 we saw how integer programming could be used to bypass this limitation, but integer programs can be difficult to solve. Second, linear programming applies only to problems that can be expressed with linear objective functions and constraints.

Dynamic programming is another approach to optimization that can readily handle nonlinear relationships and integer variables. Despite the name, many dynamic programming applications have nothing to do with time. For example, the problem of how to allocate funds to competing projects can be solved by dynamic programming. The method entails breaking down a problem into a sequence of subproblems or stages. The solutions of each subproblem are found first, and then these solutions are put together to find the solution of the overall problem.

A disadvantage of dynamic programming, in contrast with linear and integer programming, is that it does not have a standard form. Although the general approach is always the same, the actual mechanics of formulating and solving a problem depend on the specific case. Therefore, the best way to learn how to formulate and solve forest resource management problems with dynamic programming is to study many examples.

We shall start with one of the earliest applications of dynamic programming in forestry: the determination of thinning intensity in even-aged stands. From this simple example we shall be able to infer the common characteristics of problems solvable by dynamic programming as well as how to solve them.

We shall then consider two more examples: (1) the trimming of paper rolls in a paper mill to maximize value, and (2) the allocation of funds to projects to minimize the probability of extinction of a wild species.

13.2 BEST THINNING OF AN EVEN-AGED FOREST STAND

A forester in Northern California plans to manage a mixed conifer stand to maximize the total yield from thinning and the final harvest. He has already decided to do the harvest in three stages: an immediate thinning, a second thinning 20 years later, and a clear-cut 20 years after that.

However, he is undecided as to how heavily to thin the stand and will consider three possibilities at the first and second stages: no thinning, a light thinning, and a heavy thinning.

STAGES, STATES, AND DECISIONS

The different ways of managing the stand over the next 40 years can be represented by a network, as in Figure 13.1. Following dynamic programming terminology, we refer to each time a decision is made as a *stage*.

The condition of the stand at a particular stage, just before the decision, is called a *state*. In Figure 13.1 each node of the network, labeled with a letter, represents a stand state. A stand state may be defined by a single variable, such as basal area, or by a combination of variables, such as basal area and number of trees.

Each arc between nodes in the network represents a *decision*. The effect of a decision is to move the stand from a particular stage and state to a different state at the subsequent stage. The change is due in part to the thinning and in part to the growth of the timber that remains after thinning.

Multistage Decision Making with Dynamic Programming

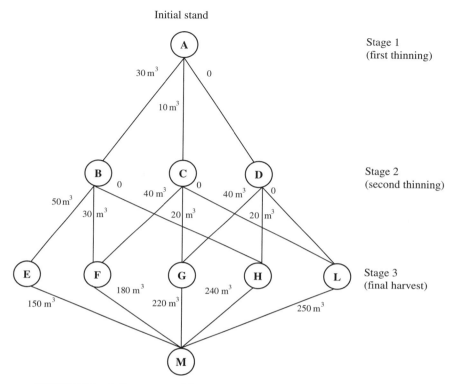

FIGURE 13.1 Different ways of thinning a forest stand (all volumes are per hectare).

The numbers along the arcs in Figure 13.1 refer to the immediate yield resulting from a decision. For example, at stage 1, a heavy thinning yields 30 m³/ha of timber while a light thinning yields 10 m³/ha.

Management Alternatives

Each path through the network represents an alternative way to manage the stand. Thus, starting from state *A* at stage 1, not thinning leads to state *D* at stage 2. At stage 2 and state *D*, a heavy thinning yields 40 m³/ha and leads to state *G* at stage 3. The stage 3 final harvest then yields 220 m³/ha. The total yield of this management regime would then be 260 m³/ha.

Clearly, the best management could be found by following all possible paths through the network and finding which one has the largest total yield. This would be a feasible method to solve this small problem. However, the number of paths becomes extremely large as we consider more stages and more options at each stage, for example, allowing thinning in any year. In such cases, the dynamic programming approach becomes very advantageous.

A Myopic Solution

One intuitive way of solving the thinning problem would be to choose, starting with stage 1, the decision that gives the largest immediate yield at each stage. This would result in thinning the stand heavily at stages 1 and 2. The total yield of the path from state A to B to E to M would then be:

$$30 + 50 + 150 = 230 \text{ m}^3/\text{ha}$$

However, this solution cannot be optimal, since the path from state A to B to F to M has a higher yield:

$$30 + 30 + 180 = 240 \text{ m}^3/\text{ha}$$

Thinning at an early stage can accelerate growth and reduce mortality in later stages, but it can also leave too little growing material. A correct solution must take into account this dependence between decisions at various stages.

Dynamic Programming Solution

Consider the possible decisions at stage t and state i. Let j refer to the decision to thin to a particular destination state. For example, $j = E$ at stage 2 means that the stand is thinned to reach state E at stage 3. Let $r(i, j)$ be the immediate yield of decision j in state i. For example, $r(B, E) = 50$ m^3/ha. Let $V_t(i, j)$ be the highest yield from stage t onward, starting in stage i and making decision j and then making the best decisions in all subsequent stages. And let $V_t^*(i)$ be the highest yield from stage t onward, starting in state i and making the best current decision j^* and the best subsequent decisions. Solving the problem involves finding this highest yield, $V_t^*(i)$, and the corresponding best decision, j^*, for all stages and states.

Dynamic programming works backwards from the last stage to the first. First we determine the highest yield at stage 3 for each state $i = E, F, G, H, L$ and the corresponding best decision:

$$V_3^*(i) \quad \text{and} \quad j^* \tag{13.1}$$

This is a trivial problem because there is only one possible decision for each state at stage 3: to harvest the remaining timber to reach state M.

Multistage Decision Making with Dynamic Programming

Next we determine the highest cumulative yield from stage 2 and stage 3 for each state at stage 2. Thus, for each state $i = B, C, D$ we find the decision j^* such that:

$$V_2^*(i) = \max_j [r(i, j) + V_3^*(j)] \tag{13.2}$$

Last, we determine the highest cumulative yield from stages 1, 2, and 3 given the initial state at stage 1. That is, for state A we find the decision j^* such that:

$$V_1^*(A) = \max_j [r(A, j) + V_2^*(j)]$$

$V_1^*(A)$ is the highest total yield that can be obtained over the 40-year planning period. The corresponding best decision, j^*, leads to the best state at stage 2. For that state, we know the best decision from having solved Equation (13.2). This best decision at stage 2 in turn leads to the best state at stage 3. For this state we know the best decision at stage 3 from having solved Equation (13.1).

The computations proceed as follows, stage by stage.

Stage 3 As shown in Figure 13.1, the stand may be in any one of five states at the time of final harvest: E, F, G, H, or L. Regardless of the state, the best (in fact, the only possible) decision is $j^* = M$. The corresponding yields are shown in Table 13.1.

Stage 2 The computations for this stage are shown in Table 13.2. The highest yield obtainable from stages 2 and 3 for a particular state i and decision j is the sum of the current yield from this decision plus the highest subsequent yield possible given that the stand will be in state j at stage 3:

$$V_2(i, j) = r(i, j) + V_3^*(j)$$

TABLE 13.1 Yields of the Final Harvest (Stage 3)

State i	Yield $V_3^*(i)$	Decision j^*
E	150	M
F	180	M
G	220	M
H	240	M
L	250	M

TABLE 13.2 Best Second Thinning as a Function of State (Stage 2)

State	Highest yield $V_2(i,j) = r(i,j) + V_3^*(j)$ Decision j				Highest yield $V_2^*(i)$	Best decision j^*
i	E	F	G	H	L	
B	$200 = 50 + 150$	$210 = 30 + 180$		$240 = 0 + 240$		240	H
C		$220 = 40 + 180$	$240 = 20 + 220$		$250 = 0 + 250$	250	L
D			$260 = 40 + 220$	$260 = 20 + 240$	$250 = 0 + 250$	260	G or H

For example, assume that the stand is in state B at stage 2 and that the decision is E. Then the highest yield obtainable from stages 2 and 3 is:

$$V_2(B, E) = r(B, E) + V_3^*(E)$$

$$= 50 + 150$$

$$= 200 \text{ m}^3/\text{ha}$$

The row of Table 13.2 for state B repeats this calculation for the two other possible decisions at stage 2, F and H. This leads to:

$$V_2(B, F) = 210 \text{ m}^3/\text{ha}$$

$$V_2(B, H) = 240 \text{ m}^3/\text{ha}$$

Therefore, the highest yield from stages 2 and 3, given state B at stage 2, is:

$$V_2^*(B) = 240 \text{ m}^3/\text{ha}$$

and the best decision at stage 2 and state B is

$$j^* = H$$

Repeating these computations for the other possible states at stage 2 leads, for state C, to:

$$V_2^*(C) = 250 \text{ m}^3/\text{ha}$$

with best decision

$$j^* = L$$

and for state D:

$$V_2^*(D) = 260 \text{ m}^3/\text{ha}$$

with best decision

$$j^* = G \text{ or } H$$

Stage 1 The calculations for stage 1, the first thinning, are shown in Table 13.3. The highest yield obtainable from stages 1, 2, and 3 for a particular state i and decision j is:

$$V_1(i, j) = r(i, j) + V_2^*(j)$$

TABLE 13.3 Best First Thinning as a Function of State (Stage 1)

State i	Highest yield $V_1(i, j) = r(i, j) + V_2^*(j)$ Decision j			Highest yield $V_1^*(i)$	Best decision j^*
	B	C	D		
A	270 = 30 + 240	260 = 10 + 250	260 = 0 + 260	270	B

There is only one possible state at stage 1, $i = A$. If the decision is $j = B$, then the highest yield obtainable from the three stages is:

$$V_1(A, B) = r(A, B) + V_2^*(B)$$
$$= 30 + 240$$
$$= 270 \text{ m}^3/\text{ha}$$

Repeating these calculations for $j = C$ and $j = D$ leads to:

$$V_1(A, C) = 260 \text{ m}^3/\text{ha}$$
$$V_1(A, D) = 260 \text{ m}^3/\text{ha}$$

Therefore, the highest yield from the three stages is:

$$V_1^*(A) = 270 \text{ m}^3/\text{ha}$$

with best decision

$$j^* = B$$

Since the best first decision is $j^* = B$, the best state at stage 2 is B. At stage 2 and state B the best decision is $j^* = H$ (see Table 13.2). This decision in turn leads to state H at stage 3. At stage 3 and state H there is but one possible decision, $j^* = M$ (see Table 13.1).

In summary, given a stand of initial state A and a final harvest in 40 years, the best management regime consists of an immediate heavy thinning but no subsequent thinning before final harvest.

Tables 13.1–13.3 can also be used to determine the best sequence of harvests starting from any stage and state. For example, assume the crew doing the first thinning did not thin enough. Twenty years later the stand might be in state C at stage 2 rather than in state B as planned. Table 13.2 shows that $j^* = L$ for $i = C$ at stage 2, so the best decision would still be not to thin the stand before final harvest.

The practical value of this particular model is limited, given the few alternatives considered. Nevertheless, the same method could be used to investigate additional thinning intensities at more frequent intervals. Besides thinning, clear-cutting could be considered as an option at every stage, thus making the rotation length a decision variable. The best rotation could then be found simultaneously with the best thinning regime. And if this were a plantation, the initial planting density could also be considered as a decision variable.

SPREADSHEET FORMULATION AND SOLUTION

Figure 13.2 shows a spreadsheet to solve the thinning problem by dynamic programming. The spreadsheet is set up to allow for five states and five decisions at each of three stages. The entries in bold are data; the other entries are the results of formulas. Cells C5:G9 contain the immediate yield from each state and decision at stage 3. The spreadsheet requires data for all 25 possible state and decision combination at each stage, so the yield data for decision M are repeated to complete the table. Cells C11:G15 contain the immediate yield from each state and decision at stage 2. Decisions that are not possible are given an arbitrarily high negative yield, -99 m^3, so that they will not be picked as best decisions. The yield data for state D are repeated to complete the stage 2 table. Cells C17:G21 contain the immediate yield from each state and decision at stage 1. The yield data for state A are repeated to complete the table.

The formulas in cells I5:M9 compute the highest yield for each state and decision at stage 3. Since this is the last stage, the highest yield is equal to the immediate yield. The formulas in cells I11:M15 compute the highest yield from stages 2 and 3 for each state and decision at stage 2. For example, the formula in cell I11 adds the immediate yield from state B, decision E, to the highest of stage 3, state E. The VLOOKUP function takes the decision in cell I10, "E," looks for its match in the first column of the range B5:B9, and returns the value in the 13th column (column N), the highest yield for state "E" at stage 3. The parameter FALSE allows the state names to be in any order. Similarly, the formulas in cells I17:M21 compute the highest yield from stages 1, 2, and 3 for each state and decision at stage 1.

The formulas in cells N5:N9 compute the highest return for each state at stage 3. The formulas in cells N11:N15 and N17:N21 repeat this for stages 2 and 1.

	A	B	C	D	E	F	G	H	I	J	K	L	M	N	O
1	BEST THINNING														
2	Stage	State	Immediate yield $r(i,j)$						Highest yield $V_t(i,j) = r(i,j) + V^*_{t+1}(j)$					Highest yield $V^*_t(i)$	Best decision j^*
3	t	i	Decision j						Decision j						
4	3		M	M	M	M	M		M	M	M	M	M		
5		E	150	150	150	150	150		150	150	150	150	150	150	M
6		F	180	180	180	180	180		180	180	180	180	180	180	M
7		G	220	220	220	220	220		220	220	220	220	220	220	M
8		H	240	240	240	240	240		240	240	240	240	240	240	M
9		L	250	250	250	250	250		250	250	250	250	250	250	M
10	2		E	F	G	H	L		E	F	G	H	L		
11		B	50	30	-99	0	-99		200	210	121	240	151	240	H
12		C	-99	40	20	-99	0		51	220	240	141	250	250	L
13		D	-99	-99	40	20	0		51	81	260	260	250	260	G
14		D	-99	-99	40	20	0		51	81	260	260	250	260	G
15		D	-99	-99	40	20	0		51	81	260	260	250	260	G
16	1		B	C	D	D	D		B	C	D	D	D		
17		A	30	10	0	0	0		270	260	260	260	260	270	B
18		A	30	10	0	0	0		270	260	260	260	260	270	B
19		A	30	10	0	0	0		270	260	260	260	260	270	B
20		A	30	10	0	0	0		270	260	260	260	260	270	B
21		A	30	10	0	0	0		270	260	260	260	260	270	B
22															
23						Key cell formulas									
24	Cell		Formula								Copied to				
25	I4	=C4							I4:M4,I10:M10,I16:M16						
26	I5	=C5							I5:M9						
27	I11	=C11+VLOOKUP(I$10,$B$5:$N$9,13,FALSE)							I11:M15						
28	I17	=C17+VLOOKUP(I$16,$B$11:$N$15,13,FALSE)							I17:M21						
29	N5	=MAX(I5:M5)							N5:N9,N11:N15,N17:N21						
30	O5	=INDEX(I$4:M$4,MATCH(N5,I5:M5,0))							O5:O9						
31	O11	=INDEX(I$10:M$10,MATCH(N11,I11:M11,0))							O11:O15						
32	O17	=INDEX(I$16:M$16,MATCH(N17,I17:M17,0))							O17:O21						

FIGURE 13.2 Spreadsheet to compute the best thinning policy.

The formulas in cells O5:O9 find the best decision corresponding to the highest yield for each state at stage 3. The formulas in cells O11:O15 and O17:O21 repeat this for stages 2 and 1. For example, the MATCH function in cell O11 finds that the highest yield in cell N11 is in the fourth column of the range I11:M11, and then the INDEX function determines that the fourth value in the range of the decision names, I10:M10, is "H."

The solution in Figure 13.2 shows that starting in state A, the highest yield of thinning and final harvest is 270 m^3/ha. The best decision, starting in state A, is to go to state B, that is, to thin heavily. From state B at stage 2, 20 years later, the best decision is go to state H, that is, not to thin at all. From state H

at stage 3, another 20 years later, the best (and only) decision is to clear-cut the stand to state *M*.

The spreadsheet shows other best solutions as well. For example, for a stand starting in state *C* at stage 2, it would be best to let the stand grow to state *L* in 20 years without thinning and then to clear-cut it. For a stand in state *D*, it would be best to thin it heavily to reach state *G* in 20 years and then to clear-cut it.

13.3 GENERAL FORMULATION OF DYNAMIC PROGRAMMING

Despite its simplicity, the thinning problem of the previous section has all the features of more complex dynamic programming problems. Problems of this kind are defined in terms of stages, states, and decisions, and they are solved recursively.

STAGES, STATES, AND DECISIONS

Problems that can be solved by dynamic programming can be divided into stages, where some decision must be made at each stage. In the thinning example, stages were points in time, but this is not always the case, as we shall see in the subsequent examples.

Each stage has one or more states associated with it. In defining the state variables it is helpful to think of what changes from stage to stage and what is affected by decisions. In the thinning problem, the state was the structure of the stand. Stand structure could be defined by a number of variables, such as number of trees, basal area, and volume.

The effect of a decision is to move from a state at one stage to another state at the next stage. There is an immediate return associated with a decision.

RECURSIVE SOLUTION

The objective of a dynamic programming problem is to maximize the returns from all decisions or to minimize the costs. The solution algorithm starts by finding the best solution for each possible state at the last stage. This is usually a trivial problem. The algorithm proceeds backwards using a recursive equation, one stage at a time. For the thinning problem, the recursive equation was:

$$V_t^*(A) = \max_j [r(i,j) + V_{t+1}^*(j)]$$

where t refers to stage, i refers to state, and j refers to decision. The function within brackets refers to the current-stage return $r(i,j)$ plus the highest return from all subsequent stages $V_{t+1}^*(j)$. In the thinning example, the immediate return depends on the state and decision, but it could also depend on the stage, as demonstrated in the endangered species example in Section 13.5.

Once the algorithm has determined the highest return for the first stage, the sequence of best decisions is found by working through the network from the first to the last stage.

13.4 TRIMMING PAPER SHEETS TO MAXIMIZE VALUE

This second example of dynamic programming differs from the thinning problem in that the stages do not correspond to points in time. The relationship between states and decisions is also different.

PROBLEM DEFINITION

The Maine pulp-making cooperative considered in Chapters 2, 3, and 10 has been so successful that it has added a paper machine to its mill. The machine produces a continuous sheet of high-quality coated book paper. The sheet is 4 m wide and must be trimmed to narrower dimensions before it is shipped to customers. All paper is sold in large rolls of the same diameter but of different width.

Currently, the selling price of paper of this quality is $400, $1,000, and $1,500 for rolls 1 m, 2 m, and 3 m wide, respectively. Note that the price is not directly proportional to the width. The premium per additional meter is higher between 1 and 2 m than between 2 and 3 m.

Because there are only a few ways of trimming the 4-m-wide sheet coming out of the paper machine to produce rolls of the desired width, one could find the combination of highest value by complete enumeration. However, the problem would become very complex with just a few more alternative widths for the rolls of paper, and dynamic programming would be needed to solve it efficiently.

Stages, States, and Decision Variables

The trimmers located at the end of a paper-making machine simultaneously cut the sheet to specified widths. However, to formulate the problem by dynamic programming it is helpful to think of the process as a sequence of stages, where each stage corresponds to choosing the width of one roll of paper to trim from the sheet, that is, of choosing the setting for one trimmer knife. Only four stages are needed, because at most four rolls can be obtained, each one 1 m wide. We shall denote the stage by t, where t varies from 1 to 4. The state i is the width of the "untrimmed" part of the sheet. The decision at each stage is the width of the roll of paper to be cut from the untrimmed width or, equivalently, the width of the untrimmed sheet to be considered at the next stage. This latter definition underlines the similarity with the stand thinning problem, where the decision was defined as a choice of the residual stocking after thinning. Therefore, in this example, if i is the state at a particular stage and j is the decision, then $i - j$ is the width of the roll trimmed from the sheet at that stage.

Description of Alternatives

The different ways of cutting the sheet of paper can be represented by a network, as in Figure 13.3. Each node corresponds to a particular state. A row of nodes is a stage. The arcs between nodes correspond to possible decisions.

Stage 1 corresponds to an untrimmed sheet 4 m wide; therefore the initial state is $i = 4$. The possible decisions are $j = 1$ if a 3-m roll worth \$1,500 is trimmed from the sheet, $j = 2$ if a 2-m roll worth \$1,000 is trimmed, and $j = 3$ if a 1-m roll worth \$400 is trimmed. Then, if $i = 3$ at stage 2 and the width of the second roll is 3 m, $j = 0$ at stage 3. The other possible decisions are $j = 1$ and $j = 2$.

The complete network in Figure 13.3 is obtained in this way by noting all possible decisions for each stage and state.

Dynamic Programming Solution

Consider the possible decisions at stage t and state i. Let $r(i, j)$ be the immediate return of decision j in state i. This is just the price of a roll of paper $i - j$ m wide. For example, $r(4, 3) = \$400$. Let $V_t^*(i)$ be the highest value of the rolls of paper obtained from stage t onward, starting in state i and making the best current decision j^* and the best subsequent decisions. That is:

$$V_t^*(i) = \max_j [r(i, j) + V_{t+1}^*(j)] \qquad (13.3)$$

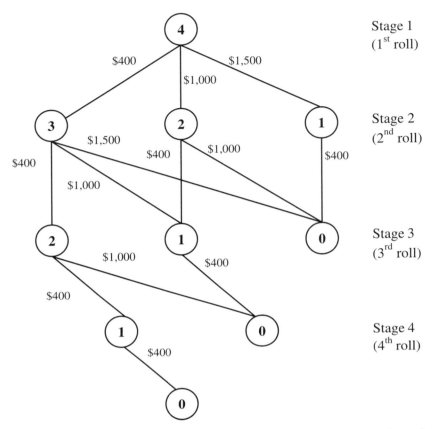

FIGURE 13.3 Different ways of trimming a 4-m-wide roll of paper into rolls 1, 2, or 3 m wide.

where the term in brackets is the value of the roll obtained at the current stage plus the highest value of the rolls obtained at subsequent stages. The objective is to determine $V_1^*(4)$, the highest value gotten by trimming a 4-m-wide sheet of paper. We do this by applying Recursive Relation (13.3), stage by stage, from stage 4 back to stage 1.

Stage 4 Figure 13.3 shows two possible states at stage 4, $i = 0$ and $i = 1$. It is possible to obtain a fourth roll of paper only if $i = 1$. In that state, the best decision is to produce a 1-m roll with a return of $400; that is:

$$V_4^*(1) = \$400$$

with the best decision

$$j^* = 0$$

Stage 3 The computations for this stage appear in Table 13.4. The highest value obtainable from stages 3 and 4 for a particular state and decision at stage 3 is:

$$V_3(i, j) = r(i, j) + V_4^*(j) \quad (13.4)$$

Figure 13.3 shows that the possible states are $i = 2$, 1, or 0. More rolls can be obtained only if $i = 2$ or 1. For example, if $i = 2$, then producing a 1-m roll would give an immediate return of \$400 and leave $j = 1$ m of untrimmed sheet at stage 4, which in turn would give a highest return of \$400 at stage 4. In terms of Equation (13.4), we have:

$$V_3(2,1) = r(2,1) + V_4^*(1)$$
$$= \$400 + \$400 = \$800$$

Still assuming $i = 2$ but producing a 2-m roll would give an immediate return of \$1,000 and leave $j = 0$ m of untrimmed sheet at stage 4, so there would be no return at stage 4. Thus:

$$V_3(2, 0) = r(2, 0)$$
$$= \$1,000$$

Therefore, the highest value obtainable from stages 3 and 4 given $i = 2$ at stage 3 is:

$$V_3^*(2) = \$1,000$$

with the best decision

$$j^* = 0$$

And for $i = 1$ at stage 3, we have only one possible decision, $j = 0$, so:

$$V_3^*(1) = \$400$$

with the best decision

$$j^* = 0$$

Stages 2 and 1 The calculations for stages 2 and 1 are similar to those for stage 3. The results for each stage appear in Tables 13.5 and 13.6.

Consider Table 13.6, which refers to stage 1. It shows that:

$$V_1^*(4) = \$2,000$$

with the best decision

$$j^* = 2$$

Thus, the highest value that can be obtained by trimming the 4-m sheet of paper is $2,000. The best decision in the first stage is $j^* = 2$, that is, to produce

TABLE 13.4 Best Width of the Third Roll of Paper as a Function of Remaining Sheet Width (Stage 3)

State i	Highest return $V_3(i, j) = r(i, j) + V_4^*(j)$ Decision j		Highest return $V_3^*(i)$	Best decision j^*
	0	1		
1	400		400	0
2	1,000	800 = 400 + 400	1,000	0

TABLE 13.5 Best Width of the Second Roll of Paper as a Function of Remaining Sheet Width (Stage 2)

State i	Highest return $V_2(i, j) = r(i, j) + V_3^*(j)$ Decision j			Highest return $V_2^*(i)$	Best decision j^*
	0	1	2		
1	400			400	0
2	1,000	800 = 400 + 400		1,000	0
3	1,500	1,400 = 1,000 + 1,000	1,400 = 400 + 1,000	1,500	0

TABLE 13.6 Best Width of the First Roll of Paper as a Function of Remaining Sheet Width (Stage 1)

State i	Highest return $V_1(i, j) = r(i, j) + V_2^*(j)$ Decision j			Highest return $V_1^*(i)$	Best decision j^*
	1	2	3		
4	1,900 = 1,500 + 400	2,000 = 1,000 + 1,000	1,900 = 400 + 1,500	2,000	2

a 2-m roll. The best state at stage 2 is therefore $i = 2$. At stage 2 and state 2, the best decision is $j^* = 0$ (see Table 13.5), that is, to produce another 2-m roll.

SPREADSHEET FORMULATION AND SOLUTION

Figure 13.4 shows a spreadsheet to solve the paper-trimming problem by dynamic programming. The spreadsheet is set up to allow up to three states and three decisions at each of four stages. The entries in bold are data; the other entries are the results of formulas. Cells C5:E7 contain the immediate return from each state and decision at stage 4. The spreadsheet requires data for all nine combinations of states and decision at each stage, so the returns from the only possible state and decision are repeated. Cells C9:E11 contain the immediate return from each state and decision at stage 3. Decisions that are not possible (starting and ending with a 1-m-wide sheet) are given an arbitrarily high negative return, $-999, so that they will not be picked for the best solution. The return data for state 2 are repeated to complete the table for stage 3. Similarly, cells C13:E15 and C17:E19 contain the immediate return from each state and decision at stages 2 and 1, respectively.

The formulas in cells G5:I7 compute the highest return for each state and decision at stage 4. Since this is the last stage, the highest return is equal to the immediate return. The formulas in cells G9:I11 compute the highest yield from stages 3 and 4 for each state and decision at stage 3. For example, the formula in cell H10 adds the immediate return from state 2 and decision 1 at stage 3 to the highest return of state 1 at stage 4. The IF function sets the highest return from state 1 at stage 4 to 0 if the roll is less than 1 m long. The VLOOKUP function takes the decision in cell H10, "1," looks for its match in the first column of the range B5:B11, and returns the value in the ninth column, the highest yield for state "1" at stage 4. Similarly, the formulas in cells G13:I15 compute the highest return from stages 3 and 4 for each state and decision at stage 2, and the formulas in cells G17:I19 compute the highest return from stages 2, 3, and 4 for each state and decision at stage 1.

The formulas in cells J5:J7 compute the highest return for each state at stage 4. The formulas in cells J9:J11, J13:J15 and J17:J19 repeat this for stages 3, 2, and 1.

The formulas in cells K5:K7 find the decision giving the highest return for each state at stage 4. The formulas in cells K9:K11, K13:K15, and K17:K19 repeat this for stages 3, 2, and 1. For example, the MATCH function in cell K9 finds that the highest yield in cell J9 is in the first column of the range G9:I9, and then the INDEX function determines that the first value in the range of the decision names, G8:I8, is "0."

The solution in Figure 13.4 shows that starting in state 4 at stage 1 (that is, with a 4-m untrimmed sheet of paper), the highest return is $2,000. The best

	A	B	C	D	E	F	G	H	I	J	K
1	BEST TRIMMING OF PAPER SHEET										
2	Stage	State	Immediate return $r(i,j)$				Highest return $V_t(i,j)=r(i,j)+V^*_{t+1}(j)$			Highest return $V^*_t(i)$	Best decision j^*
3	t	i	Decision j				Decision j				
4	4		0	0	0		0	0	0		
5		1	400	400	400		400	400	400	400	0
6		1	400	400	400		400	400	400	400	0
7		1	400	400	400		400	400	400	400	0
8	3		0	1	1		0	1	1		
9		1	400	-999	-999		400	-599	-599	400	0
10		2	1000	400	400		1000	800	800	1000	0
11		2	1000	400	400		1000	800	800	1000	0
12	2		0	1	2		0	1	2		
13		3	1500	1000	400		1500	1400	1400	1500	0
14		2	1000	400	-999		1000	800	1	1000	0
15		1	400	-999	-999		400	-599	1	400	0
16	1		3	2	1		3	2	1		
17		4	400	1000	1500		1900	2000	1900	2000	2
18		4	400	1000	1500		1900	2000	1900	2000	2
19		4	400	1000	1500		1900	2000	1900	2000	2
20											
21					Key Formulas						
22	Cell		Formula							Copy to	
23	G4		=C4							G4:I4,G8:I8,	
24										G12:I12,G16:I16	
25	G5		=C5							G5:I7	
26	G9		=C9+IF(G$8<1,0,								
27			VLOOKUP(G$8,$B$5:$J$7,9,FALSE))							G9:I11	
28	G13		=C13+IF(G$12<1,0,								
29			VLOOKUP(G$12,$B$9:$J$11,9,FALSE))							G13:I15	
30	G17		=C17+IF(G$16<1,0,								
31			VLOOKUP(G$16,$B$13:$J$15,9,FALSE))							G17:I19	
32	J5		=MAX(G5:I5)							J5:J7,J9:J11,	
33										J13:J15,J17:J19	
34	K5		=INDEX(G$4:I$4,MATCH(J5,G5:I5,0))							K5:K7	
35	K9		=INDEX(G$8:I$8,MATCH(J9,G9:I9,0))							K9:K11	
36	K13		=INDEX(G$12:I$12,MATCH(J13,G13:I13,0))							K13:K15	
37	K17		=INDEX(G$16:I$16,MATCH(J17,G17:I17,0))							K17:K19	

FIGURE 13.4 Spreadsheet to compute the best trimming of a paper sheet.

decision from state 4 at stage 1 is to go to state 2 at stage 2, that is, to produce a 2-m roll. From state 2 at stage 2, the best decision is go to state 0 at stage 3, that is, to produce another 2-m roll of paper.

13.5 MINIMIZING THE RISK OF LOSING AN ENDANGERED SPECIES

The two previous examples of dynamic programming involved additive returns. The highest returns at stage t were in both cases the highest of the immediate returns at stage t plus the highest of the possible returns at stage $t + 1$ and beyond. However, dynamic programming can be applied to return functions with other forms. The following example uses a multiplicative return function in which the highest return at stage t is the product of the immediate return at stage t and of the highest return at stage $t + 1$ and beyond. Dynamic programming can deal readily with such nonlinear functions.

PROBLEM DEFINITION

Consider a nonprofit organization whose main objective is the protection of wildlife. This organization is particularly concerned by the status of a species that is close to extinction. Three projects have been set up to try to save this species. Each project is in a different area and uses a somewhat different approach, so the probability of success or failure of a project is independent of that of the others.

The organization has $2 million to grant the three projects. It wants to allocate the money to minimize the probability that all three projects will fail, which would result in extinction of the species. Recall that the probability of the simultaneous occurrence of three independent events is the product of the probabilities of occurrence of each event. Thus, the probability of total failure is the product of the probabilities of failure of each individual project.

The organization's managers believe strongly in the effectiveness of large grants. Therefore, they will consider only three funding levels for each project: $0, $1 million, or $2 million.

The organization has hired a consultant to determine the probability that each project will fail, conditional on the amount of money granted to it. The result of the consultant's work is summarized in Table 13.7. The table shows, for example, that if project 1 is granted nothing, there is a 50% chance that it will fail. A $1 million grant will decrease that probability to 30%, and $2 million will reduce it to 20%.

TABLE 13.7 Probability of Project Failure as a Function of Funding Level

Project (stage)	Funding level (state)		
	$0	10^6	2×10^6
1	.5	.3	.2
2	.7	.5	.3
3	.8	.5	.4

There are many possible ways of allocating the $2 million to the different projects. We shall use dynamic programming to find the combination that is best, in the sense that it minimizes the probability of total failure.

STAGES, STATES, AND DECISION VARIABLES

Again, as in the paper-trimming example, it is useful to think of the allocation of money to each project as a sequence of decisions, or stages. Consequently, stages $t = 1, 2,$ and 3 refer to the allocation of money to projects 1, 2, and 3, respectively.

Let us define state i as the budget available for allocation at stage t, and decision j as the budget available for allocation at stage $t + 1$. Thus, the money allocated at stage t (that is, to project t) is $i - j$. In this problem, i and j can only take the value 0, 1, or 2.

DYNAMIC PROGRAMMING SOLUTION

Consider the possible decisions at stage t and state i. Let $p_t(i, j)$ be the probability of failure of project t if the decision is j, that is, if it is granted $(i - j)$ million dollars. In contrast with the previous examples, this probability depends not only on the state and decision but also on the stage. Let $V_t^*(i)$ be the smallest probability of total failure from stage t onward, starting in state i and making the best decision j^*. Thus:

$$V_t^*(i) = \min_j [p_t(i, j) V_{t+1}^*(j)] \qquad (13.5)$$

The objective is to determine $V_1^*(2)$, the smallest probability of total failure that can be achieved by allocating $2 million to projects 1, 2, and 3. We do this by applying Recursive Relation (13.5) stage by stage, from stage 3 to stage 1.

Stage 3 Three states are possible at stage 3: $i = 0$, $i = 1$, and $i = 2$. If $i = 0$, no money is left for the third project and the only possible decision is $j^* = 0$.

Multistage Decision Making with Dynamic Programming

The probability that the third project will fail is then $V_3^*(0) = .8$ (Table 13.7). If $i = 1$, the best decision is $j^* = 0$, that is, spending $1 million, because it minimizes the probability of failure $V_3^*(1) = .5$. If $i = 2$, the best decision is $j^* = 0$, that is, spending $2 million, because it minimizes the probability of failure, $V_3^*(2) = .4$. The computations for stage 3 are summarized in Table 13.8.

Stage 2 The computations for this stage are in Table 13.9. The three possible states are $i = \$0$, $\$1$ million, or $\$2$ million. If, for example, $i = 1$, the possible decisions are $j = 0$ or 1. If $j = 0$, then $1 million is allocated to project 2 and its probability of failure, is .5, and the state at stage 3 is 0, which in turn leads to a probability of failure of project 3 of .8 (see Table 13.8). So the lowest probability of total failure for stages 2 and 3, starting in state $i = 1$ and making the decision $j = 0$, is:

$$V_2(1,0) = p_2(1,0)V_3^*(0)$$

$$= .5 \times .8 = .4$$

If $i = 1$ and $j = 1$, then project 2 receives no money and its probability of failure is .70. One million dollars is left for the third project, so its probability of failure is .50. Thus:

$$V_2(1,1) = p_2(1,1)V_3^*(0)$$

$$= .7 \times .5 = .35$$

TABLE 13.8 Best Funding of Third Project (Stage 3)

Budget state i (10^6)	Smallest probability of failure $V_3^*(i)$	Best decision j^* (10^6)
0	.8	0
1	.5	1
2	.4	2

TABLE 13.9 Best Funding of Second Project (Stage 2)

Budget state i (10^6)	Smallest probability of failure $V_2(i,j) = p_2(i,j) V_3^*(j)$ Decision j (10^6)			Smallest probability of failure $V_3^*(i)$	Best decision j^* (10^6)
	0	1	2		
0	$.56 = .7 \times .8$			.56	0
1	$.40 = .5 \times .8$	$.35 = .7 \times .5$		.35	1
2	$.24 = .3 \times .8$	$.25 = .5 \times .5$	$.28 = .7 \times .4$	.24	0

Therefore, the smallest probability of failure for stages 2 and 3, given $i = 1$ at stage 2, is:

$$V_2^*(1) = .35$$

with the best decision

$$j^* = \$1 \text{ million}$$

Repeating these calculations for state $i = 0$ leads to:

$$V_2^*(0) = .56$$

with the best decision

$$j^* = 0$$

And for $i = 2$:

$$V_2^*(2) = 0.24$$

with the best decision

$$j^* = 0$$

Stage 1 The calculations for stage 1 follow the same procedure as for stage 2. However, there is only one state at stage 1, $i = 2$, meaning that no money has been allocated yet. The results of the computations appear in Table 13.10. They show that:

$$V_1^*(2) = .105 \quad \text{and} \quad j^* = \$1 \text{ million}$$

Thus, the lowest probability of failure that the organization can achieve is .105. Achieving this requires granting $1 million at stage 1, that is, to the first

TABLE 13.10 Best Funding of First Project (Stage 1)

Budget state i (10^6)	Smallest probability of failure $P_1(i, j) = p_1(i, j) V_2^*(j)$			Smallest probability of failure $V_1^*(i)$	Best decision j^* (10^6)
	Decision j (10^6)				
	0	1	2		
2	$.112 = .2 \times .56$	$.105 = .3 \times .35$	$.12 = .5 \times .24$	.105	1

project. This means that $i = 1$ at stage 2. Table 13.9 shows that the best decision in state 1 at stage 2 is $j^* = 1$, so no money should be granted to the second project. This leads to state $i = 1$ at stage 3. Table 13.8 shows that in state 1 at stage 3 the best decision is $j^* = 0$, that is, to allocate the remaining $1 million to project 3.

One interesting aspect of this solution is that, while project 1 has the lowest probability of failure among all projects given a $2 million grant, the best solution is to share the funding between projects 1 and 3. The lower risk achieved by sharing the budget between two projects is an illustration of the principle that diversification helps mitigate risk.

SPREADSHEET FORMULATION AND SOLUTION

Figure 13.5 shows a spreadsheet to solve the species preservation problem by dynamic programming. The spreadsheet can deal with three states and three decisions at each of three stages. The entries in bold are data; the other entries are the results of formulas. Cells C5:E7, C9:E11, and C13:E15 contain the probability of failure of projects 3, 2, and 1, respectively, by state and decision. States and decisions that are not possible are given an arbitrarily high probability of failure, 999, so that they would not be picked in the solution.

The formulas in cells G5:I7 compute the smallest probability of failure for each state and decision at stage 3. This is equal to the probability of failure of the third project for different states and decisions. The formulas in cells G9:I11 compute the smallest probability of failure from projects 2 and 3, for each state and decision for project 2. For example, the formula in cell G9 multiplies the probability of failure in state 2 and decision 1 by the highest probability of failure of state 0 at stage 1. The VLOOKUP function takes the decision in cell G8, "0," looks for its match in the first column of the range B5:J7, and returns the value in the ninth column, .80. Similarly, the formulas in cells G13:I15 compute the lowest probability of failure from stages 1, 2, and 3 for each state and decision at stage 1.

The formulas in cells J5:J7, J9:J11, and J13:J15 compute the lowest probability of failure for each state at stages 3, 2, and 1, respectively.

The formulas in cells K5:K7, K9:K11, and K13:K15 find the decision giving the smallest probability of failure for each state at stages 3, 2, and 1, respectively. For example, the MATCH function in cell K9 finds the decision for which the probability of failure is smallest.

The results in Figure 13.5 show that starting with a $2 million budget (state 2 at stage 1), the best decision is to spend $1 million on project 1. At stage 2, state 1, the best decision is 1, so nothing should be spent on the second project. Last, at stage 3 and state 1, the best decision is 0, so $1 million should be spent

	A	B	C	D	E	F	G	H	I	J	K
1	MINIMIZING PROBABILITY OF FAILURE										
2	Stage	State	$p_t(j)$				Smallest probability $P_t(i,j)=p_t(j)P^*_{t+1}(j)$			Smallest probability	Decision
3	t	i	Decision j				Decision j			$P^*_t(i)$	j^*
4	3		0	1	2		0	1	2		
5		0	0.80	999	999		0.800	999.000	999.000	0.800	0
6		1	0.50	0.80	999		0.500	0.800	999.000	0.500	0
7		2	0.40	0.50	0.80		0.400	0.500	0.800	0.400	0
8	2		0	1	2		0	1	2		
9		0	0.70	999	999		0.560	499.500	399.600	0.560	0
10		1	0.50	0.70	999		0.400	0.350	399.600	0.350	1
11		2	0.30	0.50	0.70		0.240	0.250	0.280	0.240	0
12	1		0	1	2		0	1	2		
13		0	0.50	999	999		0.280	349.650	239.760	0.280	0
14		1	0.30	0.50	999		0.168	0.175	239.760	0.168	0
15		2	0.20	0.30	0.50		0.112	0.105	0.120	0.105	1
16											
17							Key cell formulas				
18	Cell		Formula							Copied to	
19	G4		=C4							G4:I4,G8:I8,G12:I12	
20	G5		=C5							G5:I7	
21	G9		=C9*VLOOKUP(G$8,$B$5:$J$7,9,FALSE)							G9:I11	
22	G13		=C13*VLOOKUP(G$12,$B$9:$J$11,9,FALSE)							G13:I15	
23	J5		=MIN(G5:I5)							J5:J7,J9:J11,J13:J15	
24	K5		=INDEX(G$4:I$4,MATCH(J5,G5:I5,0))							K5:K7	
25	K9		=INDEX(G$8:I$8,MATCH(J9,G9:I9,0))							K9:K11	
26	K13		=INDEX(G$12:I$12,MATCH(J13,G13:I13,0))							K13:K15	

FIGURE 13.5 Spreadsheet to allocate funds to minimize the probability of failure.

on the third project. With this allocation of funds, the probability of total failure would be .105.

The results also show that if only $1 million were available, it would be best to allocate it all to project 1, leading to a total probability of failure of .168.

13.6 CONCLUSION

The examples used in this chapter illustrate the potential of dynamic programming to solve many different kinds of forest management problems. The method is attractive because it is a general way to formulate and solve optimization problems. As the examples have shown, dynamic programming can be used to solve linear as well as nonlinear problems.

The method can also handle stochastic elements. For example, the growth of the stand in the thinning example could be a function of the state of the stand and of a random variable reflecting catastrophic events. This will be studied further in Chapter 17, which will also deal with infinite time horizons and with discounted returns, in an application of dynamic programming in conjunction with Markov chains.

Although the principles of dynamic programming are general, each application is specific. In contrast with linear programming, dynamic programming models do not have a standard form. For this reason, dynamic programming is generally more difficult to use than linear programming. In addition, solving dynamic programming problems becomes much more difficult as the number of state variables increases. The difficulties increase with the number of constraints. While linear programming routinely solves problems with thousands of constraints, this is not possible with dynamic programming. In fact, constraints in dynamic programming are handled quite differently than in linear programming. They are embedded in the formulation of the problem, in the definition of the states and decisions, which is why there isn't a standard form and why so much care must be paid to problem formulation, to ensure that it does cover all the constraints.

Furthermore, many problems that can be solved by dynamic programming can also be solved by other methods. For example, as suggested in this chapter, comparing all alternatives (the so-called exhaustive search approach) may be a good way to solve even fairly large problems, regardless of how much computer time it takes.

Ultimately, an efficient method is one that gives the correct solution at minimum cost. In this cost, computer time is negligible compared to the time spent in formulating a problem, developing a model, and writing the computer program to solve it.

PROBLEMS

13.1 (a) Set up your own spreadsheet model to compute the best thinning policy for a mixed conifer stand like the one shown in Figure 13.2. Using the same data, verify that your results are the same.

(b) Assume that the objective of the forester managing the stand is to maximize the present value of the harvests instead of the volume of harvests. Assume that the real price (net of inflation) of mixed conifer stumpage is expected to remain at $50/m^3 over the next 40 years and that the real rate of interest is 3% per year. Revise the network diagram in Figure 13.1 to show the new returns from each decision at each stage.

(c) Solve the new problem by hand, using the dynamic programming method.

(d) Revise the data in the spreadsheet in Figure 13.2 to solve the new problem. Verify that the solutions you obtained in parts (c) and (d) are the same. (e) How does this solution change if the real guiding rate of interest changes to 6% per year? Explain the changes.

13.2 The table describes different ways of managing a red pine plantation, from bare land to final harvest, over the next 60 years. The plantation may be established with an initial density of 400, 800 or 1,600 trees/ha. After 10 years, the plantation may be precommercially thinned or left as is. At age 30, a commercial thinning may be done as shown in the table; the yield of the thinning depends on the state of the stand and thus on earlier decisions. Finally, when the trees are 60 years old, the stand will be clear-cut.

Age (years)	Starting state	Action	Harvest (m^3/ha)	Ending state
0	Bare land	Plant 400 trees/ha	0	B
		Plant 800 trees/ha	0	C
		Plant 1600 trees/ha	0	D
10	B	Thin	0	E
		None	0	F
	C	Thin	0	F
		None	0	G
	D	Thin	0	G
		None	0	H
30	E	Thin	35	I
		None	0	K
	F	Thin	35	J
		None	0	L
	G	Thin	40	K
		None	0	M
	H	Thin	45	L
		None	0	N
60	I	Clear-cut	200	O (Bare land)
	J	Clear-cut	225	O (Bare land)
	K	Clear-cut	235	O (Bare land)
	L	Clear-cut	275	O (Bare land)
	M	Clear-cut	245	O (Bare land)
	N	Clear-cut	240	O (Bare land)

(a) Use the data in the table to express this problem as a network diagram, with nodes corresponding to states and arcs to decisions. Show the returns from each decision at each stage.
(b) Solve this problem by hand, using the dynamic programming method.
(c) Revise the spreadsheet model in Figure 13.2 to solve this problem. Verify that the solutions you obtained in parts (b) and (c) are the same.

13.3 Assume that the objective of managing the red pine stand described in Problem 13.2 is to maximize the present value of the harvests instead of the volume of the harvests. Assume further that the real price (net of inflation) of red pine stumpage is expected to remain at $50/m^3 over the next 60 years, the real guiding rate of interest is 3% per year, the costs of site preparation and planting are $200, $300, and $400/ha for densities of 400, 800, and 1,600 trees/ha, respectively, and precommercial thinning costs are $80/ha.

(a) Revise the network diagram from Problem 13.2 to show the new returns from each decision at each stage.

(b) Solve the new problem by hand, using the dynamic programming method.

(c) Revise the data in the spreadsheet model in Figure 13.2 to solve the new problem. Verify that the solutions you obtained in parts (b) and (c) are the same.

13.4 (a) Set up your own spreadsheet model like the one shown in Figure 13.4 to compute the best way to trim a 4-m roll of paper. Using the same data, verify that your results are the same.

(b) Assume that a new market has developed for rolls of paper 2.5 m wide where they are currently selling for $1,300/roll. Revise the network diagram in Figure 13.3 to show the new possible decisions at each stage, along with their returns.

(c) Solve the new problem by hand, using the dynamic programming method.

(d) Revise the data in the spreadsheet model in Figure 13.4 to solve the new problem. Verify that the solutions you obtained in parts (b) and (c) are the same.

13.5 A logging contractor cuts and sells oak logs. To maximize profits, the contractor must pay careful attention to the bucking operation, that is, how the crews cut the trees into logs after felling them. On the local market, logs sell according to three grades based on length and small-end diameter. Grade 1 logs must be 4 m long and at least 30 cm in diameter at the small end, grade 2 logs must be 3 or 4 m long and at least 25 cm in diameter, and grade 3 logs must be 2, 3, or 4 m long and at least 15 cm in diameter. The value of logs, by grade and length, is shown in the table. Consider a fallen tree that is 8 m long with a 60-cm large-end diameter. Moving from the large to the small-end of the tree, the diameter decreases by 5 cm every meter.

Tree Value, by Log Grade and Length

Log grade	Length (m)		
	2	3	4
1	—	—	$75
2	—	$55	$60
3	$25	$30	$40

(a) Use the data in the table to express this problem as a network diagram, with nodes corresponding to states and arcs to decisions. Show the returns from each decision at each stage. (*Hint:* The formulation of this problem is similar to that of the paper-trimming problem. Be sure to consider small-end diameter in determining what bucking patterns are possible at each stage.)

(b) Solve the new problem by hand, using the dynamic programming method.

(c) Revise the spreadsheet model in Figure 13.4 to solve this problem. Verify that the solutions you obtained in parts (b) and (c) are the same.

13.6 The logging foreman for a hardwood veneer mill has been authorized to spend up to two extra hours every time she fells a highly valuable tree in order to reduce the risk of breaking it. The foreman is trying to decide how to allocate this time between three modifications of the usual felling procedure: limbing trees that might break the tree as it falls, preparing a "bed" for trees to fall in; and using greater care in skidding fallen trees.

The foreman has estimated that extra time spent on each modification would lead to the probabilities of avoiding breakage shown in the table.

Probability of Not Breaking a Tree

Modification	Time spent (hours)		
	0	1	2
Limbing trees	.80	.95	.96
Preparing a bed	.90	.92	.93
More careful skidding	.91	.96	.96

For example, if no extra time is used to limb surrounding trees, the probability of not breaking the tree when it falls is .80, but this probability increases to .95 if 1 extra hour is spent to fell or limb surrounding trees.

A tree can break while it falls, when it hits the ground, or during skidding. Thus, the probability of avoiding breakage altogether is the product of the probabilities of avoiding breakage, given how much additional time is devoted to each modification of the felling procedure.

(a) Formulate this problem as a dynamic programming problem to find how much of the two extra hours should be spent on each modification to maximize the probability of no breakage. (*Hint:* The formulation of this problem is similar to the problem of minimizing the risk of losing an endangered species discussed in Section 13.5.)

(b) Solve this problem by hand.

(c) Revise the spreadsheet model in Figure 13.5 to solve this problem. Verify that the solutions you obtained in parts (b) and (c) are the same.

ANNOTATED REFERENCES

Amidon, E.L., and G.S. Akin. 1968. Dynamic programming to determine optimum levels of growing stock. *Forest Science* 14(3):287–291. (Early application of dynamic programming in forest management.)

Anderson, D.J., and B.B. Bare. 1994. A dynamic programming algorithm for optimization of uneven-aged forest stands. *Canadian Journal of Forest Research* 24:1758–1765. (Dynamic programming model with number of trees and basal area per acre as state variables.)

Arthaud, G.J., and W.D. Klemperer. 1988. Optimizing high and low thinning in loblolly pine with dynamic programming. *Canadian Journal of Forest Research* 8:1118–1122. (Dynamic programming model with four state variables for optimizing thinning regimes.)

Brodie, J.D., D.M. Adams, and C. Kao. 1978. Analysis of economic impacts on thinning and rotation of Douglas fir using dynamic programming. *Forest Science* 24(4):513–522. (Determines optimal thinning schedule and rotation as a function of regeneration costs, initial stocking levels, site, quality premiums, and variable logging costs.)

Chang, S.J. 1998. A generalized Faustmann model for the determination of optimal harvest age. *Canadian Journal of Forest Research* 28:652–659. (Shows how the Faustmann rotation can be determined by dynamic programming.)

Chen, C.M., D.W. Rose, and R.A. Leary. 1980. *How to Formulate and Solve Optimal Stand Density over Time Problems for Even-Aged Stands Using Dynamic Programming*. U.S. Forest Service General Technical Report NC-56. North Central Forest and Range Experiment Station, St. Paul, MN. 17 pp. (Well-written tutorial.)

Clark, C.W. 1990. *Mathematical Bioeconomics: The Optimal Management of Renewable Resources*. Wiley, New York. 400 pp. (Basic reference on the application of dynamic programming to natural resource problems.)

Conrad, J.M. 1999. *Resource Economics*. Cambridge University Press, Cambridge, UK. 213 pp. (Chapter 8 uses a spreadsheet-based dynamic programming model.)

Doherty, P.F., Jr., E.A. Marschall, and T.C. Grubb, Jr. 1999. Balancing conservation and economic gain: A dynamic programming approach. *Ecological Economics* 29:349–358. (Describes a dynamic programming model linking a hypothetical forest birds' population with the silvicultural practices used in its natural range, demonstrating the value of making optimal management decisions earlier in time.)

Dykstra, D.P. 1984. *Mathematical Programming for Natural Resource Management*. McGraw-Hill, New York. 318 pp. (Chapter 10 introduces the basics of dynamic programming using optimal-thinning and log-bucking examples.)

Kourtz, P. 1989. Two dynamic programming algorithms for forest fire resource dispatching. *Canadian Journal of Forest Research* 19:106–112. (Presents a dynamic programming model for dispatching water bombers and firefighting crews to fires and compares its effectiveness to existing practice.)

Lembersky, M.R., and U.H. Chi. 1986. Weyerhaeuser decision simulator improves timber profits. *Interfaces* 16(1):6–15. (Describes a graphical simulator of the consequences of bucking and sawing decisions and compares user's decisions to optimal ones generated by a dynamic programming model.)

Martin, G.L., and A.R. Ek. 1981. A dynamic programming analysis of silvicultural alternatives for red pine plantations in Wisconsin. *Canadian Journal of Forest Research* 11(2):370–379. (Determines optimum thinning schedules for red pine.)

Pickens, J.B., A. Lee, and G.W. Lyon. 1992. Optimal bucking of northern hardwoods. *Northern Journal of Applied Forestry* 9(4):149–152. (Describes a model to optimize the bucking of trees into logs, emphasizing the sensitivity of hardwood log grades to the spatial arrangement of defects.)

Ritters, K., J.D. Brodie, and D.W. Hann. 1982. Dynamic programming for optimization of timber production and grazing in ponderosa pine. *Forest Science* 28(3):517–526. (Application of dynamic programming to optimization of a joint production problem.)

Sessions, J., E.D. Olsen, and J. Garland. 1989. Tree bucking for optimal stand value with log allocation constraints. *Forest Science* 35(1):271–276. (Describes a model to optimize the bucking of trees into logs when demand restrictions for different log lengths must be taken into account.)

Tait, D.E. 1986. A dynamic programming solution of financial rotation ages for coppicing tree species. *Canadian Journal of Forest Research* 16:799–801. (Presents a dynamic programming model for deciding when to cut and when to replant a species that reproduces by sprouting.)

Valsta, L.T. 1990. A comparison of numerical methods for optimizing even-aged stand management. *Canadian Journal of Forest Research* 20:961–969. (Compares dynamic programming to other optimization methods.)

Winston, W.L. 1995. *Introduction to Mathematical Programming*. Duxbury Press, Belmont, CA. 818 pp. (Chapter 13 discusses dynamic programming.)

CHAPTER 14

Simulation of Uneven-Aged Stand Management

14.1 INTRODUCTION

Past chapters have dealt mostly with optimization models. The common theme throughout those chapters was to find management strategies that were best for specific criteria. The criteria were represented either by an objective function, by constraints, or by both. Optimization is indeed a fundamental goal in analyzing forestry operations. Ideally, one wishes always to find a decision that is not only good, but also better than any other decision.

However, optimization methods such as linear, goal, and integer programming have limitations. They force us to design models that fit very specific forms. For example, linear programming requires that the objective function and all the constraints be linear in the variables. Heroic assumptions must sometimes be made to cast a forest management problem into such a linear form.

Simulation allows for much more flexibility. Any phenomenon that can be represented by mathematical relationships is tractable by simulation. Simulation can be described as the process of developing a model of a real system and then conducting experiments with the model. In a sense, optimality is still the goal, since by experimenting with the model we hope to discover the best way of managing a system. But in contrast with programming models, there are no general solution algorithms, like the simplex, to identify an optimal solution. In a simulation experiment, all we do is observe the consequences of a specific set of actions corresponding to a management strategy. In essence, simulation allows us to bring the real world to the laboratory for intensive study. For that reason, simulation has become one of the most powerful and versatile tools for problem solving in forest management. Given that experiments with real forests might take decades or centuries and that certain outcomes would result in public outrage, the benefits of simulation experiments that can be run in an instant are obvious.

14.2 TYPES OF SIMULATION

Simulation models may be either deterministic or stochastic. A *deterministic* simulation assumes that the future state of a forest system can be predicted exactly from knowledge of the present (which may include memory of the past). Alternatively, a deterministic simulation may be viewed as an attempt to predict only the average state of a system, but not its actual state. A *stochastic* simulation model explicitly recognizes the uncertainty of all predictions. Uncertainty may take the form of variation in a continuous process. For example, the annual growth of a tree or stand may vary due to unpredictable changes in the weather, and the price of timber may vary depending on market conditions. Uncertainty may also take the form of *discrete random events*. Such events, for example, storms and fires, are discrete in that they either do or do not occur in a given time interval, and they are random, in that their exact time of occurrence is unpredictable.

Many simulation models contain both continuous and discrete elements as well as both deterministic and stochastic elements. Treating some of the variables in a stochastic model as if they were known exactly often leads to a model that is simpler to build and use. The art of the model builder is to recognize the key features of a system, avoiding detail that only complicates the model without making it necessarily more useful. The precept that "Small is beautiful" should always guide model building and model selection.

To understand simulation, perhaps more than for any other technique, one must do it. With this in mind, we will study two simulation examples. In this chapter we will simulate uneven-aged forest management. We will use the same growth equations as in the linear programming model in Chapter 8. This example will help clarify the differences between simulation and optimization models and between deterministic and stochastic models. In Chapter 15 we will simulate even-aged forest management, starting with a deterministic model and then introducing discrete random events that simulate catastrophic storms.

14.3 DETERMINISTIC SIMULATION OF UNEVEN-AGED FOREST MANAGEMENT

As in Chapter 8, the condition of an uneven-aged stand at a particular point in time is described by the number of trees per unit area in different size classes. We continue to classify trees by size only, because size is the key monitoring parameter in uneven-aged management. This costs us nothing in generality, because both size and species classes can be handled with the same modeling principles.

The purpose of the deterministic simulation model presented in this section is to predict the condition of the stand at any future point in time, given: (1) its initial condition, (2) how the stand grows over short time intervals, and (3) possible harvests.

The simulator will also compute the present value of the returns obtained from the stand over a long time period and a nonlinear index of the tree diversity. With these indicators, we shall compare different management regimes in terms of their economic and environmental effects.

SPREADSHEET SIMULATION

The spreadsheet model in Figure 14.1 uses the deterministic growth equations for uneven-aged forests studied in Chapter 8. It is designed to simulate q-ratio management guides for different cutting cycles. The spreadsheet also shows the economic implications of this type of management and its implications for stand diversity.

q-RATIO MANAGEMENT

A common way of specifying the desired tree distribution in uneven-aged stands is with the q-ratio system. The distribution of trees by size class in uneven-aged stands typically has an inverse-J shape: the number of trees per unit area decreases as the size of the trees increases. A q-ratio distribution specifies that the ratio of the number of trees in successive size classes is a constant, q.

	A	B	C	D	E	F	G	H	I	J	K	L	M
1	Q-RATIO MANAGEMENT												
2			Stock (trees/ha)			BP_t		Cut (trees/ha)			V_t	F_t	NPV_t
3	Year	Delay	Size1	Size2	Size3	index		Size1	Size2	Size3	($/ha)	($/ha)	($/ha)
4	0	0	840	234	14	1.30		520	194	9	1888	200	1688
5	5	5	387	49	5	1.14		0	0	0	0	0	0
6	10	10	446	59	6	1.15		0	0	0	0	0	0
7	15	15	496	71	6	1.16		0	0	0	0	0	0
8	20	0	539	84	7	1.17		219	44	2	461	200	98
9	25	5	387	49	5	1.14		0	0	0	0	0	0
10	30	10	446	59	6	1.15		0	0	0	0	0	0
11	35	15	496	71	6	1.16		0	0	0	0	0	0
12	40	0	539	84	7	1.17		219	44	2	461	200	37
13	45	5	387	49	5	1.14		0	0	0	0	0	0
14	50	10	446	59	6	1.15		0	0	0	0	0	0
15	55	15	496	71	6	1.16		0	0	0	0	0	0
16	60	0	539	84	7	1.17		219	44	2	461	200	14
17	65	5	387	49	5	1.14		0	0	0	0	0	0
18	70	10	446	59	6	1.15		0	0	0	0	0	0
19	75	15	496	71	6	1.16		0	0	0	0	0	0
20	Cycle (y)		Desired stock (trees/ha)					q-ratio			NPV ($/ha)		1837
21	20		320	40	5			8					
22													
23					Key cell formulas								
24	Cell		Formula								Copied to		
25	A5		=A4+5								A5:A19		
26	B5		=IF(B4+5=A$21,0,B4+5)								B5:B19		
27	C5		=0.92*(C4-H4)-0.29*(D4-I4)-0.96*(E4-J4)+109								C5:C19		
28	D5		=0.04*(C4-H4)+0.90*(D4-I4)								D5:D19		
29	E5		=0.02*(D4-I4)+0.90*(E4-J4)								E5:E19		
30	F4		=SUM(C4:E4)/MAX(C4:E4)								F4:F19		
31	H4		=IF(B4=0,MAX(0,C4-C$21),0)								H4:J19		
32	K4		=0.3*H4+8*I4+20*J4								K4:K19		
33	L4		=IF(SUM(H4:J4)>0,200,0)								L4:L19		
34	M4		=(K4-L4)/(1.05)^A4								M4:M19		
35	M20		=SUM(M4:M19)										
36	D21		=E21*$H21								C21		

FIGURE 14.1 Spreadsheet simulation of q-ratio management in uneven-aged stands.

The growth model in Figure 14.1 deals with three size classes. The number of trees by size class that satisfy a particular q-ratio are related as follows:

$$y_1^* = q\, y_2^*$$
$$y_2^* = q\, y_3^*$$
(14.1)

Simulation of Uneven-Aged Stand Management

where y_i^* is the desired number of trees in size class i and q is a positive constant greater than 1. The q-ratio alone is not enough to define a tree distribution. We also need the number of trees in one size class. Usually, the number of trees in the largest size class is used. For example, if $y_3^* = 5$ trees/ha and $q = 8$, then, $y_2^* = 40$ trees/ha and $y_1^* = 320$ trees/ha.

One advantage of the q-ratio approach is that, regardless of the number of size classes, the desired distribution is completely defined with two numbers: the q-ratio itself and the number of trees in one size class. The disadvantage is that it is less flexible than directly specifying the number of trees desired in each size class.

A q-ratio-based harvesting rule would be to cut the trees in each size class in excess of the desired distribution. In our example:

$$h_{1t} = \max(0, y_{1t} - y_1^*)$$
$$h_{2t} = \max(0, y_{2t} - y_2^*) \qquad (14.2)$$
$$h_{3t} = \max(0, y_{3t} - y_3^*)$$

where h_{it} is the number of trees per hectare cut from size class i at time t and y_{it} is the number of trees per hectare in size class i at time t, before the cut. In Figure 14.1, the q-ratio is in cell H21, and the desired number of trees in the largest size class is in cell E21. The formulas in cells C21:D21 compute the desired number of trees in the small and medium size classes according to Equation (14.1).

VARYING THE CUTTING CYCLE

The stand growth model in Figure 14.1 is the same as in Chapter 8 (see Equations 8.10):

$$y_{1,t+5} = 0.92(y_{1t} - h_{1t}) - 0.29(y_{2t} - h_{2t}) - 0.96(y_{3t} - h_{3t}) + 109$$
$$y_{2,t+5} = 0.04(y_{1t} - h_{1t}) + 0.90(y_{2t} - h_{2t}) \qquad (14.3)$$
$$y_{3,t+5} = \qquad\qquad 0.02(y_{2t} - h_{2t}) + 0.90(y_{3t} - h_{3t})$$

where t is in years so that the model implies that the stand state is recorded every 5 years. Cells A5:A19 in Figure 14.1 contain the formulas for the current time, t, starting with $t = 0$, and cells C5:E19 contain the formulas of the growth equations. The given initial stand state is in cells C4:E4.

Harvests, however, need not occur every 5 years just because the growth model uses that time interval. In many applications, it is desirable to know the effect of different cutting cycles. For example, in Chapter 9 we showed the advantage of lengthening the cutting cycle when there were fixed harvesting costs per unit area.

To simulate cutting cycles of variable length, we need to keep track of the elapsed time since the last harvest. This *delay* starts at zero and increases by 5 years each growth period. When it is equal to the length of the cutting cycle, *delay* is reset to zero, signaling a harvesting event and related fixed costs.

In the spreadsheet in Figure 14.1, the cutting cycle is set at 20 years in cell A21. The delay at year 0 is set at 0 in cell B4. The formulas in cells B5:B19 increment the delay and reset it to zero when the delay is equal to the cutting cycle. The formulas in cells H4:J19 compute the harvest according to Equation (14.2) if the delay is 0; otherwise they set the harvest at 0.

Environmental Performance

One measure of the environmental performance of a management regime is the diversity of the stand that it leads to. In Chapters 8 and 9 we used the smallest number of trees in any size class as an index of diversity. We then used linear programming to maximize this smallest number (MaxiMin criterion).

There are many other measures of diversity, most of which are not linear functions of the number of trees. There is no perfect index of diversity, linear or nonlinear; It is just not possible to express the full complexity of a population structure with a single number. Still, it is best to use simple indices in models, unless there are strong reasons to do otherwise.

Among the nonlinear indices of diversity, one of the simplest to express and interpret is that of Berger and Parker (1970). According to Magurran (1988), this is one of the most satisfactory indices of diversity available. In our example, the Berger–Parker (BP) index of diversity of trees in the uneven-aged stand observed at time t is:

$$BP_t = \frac{y_{1t} + y_{2t} + y_{3t}}{\max(y_{1t}, y_{2t}, y_{3t})}$$

The Barger–Parker index is smallest and equal to 1 if all the trees are in one size class. It is largest and equal to the number of size classes if the number of trees is the same in all size classes. Thus, in this example the largest possible value of BP is 3.

In the spreadsheet in Figure 14.1, cells F4:F19 contain the formulas of the Berger–Parker index of diversity in each period.

Simulation of Uneven-Aged Stand Management

ECONOMIC PERFORMANCE

Assume that the average real prices of trees of different sizes are those used in Chapter 9: $0.30 per tree in the smallest size class, $8 per tree of medium size, and $20 per tree of largest size. Then the value of the harvest in year t is:

$$V_t = 0.30 h_{1t} + 8 h_{2t} + 20 h_{3t} \quad (\$/ha)$$

Furthermore, assume, as in Chapter 9, that the fixed cost of a harvest is $200/ha. This cost is incurred if and only if there is a harvest, and it is independent of the amount harvested. Thus, the expression of fixed cost in year t is:

$$F_t = \begin{cases} 200 & \text{if } h_{1t} + h_{2t} + h_{3t} > 0 \\ 0 & \text{otherwise} \end{cases}$$

The net present value of the harvest in year t, assuming an interest rate of 5% per year, is:

$$\text{NPV}_t = \frac{V_t - F_t}{1.05^t}$$

And the total net present value of all harvests is:

$$\text{NPV} = \text{NPV}_0 + \text{NPV}_5 + \cdots + \text{NPV}_T$$

where T is the length of the simulation, in years. For large values of T, NPV would approach the net present value that the stand would produce with this management over an infinite length of time, that is, the stand value inclusive of the land and of the initial trees.

In the spreadsheet in Figure 14.1, the formulas for the value of the harvest in each period are in cells K4:K19. The formulas for the fixed cost are in cells L4:L19. They indicate that the fixed cost is incurred if and only if there is a harvest. The formulas for the net present value of the periodic harvests are in cells M4:M19, and the formula for total net present value is in cell M20.

14.4 APPLICATIONS OF DETERMINISTIC SIMULATION MODEL

Figure 14.1 shows the results of an application of the simulation model for a particular set of parameters (shown in bold): a given initial stand state (y_{1t} = 840, y_{2t} = 234, y_{3t} = 14), a cutting cycle of 20 years, a q-ratio of 8, and a desired stock of 5 trees/ha in the largest size class (y_3^* = 5 trees/ha).

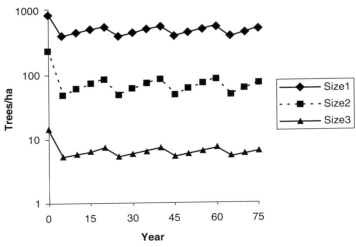

FIGURE 14.2 Stand dynamics under management.

STAND DYNAMICS

With these parameters, the simulation shows the sequence of harvests that will maintain the desired number of trees in all size classes throughout the 75 years of the simulation. The data show that a stand would quickly reach a steady state. From year 5 onward, the growth of the stand over 20 years just replaces the harvest (Figure 14.2). Therefore, the management is sustainable in the long run. However, as shown in cells F4:F19 of Figure 14.1, the index of diversity would decrease as a result of the initial harvest, although it would remain stable thereafter at about 1.15.

STAND VALUE

The financial results in Figure 14.1 show that, under the simulated q-ratio regime, the value of the first harvest, \$1,888/ha, is much larger than that of the subsequent periodic harvests, which are constant at \$461/ha. The effect of discounting reduces the contribution of future harvests to NPV even further (compare the value in cell M4 to those in cells M8, M12, and M16). If the simulation were continued beyond 75 years, the NPV would converge to a value just slightly larger than the \$1,837/ha obtained with a time horizon of 75 years. Thus, the stand value, inclusive of land and trees, is, to a close approximation, \$1,837/ha.

Simulation of Uneven-Aged Stand Management   297

	O	P	Q
2	Cutting		
3	cycle	NPV	BP
4	(y)	($/ha)	index
5		1837	1.16
6	5	1323	1.14
7	10	1700	1.14
8	20	1837	1.16
9	30	1840	1.16
10	40	1802	1.22
11			
12		*Key cell formulas*	
13	Cell	Formula	
14	P5	=M20	
15	Q5	=F19	
16	P6:Q10	see text	

FIGURE 14.3 Spreadsheet to compute the effects of the cutting cycle on net present value and stand diversity after 75 years.

SENSITIVITY TO CUTTING CYCLE

The simulation results in Figure 14.1 are for a cutting cycle of 20 years. As stated earlier, the cutting cycle is an important management decision because of its consequences for stand structure, harvest, and economic performance. We can analyze the sensitivity of the results to the length of the cutting cycle by repeating the simulation with different cutting cycles, keeping all the other parameters constant.

Figure 14.3 shows another range of the spreadsheet in Figure 14.1. It automatically performs the simulations for cutting cycles from 5 to 40 years, and records the corresponding NPV and the Berger–Parker index of diversity at the end of each 75-year simulation. To create this table in Excel, execute the following steps:

1. Set cell P5 equal to cell M20 in Figure 14.1, which contains the formula for the NPV. Set cell Q5 equal to cell F19 (in Figure 14.1), which contains the formula for the Berger–Parker index at year 75. Enter the different values of the cutting cycle to be tested in cells O6:O10.
2. Select the range of cells O5:Q10.
3. On the Data menu, click **Table**. In the **Column input cell** box, enter cell A21, which contains the cutting cycle in Figure 14.1.

The results in Figure 14.3 show that the forest value increases markedly as the cutting cycle increases from 5 to 30 years, and decreases slightly for a cutting cycle of 40 years. The Berger–Parker diversity index is slightly higher for cutting cycles of 20–30 years than for cutting cycles of 5 or 10 years. The highest diversity is obtained for a cutting cycle of 40 years.

Thus, in this case there seems to be little conflict between the economic and diversity goals. Both are best achieved with longer cutting cycles. But this result is predicated on the use of a target distribution with five trees in the largest size class, and a q-ratio of 8. How would these results change with different values of these parameters?

SENSITIVITY TO q-RATIO AND NUMBER OF LARGE TREES

To explore the sensitivity of our results to the parameters of the q-ratio distribution, we repeat the simulation with a cutting cycle of 20 years and vary first the desired number of trees in the largest size class, y_3^*, and then the q-ratio, all other parameters being held constant. Each time, we record the values of the variables of interest: the Berger–Parker diversity at age 75 years and the net present value.

Figure 14.4 shows another range of the spreadsheet in Figure 14.1. It automatically performs the simulations for different values of q and the number of trees in the largest size class. To create this table in Excel, do the following:

1. Set cell T4 equal to cell F19 in Figure 14.1, which contains the formula for the Berger–Parker index at year 75.
2. Type the different q-ratios to be tested in row cells U4:X4 and the different number of largest trees to be tested in column cells T5:T8.

	S	T	U	V	W	X
2	BERGER-PARKER DIVERSITY					
3			q-ratio			
4		1.16	2	4	8	10
5	Number of	5	1.08	1.12	1.16	1.17
6	largest	10	1.12	1.18	1.21	1.22
7	trees/ha	20	1.19	1.26	1.35	1.46
8		30	1.25	1.32	1.66	1.71
9						
10			Key cell formula			
11	Cell		Formula			
12	T4		=F19			
13	U5:X8		see text			

FIGURE 14.4 Spreadsheet to compute the effect of the q-ratio and the number of largest trees on stand diversity.

3. Select the range of cells T4:X8.
4. On the Data menu, click **Table**. In the **Row input cell** box, enter cell H21, which contains the q-ratio in Figure 14.1. In the **Column input cell** box, enter cell E21, which contains the number of trees in the largest size class in Figure 14.1.

The results in Figure 14.4 show that diversity is positively related with both the q-ratio and the number of trees in the largest size class. The highest predicted diversity would be achieved with a q-ratio of 10 and 30 trees/ha in the largest size class. But what are the economic implications?

Figure 14.5 shows another range of the spreadsheet in Figure 14.1 to compute the NPV for different q-ratios and number of trees in the largest size class. The setup is like that in Figure 14.4. The only change is in the formula in cell T17, equal to cell M20, which contains the formula of the NPV in Figure 14.1. The results in Figure 14.5 show that forest value decreases as either the q-ratio or the number of trees in the largest size class increase.

Figure 14.6 shows a scatterplot of BP diversity index values against the forest value, using the data from Figures 14.4 and 14.5. Each point corresponds to a particular combination of q-ratio and number of trees in the largest size class. All the values assume a cutting cycle of 20 years. There is a clear negative relationship between stand diversity and net present value.

However, it must be kept in mind that the net present value is expressed only in terms of timber revenues. Diversity itself has a value as real as timber, though it is more difficult to measure in monetary terms due to the absence of markets for diversity. The benefit of the approach taken here is to show clearly the trade-off between the two objectives. For example, Figure 14.5 shows

	S	T	U	V	W	X
15	NPV ($/HA)					
16			q-ratio			
17		1837	2	4	8	10
18	Number of	5	2058	1969	1837	1785
19	largest	10	1855	1676	1410	1267
20	trees/ha	20	1577	1219	544	167
21		30	1384	847	-237	-293
22						
23			Key cell formulas			
24	Cell		Formula			
25	T17		=M20			
26	U18:X21		see text			

FIGURE 14.5 Spreadsheet to compute the effect of the q-ratio and the number of largest trees on net present value.

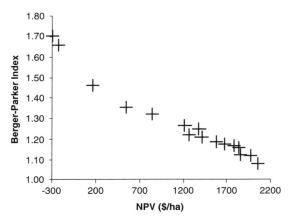

FIGURE 14.6 Trade-off between net present value and stand diversity.

that the management regime leading to the highest Berger–Parker diversity index after 75 years would have NPV = $–293/ha, while the highest NPV was $2,058/ha. Therefore, the opportunity cost of the management regime of highest diversity was $2,351/ha in terms of foregone timber value.

14.5 STOCHASTIC SIMULATION

All the simulations done so far were deterministic, that is, based on models with constant parameters. Although these deterministic models are very useful and give correct "average" results, they do not give us any information on the variability of possible outcomes. Most biological processes have a random, or stochastic, component. Similarly, economic variables such as timber prices fluctuate considerably over time. The goal of stochastic simulation is to represent this variability as fully as possible.

BIOLOGICAL RISK

In the stand growth model of Equations (14.3), the number of trees in the smallest size class, y_{1t}, depends in part on the recruitment, which can be very erratic. As a result, the number of trees in the smallest size class can be very different from the average number predicted by the deterministic equation. In comparison, the equations predicting the number of trees in the medium and large size classes are quite accurate: It is relatively easy to predict the future of trees that are already big enough to be counted in one of the three size classes.

Simulation of Uneven-Aged Stand Management

Therefore, in a stochastic version of Model (14.3), it is most useful to recognize the wide possible range of recruitment rates, independent of stand condition. To this end, rewrite the first equation in Model (14.3) as:

$$y_{1,t+5} = [(0.92(y_{1t} - h_{1t}) - 0.29(y_{2t} - h_{2t}) - 0.96(y_{3t} - h_{3t}) + 109]u_t \quad (14.4)$$

where u_t is a random variable representing the random part of recruitment. This equation means that part of the number of trees in size class 1 at $t + 5$ years can be predicted from the state of the stand after harvest at t, while part of it is unpredictable.

The object of stochastic simulation is to generate a realistic value of u_t for each period of the simulation, t. If the deterministic part of the model is any good, the mean of u_t over many years must be 1, because the deterministic model should predict correctly on average. Our ability to predict the rest of the distribution of u_t, that is, the spread of u_t around its mean of 1, and the probability of occurrence of particular values depends on our ecological and biometric knowledge.

Assume that all we have is a rough estimate of the variability around the mean. We judge that there may be up to a 25% difference between the actual number of trees in size class 1 and the number predicted by the deterministic equation. Furthermore, there is an equal probability for any value of the number of trees within this interval. Then random values of u_t can be generated by:

$$u_t = 0.50 R_t + 0.75 \quad (14.5)$$

where R_t is a random number between 0 and 1. The stochastic simulation entails drawing such a random number in each period, t. For the highest possible values, $R_t = 1$, the number of trees is 25% larger than the number predicted by the deterministic equation. For the lowest possible values, $R_t = 0$, it is 25% smaller. We assume that any value between these bounds can occur with equal probability. Note that the average value of R_t is 0.50, so on average, $0.50R + 0.75 = 1$, and the stochastic equation predicts the same number of trees as the deterministic equation. However, the actual stand dynamics may be quite different from those predicted by the deterministic model.

Similarly, we may recognize the stochastic elements in the other two growth equations of the model. They arise from the variability of mortality and growth rates of individual trees. But this is typically smaller than the variations in recruitment, because mortality rates are small and trees grow at a fairly constant rate. Thus, the stochastic part of the growth of trees in the medium and large size classes is small and can be ignored (see Problem 14.10).

Economic Risk

In the deterministic simulations, the economic implications of different management policies were calculated under the assumption that the price of timber was constant. In reality, it is well known that most timber prices vary over time. Real prices (net of inflation) may have a deterministic (predictable) tendency to increase or decrease. This deterministic trend can be dealt with by adjusting the interest rate. The interest rate is set lower if the real price is expected to increase, or it is set higher if the real price is expected to decrease (see Chapter 18).

In addition, there is unpredictable variation around the trend. High demand or low supply lead typically to prices that are higher than usual. Low demand or high supply lead instead to prices that are lower than usual. This variation can be modeled explicitly in different ways, depending on the data available on price variability.

Assume that the average real price of trees for timber is, as before, $0.30 per tree in the smallest size class, $8 per tree of medium size, and $20 per tree of the largest size. Deterministic price trends are already reflected in the interest rate, assumed to be 5% per year. However, we know from past experience that the price may be up to 50% lower or higher than the average price. Prices in this interval can occur with equal probability. With these assumptions, a price index, P_t, can be used to change timber prices:

$$P_t = 1.0 R_t + 0.5 \tag{14.6}$$

where R_t is a random number, uniformly distributed between 0 and 1. The highest possible value of the price index is $P_t = 1.5$, obtained when $R_t = 1$. This means that the price is 50% higher than average. The lowest possible value of the price index is $P_t = 0.5$, obtained when $R_t = 0$. This means that the price is 50% lower than average. The average value of P_t is 1, because the mean of R_t is 0.5.

With this expression of the price index, the stochastic value of the harvest in year t is:

$$V_t = (0.30 h_{1t} + 8 h_{2t} + 20 h_{3t}) P_t \tag{14.7}$$

where the expression in parentheses is the value of the harvest at average prices and the price index P_t randomly increases or decreases this value by up to 50%.

Spreadsheet Stochastic Simulation

Figure 14.7 shows a spreadsheet to simulate the growth of a managed stand, subject to biological and economic risk. The changes with respect to the spreadsheet in Figure 14.1 are:

Simulation of Uneven-Aged Stand Management

	A	B	C	D	E	F	G	H	I	J	K	L	M	N	O
1	MANAGEMENT UNDER RISK														
2			Stock (trees/ha)			BP		Cut (trees/ha)			V_t	F_t	NPV_t	Risk	
3	Year	Delay	Size1	Size2	Size3	index		Size1	Size2	Size3	($/ha)	($/ha)	($/ha)	Bio	Price
4	0	0	840	234	14	1.30		520	194	9	993	200	793	1.2	0.53
5	5	5	353	49	5	1.15		0	0	0	0	0	0	0.9	1.02
6	10	10	338	58	6	1.19		0	0	0	0	0	0	0.8	1.37
7	15	15	356	66	6	1.20		0	0	0	0	0	0	0.9	1.08
8	20	0	416	73	7	1.19		96	33	2	301	200	38	1.0	0.9
9	25	5	386	49	5	1.14		0	0	0	0	0	0	1.0	1.05
10	30	10	363	59	6	1.18		0	0	0	0	0	0	0.8	1.15
11	35	15	492	68	6	1.15		0	0	0	0	0	0	1.2	1.19
12	40	0	650	81	7	1.14		330	41	2	650	200	64	1.2	1.39
13	45	5	415	49	5	1.13		0	0	0	0	0	0	1.1	1.1
14	50	10	429	61	6	1.15		0	0	0	0	0	0	0.9	1.03
15	55	15	559	72	6	1.14		0	0	0	0	0	0	1.2	0.64
16	60	0	543	87	7	1.17		223	47	2	597	200	21	0.9	1.23
17	65	5	441	49	5	1.12		0	0	0	0	0	0	1.1	1.24
18	70	10	411	62	6	1.16		0	0	0	0	0	0	0.8	1.02
19	75	15	488	72	6	1.16		0	0	0	0	0	0	1.1	0.52
20	Cycle (y)		Desired stock (trees/ha)					q ratio			NPV ($/ha)		917		
21	20		320	40	5			8							
22															
23					Key cell formulas										
24	Cell		Formula											Copied to	
25	A5		=A4+5											A5:A19	
26	B5		=IF(B4+5=A$21,0,B4+5)											B5:B19	
27	C5		=(0.92*(C4-H4)-0.29*(D4-I4)-0.96*(E4-J4)+109)*N4											C5:C19	
28	D5		=0.04*(C4-G4)+0.90*(D4-H4)											D5:D19	
29	E5		=0.02*(D4-H4)+0.90*(E4-I4)											E5:E19	
30	F4		=SUM(C4:E4)/MAX(C4:E4)											F4:F19	
31	H4		=IF(B4=0,MAX(0,C4-C$21),0)											H4:J19	
32	K4		=(0.3*H4+8*I4+20*J4)*O4											K4:K19	
33	L4		=IF(SUM(H4:J4)>0,200,0)											L4:L19	
34	M4		=(K4-L4)/(1.05)^A4											M4:M19	
35	M20		=SUM(M4:M19)												
36	D21		=E21*$H21											C21	
37	N4		=RAND()*0.5+0.75											N4:N19	
38	O4		=RAND()+0.5											O4:O19	

FIGURE 14.7 Spreadsheet model to simulate uneven-aged management under biological and economic risk; cutting cycle = 20 years.

The addition of formulas of the random biological shocks in cells N4:N19 according to Equation (14.5)

The addition of formulas of the random price index in cells O4:O19 according to Equation (14.6)

The growth formulas in cells C5:C19 according to Equation (14.4).

The formulas of the value of the harvests in cells K4:K19 according to Equation (14.7)

All other parameters and assumptions are the same as in the deterministic simulation in Figure 14.1. In particular, the initial stand state is the same, and so are the cutting cycle, 20 y, the q-ratio $= 8$, and the desired number of trees in the largest size class, 5 trees/ha.

The results in Figure 14.7 are for one replication, that is, for one single sequence of random numbers. In this case, the NPV is $917/ha, which is about half the forest value obtained in the deterministic simulation in Figure 14.1. The Berger–Parker index of diversity at year 75 is 1.16, which is the same as the diversity predicted by the deterministic simulation.

However, this is only one of the many possible outcomes. A few replications obtained by pressing the F9 key, which generates a new sequence of R_t values, show that many different results are possible. Thus, we need to replicate this experiment many times to get a better picture of the variability of the results.

REPLICATING THE EXPERIMENT

The outcome of each simulation is a random observation. To make good statistical statements about the outcomes, we need many observations, that is, many replications of the simulation. Let N be the number of replications, and assume that the performance measure is the Berger–Parker index of stand diversity after 75 years, BP.

Each experiment, i, gives a random observation BP_i. The summary statistics of interest for management include the largest and smallest value of BP_i, its average value or mean, $\overline{BP}$:

$$\overline{BP} = \frac{\sum_{i=1}^{N} BP_i}{N}$$

and its standard deviation, SD, which measures the variability of BP_i around the mean:

$$SD = \sqrt{\frac{\sum_{i=1}^{N} (BP_i - \overline{BP})^2}{N-1}}$$

Furthermore, the mean itself is a random variable, because repeating the N replications several times would each time give a different mean. This variability of the mean is measured by the standard deviation of the mean, or standard error,

SE. The standard error of the mean is:

$$SE = \frac{SD}{\sqrt{N}}$$

There is a 95% probability that the true mean lies in the interval

$$\overline{BP} \pm 2SE$$

To compare the effect of two policies, say, two different cutting cycles on the mean diversity, one may compare the confidence intervals of the means under the two policies. If they do not overlap, then the probability that the two policies will result in the same mean diversity is less than 5%. The same reasoning applies to the NPV or to any other performance indicator.

Figure 14.8 shows another range of the spreadsheet in Figure 14.7, set up to make $N = 100$ replications, collect the outcomes for the Berger–Parker index and for the NPV, and calculate their summary statistics.

The replication numbers are in column Q5:Q104. They range from 1 to 100. To calculate the 100 replications with Excel, do the following:

1. Set cell R5 equal to cell M20, which contains the formula for NPV (Figure 14.7). Set cell S5 equal to cell F19, which contains the formula of the Berger–Parker index of diversity after 75 years.
2. Select the range of cells Q5:S104.
3. On the Data menu, click **Table**. In the **Column input cell** box, enter cell O1 or any other cell of the spreadsheet that does not affect the computation. The object is to induce the spreadsheet to redo the calculations for each row of the table. The results will differ for each row only because the random numbers change while all the other parameters remain constant.

The results in Figure 14.8 show that in these 100 replications, the NPV ranged from a minimum of $823/ha to a maximum of $2,836/ha. The mean NPV was $1,910/ha, with a standard error of $58/ha. The Berger–Parker index of stand diversity after 75 years ranged from 1.11 to 1.25, with a mean of 1.17 and a standard error of less than 0.5%.

The 95% confidence interval of the mean NPV is:

$$1,910 \pm 2 \times 58 = (\$1,794/ha, \$2,026/ha)$$

	Q	R	S	T	U	V	W	X
1	REPLICATIONS				SUMMARY STATISTICS			
2			Berger-				Berger-	
3		NPV	Parker			NPV	Parker	
4	#	($/ha)	index			($/ha)	index	
5	1	2373	1.17		Max	2836	1.25	
6	2	2160	1.12		Min	823	1.11	
7	3	1230	1.15		Mean	1910	1.17	
8	4	823	1.15		SD	582	0.03	
9	5	1891	1.21		SE	58	0.00	
10	6	2533	1.12					
11	7	1916	1.13			Key cell formulas		
12	8	2153	1.24		Cell	Formula		Copy to
13	9	1350	1.14		Q6	=Q5+1		Q6:Q104
14	10	924	1.15		R5	=M20		
15	11	2836	1.14		S5	=F19		
16	12	2363	1.17		R6:S104	see text		
17	13	2597	1.19		V5	=MAX(R5:R104)		V5:W5
18	14	1044	1.15		V6	=MIN(R5:R104)		V6:W6
19	15	1858	1.18		V7	=AVERAGE(R5:R104)		V7:W7
20	16	2535	1.21		V8	=STDEV(R5:R104)		V8:W8
21	17	2368	1.14		V9	=V8/10		V9:W9

FIGURE 14.8 Spreadsheet for 100 replications of net present value and stand diversity under risk; cutting cycle = 20 years.

This 95% confidence interval contains the NPV obtained by deterministic simulation, $1,837/ha (Figure 14.1). The standard error of the mean Berger–Parker index is practically zero, so the 95% upper and lower confidence limits are very close to the mean value of 1.17, which is close to the value given by the deterministic simulation.

Thus, the mean results from the stochastic simulation are not very different from those of the deterministic simulation. What has been gained from the stochastic simulation is information about the possible variability of outcomes. While the law of averages may be relevant for owners who have many stands of this type, it may not be for owners of a single woodlot. For them, any one of the outcomes of the stochastic simulation is possible, and variability in outcomes will affect property values and influence managerial decisions.

Deciding What Matters

Because it fully reflects the range of outcomes, stochastic simulation can tell us what policy choices really matter. For example, the deterministic simulation in

Simulation of Uneven-Aged Stand Management

	Q	R	S	T	U	V	W	X
1	REPLICATIONS				SUMMARY STATISTICS			
2		Stand	Berger-				Berger-	
3		value	Parker			NPV	Parker	
4	#	($/ha)	index			($/ha)	index	
5	1	822	1.17		Max	2823	1.35	
6	2	2043	1.24		Min	822	1.16	
7	3	2405	1.22		Mean	1738	1.23	
8	4	2712	1.27		SD	529	0.04	
9	5	2175	1.17		SE	53	0.00	
10	6	2106	1.26					
11	7	1657	1.19			Key cell formulas		
12	8	2381	1.20		Cell	Formula		Copy to
13	9	1410	1.30		Q6	=Q5+1		Q6:Q104
14	10	1337	1.20		R5	=M20		
15	11	1445	1.25		S5	=F19		
16	12	2265	1.25		R6:S104	see text		
17	13	970	1.23		V5	=MAX(R5:R104)		V5:W5
18	14	1588	1.23		V6	=MIN(R5:R104)		V6:W6
19	15	2086	1.19		V7	=AVERAGE(R5:R104)		V7:W7
20	16	2017	1.20		V8	=STDEV(R5:R104)		V8:W8
21	17	2384	1.28		V9	=V8/10		V9:W9

FIGURE 14.9 Spreadsheet for 100 replications of net present value and stand diversity under risk; cutting cycle = 40 years.

Figure 14.3 suggests that a cutting cycle of 40 years is economically inferior to one of 20 years. To test if this difference is truly significant, we can do 100 replications of a stochastic simulation with a cutting cycle of 40 years (Figure 14.9). The 95% confidence interval of the mean NPV is:

$$1738 \pm 2 \times 53 = (\$1{,}632/\text{ha}, \$1{,}844/\text{ha})$$

This 95% confidence interval overlaps the 95% confidence interval of the mean NPV for a cutting cycle of 20 years ($1,794/ha, $2,026/ha). Thus, the means are not statistically different; that is, there is little chance that the shorter cutting cycle would give a higher NPV on average.

The mean Berger–Parker index, however, is significantly higher for a cutting cycle of 40 years than for a cutting cycle of 20 years: 1.23 with a standard error of 0.00, against 1.17 with a standard error of 0.00.

Most owners would agree that the cutting cycle of 40 years is a better policy, because it results in greater diversity without significantly affecting the economic returns.

14.6 CONCLUSION

A deterministic forest simulator like the one presented in this chapter can be developed quickly with less knowledge than that needed for the optimization models of Chapters 8 and 9. And yet, despite its simplicity, it is a powerful tool. Numerous other management experiments can be done with this model. The general principle remains the same as for the analysis of the cutting cycle, q-ratio, and number of trees in the largest size class. To study the effect of one parameter, conduct a series of simulations in which all parameters remain the same except the one being studied. Then observe how a selected management criterion, such as net present value or stand diversity, responds to changes in the parameter.

One difficulty is that many parameters can be changed. To keep the number of experiments reasonably small, the analyst must use considerable judgment and intuition in guessing which parameters are important and in defining management policies that are likely to yield good results.

A deterministic simulation can be made more realistic by introducing random variables. These may reflect biological risk or economic risk. Replicating the simulation with different strings of random numbers gives a set of random observations of the measures of performance, such as net present value and tree diversity. The performance can then be analyzed statistically to show their average behavior and the variations in performance around the average. As a result it is possible to tell which policy choices really make a difference, given the variability of the actual outcomes in real life.

PROBLEMS

14.1 (a) Set up your own spreadsheet model like the one shown in Figure 14.1 to simulate q-ratio management in an uneven-aged stand. Using the same data, verify that your results are the same.

(b) Use this model to simulate the effect of cutting cycles of 10, 20, and 30 years. What changes do you observe in the stand dynamics?

(c) Change the cutting cycle back to 20 years, and then apply the model with q-ratios of 6, 8, and 10. What changes do you observe in the stand dynamics?

(d) Change the q-ratio back to 8, and then use the model with a desired number of trees in the largest size class of 5, 10, and 15 trees/ha. What changes do you observe in the stand dynamics?

14.2 The stand simulated in the spreadsheet model in Figure 14.1 moves quickly to a stable steady state, given the particular harvesting rule used.

Apply this model to predict the growth of a stand that starts in the same initial condition but is never cut. (*Hint:* Make the desired number of trees in the largest age class very high.) Graph the number of trees in each size class over time. What do you observe relative to the results in Figure 14.2?

14.3 Repeat the simulation of the growth of a stand that is never cut, as in Problem 14.2, but starting in two very different states:

	Trees per Hectare		
	Size 1	Size 2	Size 3
Initial state 1	0	0	30
Initial state 2	300	0	0

(a) What kind of stand does each initial state describe?
(b) Simulate the growth of the stand given each initial state. How do the results compare with those from Problem 14.2?
(c) Based on these results, how does the initial state of a stand seem to influence its subsequent growth and steady state?

14.4 Consider the stand simulated by the spreadsheet model in Figure 14.1. Suppose the owners want to try the following management regime: Cut the stand every 10 years, and take half of the trees in each size class.

(a) Modify the spreadsheet model in Figure 14.1 to simulate this policy. Use the spreadsheet to predict the present value of the returns to the owners and the Berger–Parker diversity of the stand after 100 years. (Use the prices, guiding rate of interest, and cutting cycle shown in Figure 14.1.)
(b) Considering net present value and diversity, which management regime would you choose, this new one or the one simulated in Figure 14.1? Why?

14.5 Consider the stand simulated by the spreadsheet model in Figure 14.1. Assume that the owners have been offered $50.00 ha/y to allow cattle to graze on their land. But the cattle would cause many young trees to die, thus reducing recruitment. Suppose that this damage would reduce the number of trees in size class 1 by about 90 trees/ha every 5 years.

(a) Modify the spreadsheet model to reflect this damage. Use the spreadsheet to predict the net present value of the returns to the owners and the Berger–Parker diversity of the stand after 100 years. (Use the prices, guiding rate of interest, and cutting cycle shown in Figure 14.1.)
(b) Compare the net present value of returns with and without grazing. Should the owners allow cattle grazing for the $50/ha fee that they have been offered? If not, what minimum price should they charge?

14.6 (a) Set up your own spreadsheet model, linked to the model shown in Figure 14.1, to predict the effects of the length of the cutting cycle on net present value and the Berger–Parker index of diversity, as in Figure 14.3. Using the same data, verify that your results are the same.

(b) What cutting cycle gives the highest NPV if the guiding rate of interest is 3%, 5%, and 10%?

(c) Set the guiding rate of interest at 5% per year. What is the best cutting cycle if there is a fixed harvesting cost of $100/ha, $200/ha, and $400/ha?

14.7 Consider the stand simulated by the spreadsheet model in Figure 14.1. Assume that a veneer mill has just opened in the vicinity of the stand. As a result, the values of trees in the largest size class have increased to $40 per tree.

(a) Modify the spreadsheet in Figure 14.1 to reflect this change. What is the effect of the higher price on the best economic cutting cycle? (Use the prices for trees in other size classes, guiding rate of interest, and cutting cycle shown in Figure 14.1.)

(b) What effect has the opening of the mill had on the value of the stand, inclusive of land and trees?

14.8 Assume that a new pathogen starts to cause heavy mortality in the stand described by the model in Figure 14.1, especially in the largest trees. A biometrician finds that the effect of the disease can be simulated by changing the last growth equation in Equations (14.3) to:

$$y_{3,t+5} = 0.02(y_{2t} - h_{2t}) + 0.80(y_{3t} - h_{3t})$$

(a) According to this equation, what fraction of the trees alive and in size class 3 at time t are still alive at time $t + 5$? (*Hint:* You may want to refer back to Chapter 8.)

(b) Change the spreadsheet in Figure 14.1 to reflect this pathogen. How much does the damage cost the stand's owner?

(c) Does this pathogen affect the economic cutting cycle?

14.9 Set up your own spreadsheet model to simulate uneven-aged management under biological and economic risk, as in Figure 14.7.

(a) With the same data and equations, the simulation results displayed in your spreadsheet will not be the same as in Figure 14.7. Why?

(b) Add a section to your spreadsheet to collect the outcomes of repeated simulations, as in Figure 14.8. Run 100 replications. How do the predictions for net present value and for the Berger–Parker diversity index compare to the results in Figure 14.8?

(c) Can you say with confidence that your spreadsheet model is the same as the one shown in Figure 14.7?

14.10 (a) Modify the spreadsheet model in Figure 14.7 to reflect uncertainty in the equations that predict the number of trees in the medium and large size classes. Assume that there may be up to a 5% difference between the actual number of trees in size classes 2 and 3 and the number predicted by the deterministic equations. Assume also that there is equal probability that the actual number of trees lies in this interval.

(b) Run 100 replications of the simulation with a spreadsheet table as in Figure 14.8.

(c) How are the means and the standard deviations of the results affected by this uncertainty?

14.11 (a) Modify the spreadsheet model in Figure 14.7 to increase the desired number of trees in the largest size class from 5 to 10 trees/ha, keeping other things equal.

(b) Run 100 replications of the simulation with a spreadsheet table as in Figure 14.8.

(c) Has the change in management affected significantly the NPV or the diversity index?

ANNOTATED REFERENCES

Baker, F.A., D.W. French, and D.W. Rose. 1982. DMLOSS: A simulator of losses in dwarf mistletoe-infested black spruce stands. *Forest Science* 28(3):590–598. (Deterministic simulation model of damage caused by a forest pathogen.)

Belcher, D.W., M.R. Holdaway, and G.J. Brand. 1982. *A Description of STEMS: The Stand and Tree Evaluation and Modeling System*. U.S. Forest Service General Technical Report NC-79. North Central Forest Experiment Station, St. Paul, MN. 18 pp. (Deterministic individual tree simulation model for Lake States forests.)

Berger, W.H., and F.L. Parker. 1970. Diversity of planktonic Foraminifera in deep sea sediments. *Science* 168:2345–2347. (Source of the Berger–Parker index of diversity used in this chapter.)

Boothby, R.D., and J. Buongiorno. 1985. *UNEVEN: A Computer Model of Uneven-Aged Forest Management*. Agricultural Research Bulletin R3285. University of Wisconsin, Madison. 62 pp. (Basis of the deterministic simulation model used in this chapter.)

Boyce, S.G. 1985. *Forestry Decisions*. U.S. Forest Service General Technical Report SE-35. Southeastern Forest Experiment Station, Asheville, NC. 318 pp. (Compares the merits of simulation and optimization models.)

Buongiorno, J., A. Kolbe, and M. Vasievich. 2000. Economic and ecological effects of diameter-limit and BDq management regimes: Simulation results for northern hardwoods. *Silva Fennica* 34(3):223–235. (Deterministic simulation modeling for multispecies stands.)

Crandall, D.A., and R.J. Luxmoore. 1982. Simulated water budgets for an irrigated sycamore phytomass farm. *Forest Science* 28(1):17–30. (Deterministic simulation model to manage irrigation of a biomass plantation.)

Grant, W.E. 1986. *Systems Analysis and Simulation in Wildlife and Fisheries Sciences*. Wiley, New York. 338 pp. (Describes a variety of deterministic simulation models of wildlife populations.)

Hansen, G.D. 1984. *A Computer Simulation Model of Uneven-Aged Northern Hardwood Stands Maintained Under the Selection System*. School of Forestry Miscellaneous Publication No. 3 (ESF 84-017). SUNY College of Environmental Science and Forestry, Syracuse, NY. 21 pp. (Describes a multispecies simulation model.)

Law, A.M., and W.D. Kelton. 2000. *Simulation Modeling and Analysis*. McGraw-Hill, New York. 760 pp. (Basic reference on simulation modeling.)

Magurran, A.E. 1988. *Ecological Diversity and its Measurement*. Princeton University Press, Princeton. 179 pp.

Nyland, R.D. 2002. *Silviculture: Concepts and Applications*. McGraw-Hill, New York. 682 pp. (Chapter 10 discusses q-ratios.)

Ragsdale, C.T. 1998. *Spreadsheet Modeling and Decision Analysis: A Practical Introduction to Management Science*. South-Western College Publishing, Cincinnati, OH. 742 pp. (Chapter 12 shows how to use spreadsheets to formulate and run stochastic simulation models.)

Schulte, B., and J. Buongiorno. 1998. Effects of uneven-aged silviculture on the stand structure, species composition, and economic returns of loblolly pine stands. *Forest Ecology and Management* 111:83–101. (Deterministic simulation modeling to evaluate q-ratio management policies, diameter-limit harvests, and high grading.)

Schulte, B., J. Buongiorno, C.R. Lin, and K. Skog. 1998. *SOUTHPRO, a Computer Program for Managing Uneven-Aged Loblolly Pine Stands*. USDA Forest Service, Forest Products Laboratory, General Technical Report FPL-GTR-112, Madison, WI. 47 pp. (Spreadsheet-based deterministic simulator of multispecies stand management, with tabular and graphic output of stand dynamics and economic and ecological indices.)

Silvert, W. 1989. Modeling for managers. *Ecological Modeling* 47:53–64. (Spreadsheet-based deterministic simulation model of the Great Bustard (*Otis tarda*) populations.)

Starfield, A.M., and A.L. Bleloch. 1986. Building models for conservation and wildlife management. Macmillan, New York. 253 pp. (Chapter 2 contains an example of a deterministic simulation model.)

Wiström, P. 2000. A solution method for uneven-aged management applied to Norway spruce. *Forest Science* 46(3):452–463. (Describes a nonoptimization approach to q-ratio management.)

Volin, V.C., and J. Buongiorno. 1996. Effects of alternative management regimes on forest stand structure, species composition, and income: A model for the Italian Dolomites. *Forest Ecology and Management* 87:107–125. (Deterministic simulation model for spruce-fir-beech-larch uneven-aged stands.)

CHAPTER **15**

Simulation of Even-Aged Forest Management

15.1 INTRODUCTION

This chapter presents a second set of deterministic and stochastic simulations, but in an even-aged management context. Some of the specific objectives of the simulation models that we shall develop are to:

Describe the evolution of the forest over time when it is managed according to a variant of area control.

Predict the effects of changing the allowable cut on both the value of the timber production and the capacity of the forest to serve as a carbon sink.

Underline the importance of context in defining best policies. In particular, we will see how economic rotations may vary when the harvest influences timber prices.

Simulate uncertain prices and catastrophic events such as fires and storms and develop best management policies in this context.

As in Chapters 6 and 7, the forest areas in each age class describe the state of the forest at a specific point in time. The silvicultural regime consists of clear-cutting followed by immediate artificial regeneration. The volume of timber in a particular age class is assumed to be strictly a function of the age of the stand. In addition to timber production, we are interested in the environmental roles of the forest, in particular in carbon sequestration.

We shall assume throughout that the management policy is a variant of area control. Specifically, the manager fixes the rotation age, and this determines the allowable cut during each time interval. Simulation models will be used to investigate the effect of the choice of the rotation on timber revenues and on carbon sequestration.

The first part of the chapter deals with a deterministic simulation model, which assumes constant parameters and constant prices. With this model we show how to simulate the case where changes in the harvest influence timber prices.

The second part of the chapter deals with a stochastic version of the same model that simulates both rare random events, such as catastrophic storms, and continuous random variations, such as those observed in prices.

15.2 DETERMINISTIC SIMULATION OF EVEN-AGED MANAGEMENT

As in Chapters 6 and 7, we will classify stands according to their age, because age is the key monitoring parameter in even-aged management. However, the same modeling principles would apply if the stands were also distinguished by species, at the cost of some complication. As in Chapter 6, the time unit is a decade, and stands are classified in 10-year age classes. The variable A_{it} refers to the area in age class i in year t, and the variable X_{it} refers to the area harvested and reforested in age class i from year t to year $t + 10$.

The purpose of the simulation model is to predict the condition of the forest at any future point in time, given its initial condition, the laws that predict how the forest grows, and a particular harvesting policy.

The model will also compute the net present value of the returns obtained from the stand over a long time period and a measure of the amount of carbon stored in the forest. With these indicators we shall compare different management regimes in terms of some of their economic and environmental effects.

INPUT DATA

The spreadsheet in Figure 15.1 shows a simulation model for a forest with four age classes. Age class 1 is covered with trees 1–10 years old, age class 2 with trees 11–20 years old, and so on up to age class 4, which is covered with trees 31 years old and older. The numbers in bold characters in Figure 15.1 show the input data for this simulation. Cells B4:E4 contain the initial condition of the

Simulation of Even-Aged Forest Management

	A	B	C	D	E	F	G	H	I	J	K	L	M	N	O
1	TIMBER AND CARBON														
2		Stock (ha) in class:					Cut (ha) in class:					H_t	N_t	NPV_t	Carbon
3	Year	1	2	3	4		1	2	3	4		$(10^3 m^3)$	$(\$10^3)$	$(\$10^3)$	$(10^3 t)$
4	0	50	50	100	300		0	0	0	83		33	1625	1625	100.4
5	10	83	50	50	317		0	0	0	83		33	1625	998	97.4
6	20	83	83	50	283		0	0	0	83		33	1625	612	91.3
7	30	83	83	83	250		0	0	0	83		33	1625	376	88.3
8	40	83	83	83	250		0	0	0	83		33	1625	231	88.3
9	50	83	83	83	250		0	0	0	83		33	1625	142	88.3
10	60	83	83	83	250		0	0	0	83		33	1625	87	88.3
11	70	83	83	83	250		0	0	0	83		33	1625	53	88.3
12	80	83	83	83	250		0	0	0	83		33	1625	33	88.3
13	90	83	83	83	250		0	0	0	83		33	1625	20	88.3
14	100	83	83	83	250		0	0	0	83		33	1625	12	88.3
15	(m^3/ha)	50	120	260	400		$(\$/m^3)$	50				NPV$(\$10^3)$	4189		
16	$(\$/ha)$	500	Interest (y)	0.05	Rotation (y)			60				Allowable cut (ha)	83.3		
17															
18							Key cell formulas								
19	Cell		Cell formula										copied to		
20	A5		=A4+10										A5:A14		
21	B5		=SUM(G4:J4)										B5:B14		
22	C5		=B4-G4										C5:D14		
23	E5		=D4-I4+E4-J4										E5:E14		
24	J4		=MIN(E4,$N$16)										J4:J14		
25	G4		=MIN(B4,$N$16-SUM(H4:$J4))										G4:I14		
26	L4		=SUMPRODUCT(B$15:E$15,G4:J4)/1000										L4:L14		
27	M4		=L4*$H$15-SUM(G4:J4)*$B$16/1000										M4:M14		
28	N4		=M4/(1+$E$16)^A4										N4:N14		
29	N15		=SUM(N4:N14)												
30	N16		=(SUM(B4:E4)/I16)*10												
31	O4		=0.65*SUMPRODUCT(B$15:E$15,B4:E4)/1000										O4:O14		

FIGURE 15.1 Spreadsheet model to simulate the effects of even-aged management on timber production and carbon storage.

forest, that is, the area in each of the four age classes at time 0. The other inputs are the volume per hectare in each age class (in cells B15:E15), the timber price (in cell H15), the reforestation cost per hectare (in cell B16), the yearly interest rate (in cell E16), and the rotation age (in cell I16).

AREA-CONTROL MANAGEMENT

For even-aged management, area control consists of harvesting no more than a prespecified area from the forest per unit of time. The rotation age, R, and the forest area determine the allowable cut. The rotation is the time it takes to cut the entire forest. Thus, the allowable cut in any single year is the fraction 1/R of the entire forest. Then the allowable cut per decade is:

$$AC = \frac{10(A_{1,0} + A_{2,0} + A_{3,0} + A_{4,0})}{R} \qquad (15.1)$$

The simulation assumes that the allowable cut is taken first from the oldest age classes. Thus, the harvest in each age class and decade is given by these equations:

$$\begin{aligned} X_{4t} &= \min(A_{4t}, AC) \\ X_{3t} &= \min(A_{3t}, AC - X_{4t}) \\ X_{2t} &= \min(A_{2t}, AC - X_{4t} - X_{3t}) \\ X_{1t} &= \min(A_{1t}, AC - X_{4t} - X_{3t} - X_{2t}) \end{aligned} \qquad (15.2)$$

The first equation states that the area cut in the oldest age class is either the area in that class or the allowable cut, whichever is smaller. The second equation states that the area cut from the second oldest age class is either the area in that class or the remaining allowable cut after the cut in the oldest age class, whichever is smaller. The last two equations are similar.

In the spreadsheet in Figure 15.1, the allowable-cut formula is in cell N16. The formulas that define the harvest are in cells G4:J14.

Forest Growth

The forest growth equations used by the spreadsheet in Figure 15.1 are similar to those in Chapter 6:

$$\begin{aligned} A_{1,t+10} &= X_{1t} + X_{2t} + X_{3t} + X_{4t} \\ A_{2,t+10} &= A_{1t} - X_{1t} \\ A_{3,t+10} &= A_{2t} - X_{2t} \\ A_{4,t+10} &= (A_{3t} - X_{3t}) + (A_{4t} - X_{4t}) \end{aligned} \qquad (15.3)$$

The first equation states that the area in age class 1 at the end of each decade is equal to the area that has been cut during that decade. This assumes that

Simulation of Even-Aged Forest Management

reforestation follows harvest immediately. The second equation states that the area in age class 2 at the end of each decade is the area that was in age class 1 a decade earlier minus what was cut from it during that decade. The third equation is similar to the second. The last equation adds the fact that any uncut area of age class 4 stays in age class 4.

In the spreadsheet in Figure 15.1, the current year, t, is calculated by the formula in cells A5:A14, starting with $t = 0$. The growth equations are expressed by the formulas in cells B5:E14.

FINANCIAL PERFORMANCE

In this model, the financial performance of a particular management policy is judged by: the periodic volume of the harvest, the monetary value of the periodic harvest, net of the reforestation cost, and its net present value at time 0 for the assumed interest rate.

The volume of the periodic harvest from year t to year $t + 10$ is:

$$\underset{(m^3)}{H_t} = \sum_{i=1}^{4} \underset{(m^3/ha)}{v_i} \underset{(ha)}{X_{it}} \qquad (15.4)$$

where v_i is the volume per unit area in age class i.

The net income from this periodic harvest is:

$$\underset{(\$)}{N_t} = \underset{(\$/m^3)}{p} \underset{(m^3)}{H_t} - \underset{(\$/ha)}{c} \sum_{i=1}^{4} \underset{(ha)}{X_{it}} \qquad (15.5)$$

where p is the timber price per unit of volume and c is the reforestation cost per unit area.

The net present value, at time 0, of this future income is:

$$\underset{(\$)}{\mathrm{NPV}_t} = \frac{N_t}{(1+r)^t} \qquad (15.6)$$

where r is the yearly interest rate and the net income is accounted for at the beginning of each decade.

The total present value of all harvests is:

$$\underset{(\$)}{\mathrm{NPV}} = \mathrm{NPV}_0 + \mathrm{NPV}_{10} + \cdots + \mathrm{NPV}_T \qquad (15.7)$$

where T is the length of the simulation, in years. For large values of T, the NPV would approach the net present value that the forest would produce with this management over an infinite length of time, that is, the forest value inclusive of the land and of the initial trees.

In the spreadsheet in Figure 15.1, the formulas in cells L4:L14 calculate the periodic production, in 1,000 m^3, according to Equation (15.4), using the yields per unit area in cells B15:E15.

The formulas in cells M4:M14 compute the net value of the harvest according to Equation (15.5), in $1,000, with the reforestation cost in cell B16 and the timber price in cell H15.

The formulas in cells N4:N14 compute the net present value of the harvest in each decade according to Equation (15.6), in $1,000, using the interest rate in cell E16.

The formula in cell N15 computes the total net present value of all the harvests over $T = 100$ years according to Equation (15.7).

Carbon Storage

In contrast to the timber revenues of the forest, which depend on periodic harvests, carbon sequestration/storage in the forest depends on the amount of timber that is left standing.

If there were a market for carbon storage, with a well-defined price per unit of carbon stored, carbon storage could be treated similarly to timber production to arrive at a global measure of economic performance. We would then seek the best combination of timber production and carbon storage. However, as long as markets for environmental services are not well developed, one is forced to deal separately with the economic performance (for timber production only) and with the carbon storage. The simulation will help understand the trade-offs between the two functions.

The amount of carbon stored in trees is closely related to the biomass of wood, inclusive of roots and branches. This biomass in turn is closely related to the volume of the standing trees. In this application we shall assume that each cubic meter of wood contains about 0.65 tons of carbon. Then the amount of carbon stored in year t is:

$$C_t = 0.65 \sum_{i=1}^{4} v_i \, A_{it} \qquad (15.8)$$
$$\text{(t)} \quad \text{(t/m}^3\text{)} \quad \text{(m}^3\text{/ha) (ha)}$$

In the spreadsheet in Figure 15.1, the formulas in cells O4:O14 calculate the amount of carbon stored in the forest stock at 10-year intervals.

15.3 APPLICATIONS OF DETERMINISTIC SIMULATION

Figure 15.1 shows the result of one simulation, projecting the evolution of the even-aged forest over 100 years, starting from a specific initial condition, and applying area-control management with a rotation age of 60 years.

Forest Dynamics and Carbon Storage

With this 60-year rotation, the allowable cut is 83.3 ha/decade. After 30 years, the forest has reached a steady state, with 83 ha in each of the three youngest age classes and 250 ha in the oldest age class. The amount of carbon stored in the forest decreases during the first 30 years and then remains stable at 88,300 tons.

Because of the large initial area in the oldest age class, the cut always occurs in that age class; and because the area cut is constant, the production is also constant, at 33,000 m^3 of timber per decade.

Forest Value

At constant prices, this policy produces a constant periodic net income of about $1.6 million per decade. However, due to discounting at 5% per year, the net present value of the periodic income decreases rapidly over time, reaching negligible values after 100 years. The NPV of the harvests would therefore change little if the simulation extended beyond 100 years. Thus, the forest value for timber production, the value of the land and the initial stock of trees, is about $4.2 million under this particular management policy. This forest value is determined largely by the initial forest condition.

Sensitivity to Rotation Age

The simulation results in Figure 15.1 are for a rotation age of 60 years. With this area-control management, the allowable cut, and therefore the rotation age, is the key determinant of the results. The rotation age has consequences for the forest structure, the harvest, the overall financial performance, and the carbon storage. An analysis of the sensitivity of the results to the rotation involves repeated simulations with different rotation ages while keeping other parameters constant.

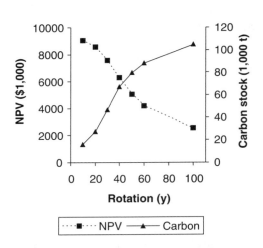

	R	S	T
3	Rotation	NPV	Carbon
4	(y)	($10³)	(10³t)
5		4189	88.29
6	10	9053	16.25
7	20	8549	27.63
8	30	7562	46.58
9	40	6284	67.44
10	50	5027	79.95
11	60	4189	88.29
12	100	2514	104.98
13			
14		Key cell formulas	
15	Cell	Cell formula	
16	S5	=N15	
17	T5	=O14	
18	S6:T12	see text	

FIGURE 15.2 Spreadsheet to compute the effects of the rotation on the NPV and carbon stock, with corresponding chart.

Figure 15.2 shows another range of the spreadsheet shown in Figure 15.1. Its purpose is to automatically do the simulations for rotations of 10–100 years and to record the corresponding NPV and the amount of carbon in the forest stock at year 100. To create the Table in Figure 15.2 in Excel, execute the following steps:

1. Set cell S5 equal to cell N15, which contains the formula for the NPV (Figure 15.1). Set cell T5 equal to cell O14, which contains the formula for the carbon in the forest stock at year 100. Type the different values of the cutting cycle to be tested in cells R6:R12.
2. Select the range of cells R5:T12.
3. On the Data menu, click **Table**. In the **Column input cell** box, enter cell I16 (Figure 15.1), which contains the rotation.

The results in Figure 15.2 show the trade-off between carbon storage and economic returns from harvesting. From a purely financial viewpoint, the best rotation is just 10 years, leading to an NPV of about $9 million. Note that 10 years is not the best rotation according to Faustmann's formula studied in Chapter 7. If you check this with the spreadsheet in Figure 7.2 you should find that the rotation that maximizes the land expectation value is 20 years.

The shorter financial rotation found in this simulation is due to the cutting policy. With area control, a rotation of 10 years allows cutting more of the valuable

Simulation of Even-Aged Forest Management

oldest timber than a rotation of 20 years, thus leading to the higher NPV, given the initial conditions of this particular forest.

This result illustrates a general principle. The best policy, in terms of rotation or some other management parameter, depends on the *context*. The context includes the initial forest condition, the management regime, and the economic and social environments. The latter point is worth stressing, since short rotations and area control would result in quick liquidation of the oldest age classes.

For environmental reasons, apart from and in addition to carbon storage, this would often be unacceptable. In choosing the rotation length, a manager would want to weigh carefully the financial gain resulting from a shorter rotation against the loss of other values that, although harder to quantify, are no less real.

The carbon storage and NPV results plotted in Figure 15.2 would help in this decision. They show carbon stock increasing steadily with rotation length while NPV decreases. The opportunity cost of a 60-year rotation compared to a 10-year rotation would be about $5 million. This would be very useful information to help stakeholders decide whether the carbon storage and the other environmental benefits of maintaining the old age classes resulting from a 60-year rotation are important enough to forego this financial gain.

DOWNWARD-SLOPING DEMAND

As another example of the importance of context in choosing a management policy, let us consider how variations in prices may influence the choice of a rotation, still within the area-control framework that we have assumed up to now. Specifically, we will assume that the amount of timber sold from the forest influences price. This is often referred to as the case of the "downward-sloping demand curve," since it implies that as the volume harvested increases, the price received by the owner declines. This may happen if there is limited demand for the timber from a forest, for example, if it is in an isolated area.

In the next simulation, we will assume that the demand equation has the following form:

$$p_t = aH_t^{-0.7}$$

where p_t is the price received for the timber, in $/ m^3$, and H_t is the volume of timber cut, in 1,000 m^3, from year t to year $t + 10$. This equation implies that an increase of the volume sold of 1% leads to a decline in price of 0.7%. Symmetrically, a rise in price of 1% causes a decline in demand of $1/0.7 = 1.43$%. Since the relative change in quantity is greater than the relative change in price, in absolute value, the demand for timber is said to be elastic with respect to price. The elasticity of demand with respect to price is precisely −1.43.

To define the demand equation completely, we need to estimate the coefficient a. This can be done if we know one point of the demand curve. Assume that when the volume sold from the forest is 30 (1,000 m³/decade) the price received is on average $50/m³. Then:

$$50 = a30^{-0.7}$$

That is:

$$a = 50 \times 30^{0.7} = 541$$

So the demand equation is:

$$p_t = 541H_t^{-0.7} \tag{15.9}$$

The geometric representation of this demand equation in Figure 15.3 shows that the price may be as high as $115/m³, when the volume offered is near 10,000 m³/decade, and as low as $40/m³, when the volume offered is 50,000 m³/decade.

Figure 15.4 shows a modification of the spreadsheet model in Figure 15.1 to introduce this downward-sloping demand feature. The fixed price in cell H15 has been replaced by the formulas in cells P4:P14, to give output-dependent

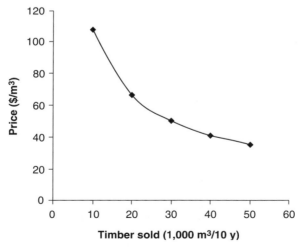

FIGURE 15.3 Demand curve for timber with a price elasticity of −1.43.

Simulation of Even-Aged Forest Management

	A	B	C	D	E	F	G	H	I	J	K	L	M	N	O	P
1	ELASTIC DEMAND															
2		Stock (ha) in class:					Cut (ha) in class:					H_t	N_t	NPV_t	Carbon	Price
3	Year	1	2	3	4		1	2	3	4		$(10^3 m^3)$	$(\$10^3)$	$(\$10^3)$	$(10^3 t)$	$(\$/m^3)$
4	0	50	50	100	300		0	0	0	83		33	1507	1507	100.4	46.5
5	10	83	50	50	317		0	0	0	83		33	1507	925	97.4	46.5
6	20	83	83	50	283		0	0	0	83		33	1507	568	91.3	46.5
7	30	83	83	83	250		0	0	0	83		33	1507	349	88.3	46.5
8	40	83	83	83	250		0	0	0	83		33	1507	214	88.3	46.5
9	50	83	83	83	250		0	0	0	83		33	1507	131	88.3	46.5
10	60	83	83	83	250		0	0	0	83		33	1507	81	88.3	46.5
11	70	83	83	83	250		0	0	0	83		33	1507	50	88.3	46.5
12	80	83	83	83	250		0	0	0	83		33	1507	30	88.3	46.5
13	90	83	83	83	250		0	0	0	83		33	1507	19	88.3	46.5
14	100	83	83	83	250		0	0	0	83		33	1507	11	88.3	46.5
15	(m^3/ha)	50	120	260	400							NPV ($\$10^3$)	3886			
16	(\$/ha)	500	Interest (/y)	0.05			Rotation (y)	60				Allowable cut (ha)	83.33			
17																
18							Key cell formulas									
19	Cell	Cell formula											copied to			
20	A5	=A4+10											A5:A14			
21	B5	=SUM(G4:J4)											B5:B14			
22	C5	=B4-G4											C5:D14			
23	E5	=D4-I4+E4-J4											E5:E14			
24	J4	=MIN(E4,$N$16)											J4:J14			
25	G4	=MIN(B4,$N$16-SUM(H4:$J4))											G4:I14			
26	L4	=SUMPRODUCT(B$15:E$15,G4:J4)/1000											L4:L14			
27	M4	=L4*P4-SUM(G4:J4)*$B$16/1000											M4:M14			
28	N4	=M4/(1+$E$16)^A4											N4:N14			
29	N15	=SUM(N4:N14)														
30	N16	=(SUM(B4:E4)/I16)*10														
31	O4	=0.65*SUMPRODUCT(B$15:E$15,B4:E4)/1000											O4:O14			
32	P4	=541*L4^-0.7											P4:P14			

FIGURE 15.4 Spreadsheet simulation model with a downward-sloping demand curve.

prices based on Equation (15.9). The formulas for the net value of the harvests in cells M4:M14 have accordingly been changed to use these prices.

We can use this new spreadsheet to simulate the effect of different rotations on the value of the forest, given a downward-sloping demand curve (see Problem 15.6). The results are shown in Figure 15.5. Two things can be observed from this figure. First, with a price-elastic demand, the NPV is lower at short rotation ages. Second, the economic rotation increases from 10 to 30 years with a price-elastic demand.

The reason for the difference in the NPV is that for a given rotation, the volume sold is the same, regardless of the assumption on demand, but unit prices tend to be lower when demand is price elastic. For example, with a rotation

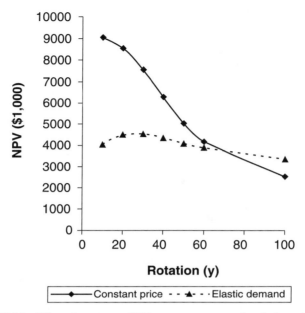

FIGURE 15.5 Effect of rotation on NPV at constant price and with elastic demand.

of 40 years, the production is 1,687,000 m³/decade, leading to a price of only $35/m³ with an elastic demand. This explains also why a rotation of 30 years gives a higher NPV when the demand is price responsive. Although a longer rotation results in lower production, it also results in higher prices. The net result is, for the price elasticity used here, an increase in periodic net revenues and thus in NPV.

15.4 SIMULATING CATASTROPHIC EVENTS

The even-aged simulation models described so far are deterministic. As time goes by, the stands move regularly from one age class to another, and the price of timber remains constant or varies with the volume produced in a fully predictable way.

As for the uneven-aged management model in the last chapter, the even-aged management model can be made more realistic by introducing random variables to reflect the real-world variability of forest growth and timber prices.

Simulation of Even-Aged Forest Management

Some random variables change continuously, such as price shocks and tree mortality. Others happen rarely, such as catastrophic events. Though infrequent, they have important effects on the state of the forest and its performance, both for timber production and carbon storage. Because of their infrequent occurrences, catastrophic events must be modeled differently.

MODELING STORMS

Assume that the forest under consideration is in an area prone to violent wind storms. The objective is to realistically represent the time of occurrence of a storm and its effects on the forest.

Storm Frequency

Although the timing of storms is random, this random process can be described with probalistic models. The models are necessarily simple, but they are consistent with the scarce information that is usually available.

In the case of catastrophic storms, meteorological records or oral history may lead to statements of the following sort: "In this area, disastrous storms occur once every 25 years." This of course does not mean that a storm will occur regularly every 25 years but that over a very long time the average period between storms has been 25 years. Some storms may actually have occurred at 5-year intervals, though 50 years may sometimes elapse without a storm.

One thing is certain: As the length of time since the last storm increases, there is a higher probability that a storm will occur. A simple model that expresses this fact quantitatively is the *exponential probability distribution*. According to the exponential model, if a is the average interval between storms, in years, then the probability that a storm will occur within T years, $P(T)$, is given by this exponential law:

$$P(T) = 1 - e^{-(T/a)} \tag{15.10}$$

where the constant $e \approx 2.718$ is the base of natural logarithms. Equation (15.10) states that for a given value of the parameter a, the probability that a storm will occur within T years approaches 1 as T increases. The exponential model assumes that the average interval between storms, a, or its inverse, the average frequency of a storm $1/a$ ($1/25$ per year in our example), is constant. That is, after a storm occurs, the probability of another storm does not change, an intuitively plausible assumption.

Figure 15.6 shows the graph of the exponential distribution for an average interval between storms $a = 25$ years. The graph shows that the probability that a storm will occur increases at a decreasing rate as T increases. The probability

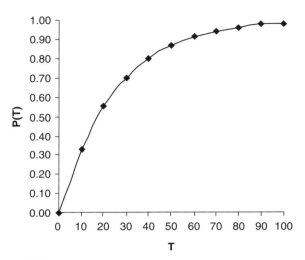

FIGURE 15.6 Probability that a storm occurs within T years, given an average time between storms of $a = 25$ years.

that a storm will occur within an interval of 10 years is about .33. Within a 25-year interval it is about .63. It is nearly certain that a storm would occur within 100 years.

In the context of the simulation model developed so far, time is incremented by decades. The probability that a storm occurs within a decade is:

$$P(10) = 1 - e^{-(10/a)}$$

Stochastic simulation entails generating, every decade, a random number R_t between 0 and 1. Then a storm occurs within the current decade if and only if R_t is greater than or equal to $P(10)$; that is:

$$R_t \geq 1 - e^{-(10/a)} \qquad (15.11)$$

Storm Intensity

When a storm hits the forest, the amount and location of the damage are assumed to be distributed randomly throughout the age classes. Specifically, the uncut area destroyed by a storm is given by the following equation:

$$L_{it} = U_{it}(A_{it} - X_{it}) \qquad (15.12)$$

where L_{it} is the forest area lost to the storm in age class i in the decade t to $t + 10$ and U_{it} is a random number between 0 and 1, different for each age class. The equation assumes that all age classes have the same probability of being destroyed.

STORM EFFECTS

A storm has essentially the same effect on forest dynamics as a harvest. The timber blown down is salvaged, if possible. Then the area is cleaned up and reforested as if it had been normally harvested. Thus, Growth Equations (15.3) become:

$$A_{1,t+10} = X'_{1t} + X'_{2t} + X'_{3t} + X'_{4t}$$
$$A_{2,t+10} = A_{1t} - X'_{1t}$$
$$A_{3,t+10} = A_{2t} - X'_{2t} \tag{15.13}$$
$$A_{4,t+10} = (A_{3t} - X'_{3t}) + (A_{4t} - X'_{4t})$$

where $X'_{it} = X_{it} + L_{it}$. That is, in each equation the area cut is replaced by the sum of the area cut and the area destroyed by the storm.

Because the carbon stored is related to the forest area in each age class by Equation (15.8), the amount of carbon stock will also be affected by the storms.

Some of the timber blown down by a storm can usually be salvaged. Let the salvage rate be a fraction, s. Then the equation of the total periodic harvest from year t to year $t + 10$ becomes:

$$H_t = \sum_{i=1}^{4} v_i X_{it} + s \sum_{i=1}^{4} v_i L_{it} \tag{15.14}$$

The areas destroyed by storms need to be reforested after cleaning up the downed trees. This will increase the reforestation costs, so the equation of the periodic net income becomes:

$$\underset{(\$)}{N_t} = \underset{(\$/m^3)}{p_t} \underset{(m^3)}{H_t} - \underset{(\$/ha)}{c} \sum_{i=1}^{4} \underset{(ha)}{X_{it}} - \underset{(\$/ha)}{c'} \sum_{i=1}^{4} \underset{(ha)}{L_{it}} \tag{15.15}$$

where c' is the cost of cleaning up and reforesting the blown-down areas.

PRICE RISK

The random occurrence of catastrophic storms has a direct economic effect, because, given an elastic demand for timber, the unusually high volume harvested after a storm will depress prices. This effect is already taken into account in deterministic Price Equation (15.9).

In addition, as discussed in the previous chapter, prices may vary over time, irrespective of what happens in the forest. For example, assume that while prices are well predicted on average by Price Equation (15.9), they can be 25% above or below that average, with equal probability of being anywhere in that range. Then the actual price level is given by the following stochastic equation:

$$p_t = (0.50R_t + 0.75)541H_t^{-0.7} \qquad (15.16)$$

where R_t is a random number between 0 and 1. When $R_t = 1$, the simulated price is 25% above the average price predicted by deterministic Price Equation (15.9). When $R_t = 0$, the simulated price is 25% below. Prices between those extremes have an equal probability of occurring. Because the average value of R_t is 0.5, the average simulated price is equal to the price predicted by the deterministic equation.

15.5 SPREADSHEET STOCHASTIC SIMULATION

Figure 15.7 shows a spreadsheet to simulate the growth of an even-aged forest with stochastic storms and prices. The model extends the deterministic model in Figure 15.4, which includes a price-elastic demand to reflect the fact that large timber harvests following a storm may indeed depress prices. The following changes have been made:

Cells Q4:Q14 contain the formulas that return the value 1 if a storm occurs in the current decade, 0 otherwise, according to Rule (15.11). These formulas use the function RAND() to generate random numbers between 0 and 1 and the mean interval between storms, $a = 30$ years, in cell U15.

The formulas in cells S4:V14 calculate the losses according to Equation (15.12) if a storm occurs and set them to zero if there is no storm.

The formulas in cells B5:E14 add the blown-down areas to the areas cut to predict forest growth according to Equations (15.13).

The formulas in cells L4:L14 add the salvaged volume to the regular harvest according to Equation (15.14). The salvage rate, $s = 0.25$, is in cell S16.

The formulas in cells M4:M14 account for the cost of reforesting the blown-down areas. The cost per hectare on blown-down areas is assumed to be double the regular reforestation cost.

The formulas in cells P4:P14 calculate random prices according to Equation (15.16).

Simulation of Even-Aged Forest Management

	A	B	C	D	E	F	G	H	I	J	K	L	M	N	O	P	Q	R	S	T	U	V
1	CATASTROPHIC STORMS																					
2		Stock (ha) in class:					Cut (ha) in class:					H_t	N_t	NPV$_t$	Carbon	Price			Loss (ha) in class:			
3	Year	1	2	3	4		1	2	3	4		(10^3m^3)	(10^3)	(10^3)	(10^3t)	($/m^3)	Storm		1	2	3	4
4	0	50	50	100	300		0	0	0	83		33	1373	1373	100.4	42.4	0.0		0	0	0	0
5	10	83	50	50	317		0	0	0	83		33	1439	883	97.4	44.4	0.0		0	0	0	0
6	20	83	83	50	283		0	0	0	83		33	1301	490	91.3	40.3	0.0		0	0	0	0
7	30	83	83	83	250		0	0	0	83		33	1359	314	88.3	42.0	0.0		0	0	0	0
8	40	83	83	83	250		0	0	0	83		43	1141	162	88.3	31.4	1.0		43	32	49	52
9	50	259	41	51	149		0	0	0	83		33	1176	103	59.0	36.5	0.0		0	0	0	0
10	60	83	259	41	117		0	0	0	83		33	1311	70	60.2	40.6	0.0		0	0	0	0
11	70	83	83	259	74		0	0	0	83		42	1202	40	72.3	34.6	1.0		53	37	129	0
12	80	303	30	46	121		0	0	0	83		33	1157	23	51.4	36.0	0.0		0	0	0	0
13	90	83	303	30	83		0	0	0	83		43	1085	13	53.1	33.8	1.0		62	222	27	0
14	100	394	21	81	3		0	0	80	3		22	1142	9	29.1	53.4	0.0		0	0	0	0
15	(m^3/ha)	50	120	260	400							NPV (10^3)		3481		Mean storm interval (y)				30		
16	($/ha)	500	Interest (/y)	0.05	Rotation (y)		60					Allowable cut (ha)		83.33		Salvage rate	0.25					
17																						
18							Key cell formulas															
19	Cell	Cell formula										copied to										
20	A5	=A4+10										A5:A14										
21	B5	=SUM(G4:J4)+SUM(S4:V4)										B5:B14										
22	C5	=B4-G4-S4										C5:D14										
23	E5	=D4-I4-U4+E4-J4-V4										E5:E14										
24	J4	=MIN(E4,$N$16)										J4:J14										
25	G4	=MIN(B4,$N$16-SUM(H4:$J4))										G4:I14										
26	L4	=SUMPRODUCT(B$15:E$15,G4:J4)/1000																				
27		+S$16*SUMPRODUCT(B$15:E$15,G4:J4)/1000										L4:L14										
28	M4	=L4*P4-SUM(G4:J4)*$B$16/1000																				
29		-SUM(S4:V4)*2*$B$16/1000										M4:M14										
30	N4	=M4/(1+$E$16)^A4										N4:N14										
31	N15	=SUM(N4:N14)																				
32	N16	=(SUM(B4:E4)/I16)*10																				
33	O4	=0.65*SUMPRODUCT(B$15:E$15,B4:E4)/1000										O4:O14										
34	P4	=(0.50*RAND()+0.75)*541*L4^-0.7										P4:P14										
35	Q4	=IF(RAND()<=1-EXP(-10/$U$15),1,0)										Q4:Q14										
36	S4	=$Q4*RAND()*(B4-G4)										S4:V4										

FIGURE 15.7   Spreadsheet simulation model with stochastic storms and prices.

All other parameters and assumptions are the same as in the deterministic simulation in Figure 15.4. In particular, the initial forest condition is the same, and so is the rotation, equal to 60 years.

The results in Figure 15.7 are for one replication, that is, for one single sequence of random numbers. In this case, storms occur between years 40 and 50, between years 70 and 80, and between years 80 and 90. As a result, the NPV is $3.5 million, which is about $400,000 less than the NPV obtained in the deterministic simulation in Figure 15.4. And there are only 29,000 tons of carbon stored in the forest by year 100, compared to 88,000 tons in the deterministic simulation.

However, this is only one of the many possible outcomes. A few replications obtained by pressing the F9 key show that many different results are possible. We need to replicate this simulation many times to get a better picture of the variability of the results.

Effect of Rotation Length on Stochastic Carbon Stock and Net Present Value

In even-aged management, one key decision variable is the rotation age. To make good statistical statements regarding the effect of the rotation age, we need many replications of the simulation at different rotation ages, holding all parameters constant except for the string of random numbers.

Figure 15.8 shows another range of the spreadsheet in Figure 15.7. It performs automatically 100 replications of the simulation for rotations of 10–100 years and produces summary statistics for carbon stock after 100 years.

The replication numbers are in columns X4:X103. The alternative rotation ages are in cells Y3:AD3. To calculate the 100 replications for each rotation age with Excel, do the following:

1. Set cell X3 equal to the cell O14 in Figure 15.7, which contains the formula for the carbon stock at year 100.
2. Select the range of cells X3:AD103.
3. On the Data menu, click **Table**. In the **Row input cell** box, enter cell I16, which contains the rotation age in Figure 15.7. In the **Column input cell** box, enter cell P1 or any other cell of the spreadsheet that

	X	Y	Z	AA	AB	AC	AD	AF	AG	AH	AI	AJ	AK	AL
1	REPLICATIONS (CARBON 10^{3}t)							SUMMARY STATISTICS (CARBON 10^{3}t)						
2		Rotation (y)							Rotation (y)					
3	29	10	20	30	40	50	100		10	20	30	40	50	100
4	1	16	28	37	35	25	105	Max	16	28	47	67	80	105
5	2	16	20	31	40	52	105	Min	16	17	17	22	22	24
6	3	16	21	31	30	26	83	Mean	16	26	40	51	58	86
7	4	16	28	47	33	80	34	SD	0	3	9	15	19	23
8	5	16	28	31	40	80	105	SE	0	0	1	1	2	2
9	6	16	28	32	33	48	105	2*SE	0	1	2	3	4	5
10	7	16	28	31	67	59	105							
11	8	16	21	47	36	57	55		Key cell formulas					
12	9	16	28	47	56	80	105	Cell	Formula			Copy to		
13	10	16	28	47	53	31	105	X3	=O14					
14	11	16	28	26	56	80	105	X5	=X4+1			X5:X103		
15	12	16	28	47	25	31	105	Y4:AD103	see text					
16	13	16	28	17	67	80	91	AG4	=MAX(Y4:Y103)			AG4:AL4		
17	14	16	27	21	54	48	105	AG5	=MIN(Y4:Y103)			AG5:AL5		
18	15	16	28	47	24	56	76	AG6	=AVERAGE(Y4:Y103)			AG6:AL6		
19	16	16	28	47	47	39	48	AG7	=STDEV(Y4:Y103)			AG7:AL7		
20	17	16	28	47	59	62	105	AG8	=AG7/10			AG8:AL8		

FIGURE 15.8 Spreadsheet to calculate 100 replications of the carbon stock under risk with different rotations.

Simulation of Even-Aged Forest Management

does not affect the computation. The goal is to induce the spreadsheet to redo the calculations for each row of the table. The results will vary by row only, because the random numbers generated are different while all the other parameters remain constant.

The summary statistics in Figure 15.8 have been computed with the formulas discussed in Chapter 14. For each rotation they give the largest and smallest carbon stock observed in 100 replications, the mean stock, the standard deviation of the stock, and the standard error of the mean.

The same spreadsheet can be used to make 100 replications and collect the data on NPV for different rotations (see Problem 15.7). We only need to reset cell X3 equal to cell N15, which contains the formula of the NPV in the simulation spreadsheet in Figure 15.7.

To better document the trade-off between carbon storage and NPV, the mean of carbon storage and the mean NPV have been plotted against rotation age in Figure 15.9. The 95% confidence interval of the mean is also shown on the chart. The confidence interval is two standard errors above and below the mean. There is a 95% probability that the true mean lies in that interval. The variability of actual outcomes is much wider, as documented by the minimum, maximum, and standard deviation of the carbon stock in Figure 15.8.

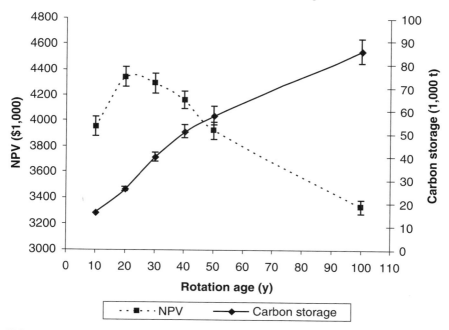

FIGURE 15.9 Effects of rotation age on carbon storage and NPV for timber, with stochastic storms and prices.

The chart in Figure 15.9 shows a steady increase of carbon stock as the rotation length increases. The variability of the carbon stock also increases with the rotation, because more catastrophic storms may impact a particular area as the rotation increases.

The mean NPV increases with rotation age up to a 30-year rotation and then decreases with longer rotations. This result is consistent with the deterministic analysis (see Figure 15.5). However, the variability of the mean NPV, shown by the standard errors, is quite large. As the confidence intervals of the mean NPV at rotations of 20–30 years overlap, it is uncertain whether different rotations truly make a difference within that range. In fact, given the relatively certain gain in carbon storage, one might very well be willing to prescribe a rotation as long as 40 years. Whether one would go beyond that depends on the willingness of the stakeholders to give up more timber income to gain carbon storage.

15.6 CONCLUSION

The simulation model of this chapter offers a straightforward approach to evaluating some management strategies for even-aged forest management. It has illustrated the importance of context in choosing management parameters, such as rotation age. For example, the deterministic version of the model has shown that under area-control management, the economic rotation tends to increase when the demand for timber is elastic with respect to its price.

Although the model dealt only with area-control management, it is not hard to see how it could be adapted to volume control by changing the allowable cut from a specified area to a volume. Other features could be added to the model in the same way that we added downward-sloping demand curves.

The stochastic version of the model illustrates the flexibility of the simulation approach in terms of adding realism to a basic deterministic model. Both continuous stochastic shocks, such as price changes, and rare discrete events, such as natural catastrophes can be handled.

There is a natural desire to add many refinements to simulation models to make them as "realistic" as possible, but this must be resisted. Added features can make models more difficult to understand and more prone to error. Paradoxically perhaps, detailed realism is not a characteristic of good models. The best models have a level of abstraction that captures only the key characteristics of the system being investigated.

PROBLEMS

15.1 (a) Set up your own spreadsheet model, like the one in Figure 15.1, to simulate the effects of even-aged management on timber production and carbon storage. Using the same data, verify that you get the same results.

(b) Set up a section of this spreadsheet, like that shown in Figure 15.2, to show how net present value and carbon storage vary with the rotation age. Using the same data, verify that your results are the same as in Figure 15.2.
(c) Reduce the interest rate from 5% to 2% per year. How does this affect the NPV at different rotations?
(d) How does the change of interest rate from 5% to 2% change the best economic rotation?
(e) How does this new best economic rotation change the amount of carbon stored after 100 years?

15.2 The forest described in the spreadsheet model in Figure 15.1 had an initial state in which much of the forest was in the oldest age class.
(a) Change the initial state to describe a forest that has recently been cut over, with 300 acres in age class 1, 200 acres in age class 2, and nothing in the two oldest age classes. Leave the other data as in Figure 15.1.
(b) Set up a section of this spreadsheet, similar to that in Figure 15.2, to find the NPV and the carbon stored after 100 years for different rotations. Compare your results to those in Figure 15.1.
(c) How important is the initial state in determining NPV?
(d) How does the initial state affect the economic rotation?
(e) How important is the initial state in determining long-term carbon sequestration?

15.3 (a) Modify the yield data for the forest shown in Figure 15.1 to reflect a less productive site with the anticipated yields shown in the table. Keep all other data the same.

	Age class			
	1	2	3	4
Site productivity (m^3/ha)	40	110	230	300

(b) Set up a section of this spreadsheet, similar to that in Figure 15.2, to find the NPV and the carbon stored after 100 years for different rotations. Compare your results to those in Figure 15.1.
(c) How important is site productivity in determining NPV?
(d) How does site productivity affect the economic rotation?
(e) How does site productivity affect long-term carbon sequestration?

15.4 The spreadsheet model to simulate management of an even-aged forest shown in Figure 15.1 can be modified to simulate a volume-control policy rather than area control. Assume that the allowable cut is based on the initial volume of the forest. The rule is still to take the allowable cut first from the oldest age classes.

(a) Change the area-control formula in cell N16 of the spreadsheet to a volume-control formula. (*Hint:* Every year, we can cut at most 1/R of the initial standing volume.)

(b) Set up a section of this spreadsheet, similar to that in Figure 15.2, to find the NPV and the carbon stored after 100 years for different rotations under volume control.

(c) How does the NPV at different rotations compare with the NPV obtained with area control?

(d) How does the best economic rotation with volume control compare to the best economic rotation with area control?

(e) If carbon sequestration is the goal, is volume control or area control the better policy? Why?

15.5 (a) Set up your own spreadsheet model, like the one in Figure 15.4, to simulate the management of an even-aged forest facing a downward-sloping demand curve for timber. Using the same data, verify that your results are the same as in Figure 15.4. (Note that this spreadsheet calculates timber price according to Demand Equation (15.9), where $a = 541$ and the exponent of H_t is -0.7, which implies an elasticity of demand with respect to price of $-1/0.7 = -1.43$.)

(b) Set $a = 250$ in the demand equation. Does this imply stronger or weaker demand for timber? How does this affect the NPV and the pattern of harvesting over time?

(c) Set $a = 750$ in the demand equation. Does this imply stronger or weaker demand for timber? How does this affect the NPV and the pattern of harvesting over time?

(d) Change the exponent of H_t to -0.35, keeping all other data as in Figure 15.4. What elasticity of demand with respect to price does this imply? How does this change in elasticity affect the NPV and the pattern of harvesting over time?

(e) Change the exponent of H_t in the demand equation to -1.0, keeping all other data as in Figure 15.4. What elasticity of demand with respect to price does this imply? How does this change affect the NPV and the pattern of harvesting over time?

15.6 Consider the spreadsheet model in Figure 15.4 to simulate management of an even-aged forest facing a downward-sloping demand curve for timber.

(a) Add a section to this spreadsheet, like the one in Figure 15.2, showing how net present value and carbon storage vary with the rotation age. Using the same data, verify that your results are the same as in Figures 15.4 and 15.5.

(b) Assume that population and income growth in the region where the forest is located have led to an increase in the demand for houses and hence for timber. This in turn has caused a permanent shift in the demand for timber. The forest can now sell a given amount of timber at a higher price; for example, the forest could now sell 30,000 m³/decade of wood for $60/m³ rather than $50/m³. What new value should be used for the constant a in the price equation, $P_t = aH_t^{-0.7}$?

(c) How does this change in demand affect the best economic rotation?

15.7 Set up your own spreadsheet model, like the one shown in Figure 15.7, to simulate the management of an even-aged forest given stochastic storms and prices.

(a) With the same data and equations, the simulation results displayed in your spreadsheet will not be the same as in Figure 15.7. Why?

(b) Set up a section of this spreadsheet, like that shown in Figure 15.8, to summarize the carbon stock data for different rotations. Run 100 replications. How do the predictions for carbon stock compare to those in Figure 15.8? Should they be the same?

(c) Can you say with confidence that your spreadsheet model is the same as the one shown in Figure 15.7?

(d) Modify your spreadsheet model to summarize the NPV data for different rotations. (*Hint:* Change the cell reference in X3 from the cell that contains the formula for carbon stock to the one with the formula for NPV.) Make a chart of the carbon stock and the NPV as a function of rotation age. Check that your results correspond to those in Figure 15.9.

(e) Modify your spreadsheet so that the average time between storms is only 10 years. Run 100 replications, and make a chart of your results like that in part (d). Then modify your spreadsheet so that the average time between storms is 50 years. Run 100 replications, and make a chart of these results. How does the average time between storms affect the economic rotation and the carbon sequestration?

ANNOTATED REFERENCES

Barrett, T.M., J.K. Gilless, and L.S. Davis. 1998. Economic and fragmentation effects of clearcut restrictions. *Forest Science* 44(4):569–577. (Simulation of the effectiveness of clear-cutting restrictions and adjacency constraints on the forest landscape.)

Betters, D.R. 1975. A timber–water simulation model for lodgepole pine watersheds in the Colorado Rockies. Water Resources Research. 11(6):903–908. (Describes a deterministic model for evaluating the impacts of alternative management activities on water and timber outputs.)

Cartwright, T.J. 1993. *Modeling the World in a Spreadsheet*. Johns Hopkins, Baltimore. (Chapter 1 contains a simulation model of an even-aged forest with stochastic fires.)

Clements, S.E., Dallain, P.L., and M.S. Jamnick. 1990. An operationally, spatially constrained harvest-scheduling model. *Canadian Journal of Forest Research* 20:1438–1447. (Uses simulation to generate spatially feasible harvest plans subject to harvest flow and adjacency constraints.)

Gould, E.M., Jr. 1977. Harvard forest management game. *Journal of Forestry* 75(9):587–589. (A noncomputerized version of Gould and O'Regan's simulation model.)

Gould, E.M., Jr., and W.G. O'Regan. 1965. *Simulation, a Step Toward Better Forest Management.* Harvard Forest Paper 13, Cambridge, MA. 86 pp. (Early application of simulation modeling in forestry.)

Grant, W.E. 1986. *Systems Analysis and Simulation in Wildlife and Fisheries Sciences.* Wiley, New York. 338 pp. (Describes a variety of stochastic simulation models of wildlife populations.)

Hoen, H.F., and B. Solberg. 1994. Potential and economic efficiency of carbon sequestration in forest biomass through silvicultural management. *Forest Science* 40(3):429–451. (Evaluates the potential for carbon sequestration in a managed forest using a linear programming model.)

Klemperer, W.D. 1996. *Forest Resource Economics and Finance.* McGraw-Hill, New York. 551 pp. (Chapter 2 discusses demand and related concepts.)

Law, A.M., and W.D. Kelton. 2000. *Simulation Modeling and Analysis.* McGraw-Hill, New York. 760 pp. (Basic reference on simulation modeling.)

Leefers, L.A., and J.W. Robinson. 1990. FORSOM: A spreadsheet-based forest planning model. *Northern Journal of Applied Foresty* 7(1):46–47. (The model can be used in simulation mode to predict the effects of a management option or in optimization mode to find the best option.)

Newell, R.G., and R.N. Stavins. 2000. Climate change and forest sinks: Factors affecting the costs of carbon sequestration. *Journal of Environmental Economics and Management* 40:211–235. (Sensitivity analysis of carbon sequestration costs to several factors, including silvicultural regime.)

Ragsdale, C.T. 1998. *Spreadsheet Modeling and Decision Analysis: A Practical Introduction to Management Science.* South-Western College Publishing, Cincinnati, OH. 742 pp. (Chapter 12 shows how to use spreadsheets to formulate and run stochastic simulation models.)

Reynolds, M.R., Jr., H.E. Burkhart, and R.F. Daniels. 1981. Procedures for statistical validation of stochastic simulation models. *Forest Science* 27(2):349–364. (Discusses validation of simulation models of natural resource systems.)

Schmidt, J.S., and P.L. Tedder. 1981. A comprehensive examination of economic harvest optimization simulation methods. *Forest Science* 27(3):523–536. (Review of harvest-scheduling models that incorporate some form of downward-sloping demand function.)

Smith, V.K., and J.V. Krutilla. 1974. A simulation model for management of low-density recreational areas. *Journal of Environmental Economics and Management* 1:178–201. (Describes a stochastic simulation model of dispersed recreation use patterns.)

Starfield, A.M., and A.L. Bleloch. 1986. *Building Models for Conservation and Wildlife Management.* Macmillan, New York. 253 pp. (Chapters 3 and 4 describe stochastic simulation models of wildlife populations.)

Winsauer, S.A. 1982. *Simulation of Grapple Skidders and a Whole-Tree Chipper.* U.S. Forest Service Research Paper NC-221. North Central Forest Experiment Station, St. Paul, MN. 42 pp. (Describes a stochastic simulation model of a logging system.)

CHAPTER 16

Projecting Forest Landscape and Income Under Risk with Markov Chains

16.1 INTRODUCTION

The last two chapters have shown how to model biological or economic risk in forests with simulation methods. The simulation approach consists in building a deterministic model of the system of interest, for example, the growth of a stand of trees (Chapter 14), or the dynamics of an entire forest (Chapter 15), and then adding stochastic elements.

Chapter 14 dealt with examples of simulation with continuous stochastic processes, such as tree regeneration and prices, which vary constantly over time. Chapter 15 gave an example of discrete event simulation, such as the occasional occurrence of major disturbances such as storms and fires.

The power of the simulation approach lies in its flexibility. Its drawback is that it lacks a standard structure. Every simulation model is different, so there are very few rules governing how to build simulation models or how to evaluate their output. Lack of a common structure makes it particularly difficult to identify best management policies. The number of possible policies is usually infinite, and comparing just a few in a stochastic environment can be cumbersome, as exemplified by the relatively simple examples shown in the previous two chapters.

In contrast to simulation, there are more structured ways of dealing with stochastic processes. Markov chains, in particular, allow problems to be formulated in a specific, refreshingly simple form. A Markov chain consists of a set of states and a table of the probabilities of moving from one state to another in a period of time. This simple form turns out to be quite powerful to predict the evolution of forests over time and to find the best way to manage them while realistically considering the role of risk. This chapter concentrates on prediction of forest landscape and income with Markov chains; the next chapter will deal with the optimization of forest systems modeled with Markov chains.

16.2 NATURAL FOREST GROWTH AS A MARKOV CHAIN

A forest may be viewed as a mosaic of stands of trees. The stands are usually in different states. A stand state may be classified by various characteristics, such as its age (if it is even aged), the species of the trees, the volume per unit area, the diameter of the trees, and the number of trees per unit area. Thus, the forest may be entirely even aged or uneven aged or a mix of the two.

STAND STATES

In practice, it is best to use a small number of characteristics to define stand states. For example, assume that there are three possible stand states and that the key characteristic for each state is the volume per unit area, as shown in Table 16.1. State L refers to a stand with low volume, less than 400 m^3/ha, state M refers to medium volume, from 400 to 700 m^3/ha; and state H refers to high volume, above 700 m^3/ha. A stand may move from one state to another due to biological growth and environmental shocks such as storms and fires or due to silvicultural treatments. In either case, the Markov model recognizes that the stand does not move from one state to another with certainty but, rather, in a probabilistic fashion.

TABLE 16.1 Volume per Unit Area, by Stand State

State i	Volume (m^3/ha)
L	<400
M	400–700
H	>700

TABLE 16.2 20-Year Transition Probabilities Without Management

Begin state i	End state j		
	L	M	H
L	.40	.60	
M		.30	.70
H	.05	.05	.90

TRANSITION PROBABILITIES

Table 16.2 shows an example of transition probabilities between stand states over a period of 20 years and with no intervention. The first row of the table indicates that for a stand that starts in state L and grows naturally, there is a .40 probability that the stand will still be in the same state after 20 years (the stand may have changed, but it would still have less than 400 m^3/ha). For the same initial stand state L, there is also a .60 probability that it will be in state M, i.e., with a volume between 400 and 700 m^3/ha. The second row indicates that for a stand initially in state M, there is a .30 probability that it will still be in that state after 20 years, while there is a .70 probability that it will be in state H, with more than 700 m^3/ha. The third row indicates that for a stand in state H, there is a .90 probability that it will be in the same state after 20 years. Due to disturbances, such as storms or fires, there is also a .05 probability that the stand will be in state M and the same probability that it will be in state L.

STAND DYNAMICS

The data in Table 16.2 define a one-period transition probability matrix, **P**, which provides a means to predict the probability that a stand that starts in a particular state would be in another state after a certain amount of time. For example, assume that we start with a low-volume stand (state L); this initial state is symbolized by the following state vector:

$$\mathbf{p}_0 = [1 \quad 0 \quad 0]$$

which means that the stand is initially in state L with probability 1 and in the other states with probability 0. Then the probability distribution of the stand state after 1 period (20 years), $\mathbf{p}_1$, is obtained by postmultiplying the initial probability distribution $\mathbf{p}_0$ by the transition probability matrix **P** (Appendix B shows how to multiply matrices):

$$\mathbf{p}_1 = \mathbf{p}_0 \mathbf{P} = [1 \quad 0 \quad 0] \begin{bmatrix} .40 & .60 & 0 \\ 0 & .30 & .70 \\ .05 & .05 & .90 \end{bmatrix} = [.40 \quad .60 \quad 0]$$

Thus, after 20 years there is a 40% chance that the stand will be still in state L and a 60% chance that it will be in state M. The probability distribution of the stand after two periods (40 years) is obtained by postmultiplying the probability distribution after one period, $\mathbf{p}_1$, by the transition probability matrix $\mathbf{P}$:

$$\mathbf{p}_2 = \mathbf{p}_1 \mathbf{P} = [.40 \quad .60 \quad 0] \begin{bmatrix} .40 & .60 & 0 \\ 0 & .30 & .70 \\ .05 & .05 & .90 \end{bmatrix} = [.16 \quad .42 \quad .42]$$

Thus, after 40 years there is a 16% chance that the stand that started initially in state L will still be in state L, a 42% chance that it will be in state M, and a 42% chance that it will be in state H.

In general, the probability distribution of the stand state in period t, $\mathbf{p}_t$, is obtained by postmultiplying the probability distribution of the stand state in the previous period, $\mathbf{p}_{t-1}$, by the transition probability matrix $\mathbf{P}$:

$$\mathbf{p}_t = \mathbf{p}_{t-1} \mathbf{P} \qquad \text{for } t = 1, \ldots, T \tag{16.1}$$

where T is the number of periods in a projection. As T increases to infinity, the vector $\mathbf{p}_T$ converges to a vector of steady-state probabilities, $\mathbf{p}^*$, which is independent of the initial state $\mathbf{p}_0$. In our example:

$$\mathbf{p}^* = [.07 \quad .12 \quad .82]$$

which means that after a long time, regardless of the initial stand state, there are .07, .12, and .82 probabilities of finding the stand in states L, M, and H, respectively. This is similar to what we found earlier for the steady state of an uneven-aged stand (Chapter 8). In both cases, the steady state depends only on the growth process, described here by the transition probabilities. The initial stand condition of a very long time ago does not matter.

SPREADSHEET IMPLEMENTATION

Figure 16.1 shows a spreadsheet that applies Recursive Equation (16.1) to predict the future probable state of a forest stand that starts in state L (volume less than 400 m³/ha) with the transition probabilities in Table 16.2. The bold

	A	B	C	D	E	F	G	H	I	J	K
1	PROBABILITY OF STATES WITHOUT MANAGEMENT									
2		State				BP			20-year probability		
3	Year	L	M	H		index		P	L	M	H
4	0	1.00	0.00	0.00		1.0		L	0.40	0.60	0.00
5	20	0.40	0.60	0.00		1.7		M	0.00	0.30	0.70
6	40	0.16	0.42	0.42		2.4		H	0.05	0.05	0.90
7	60	0.09	0.24	0.67		1.5					
8	80	0.07	0.16	0.77		1.3					
9	100	0.07	0.13	0.81		1.2					
10	120	0.07	0.12	0.82		1.2					
11	140	0.07	0.12	0.82		1.2					
12	160	0.07	0.12	0.82		1.2					
13	180	0.07	0.12	0.82		1.2					
14	200	0.07	0.12	0.82		1.2					
15	220	0.07	0.12	0.82		1.2					
16	240	0.07	0.12	0.82		1.2					
17											
18					Key cell formulas						
19	Cell		Formula							Copied to	
20	A5		=A4+20							A5:A16	
21	B5:D5*		{=MMULT(B4:D4,I$4:K$6)}							B5:D16	
22	F4		=SUM(B4:D4)/MAX(B4:D4)							F4:F16	
23	* Select B5:D5, enter formula =MMULT(B4:D4,I$4:K$6),										
24	then press CTRL+SHIFT+ENTER. Excel adds the brackets.										

FIGURE 16.1 Spreadsheet to predict the probability of stand states and the landscape diversity index, without management.

characters indicate the input data: the initial probability distribution of stand states in cells B4:D4 and the matrix of 20-year transition probabilities in cells I4:K6.

The formula in cells B5:D5 obtains the probability distribution of stand states after 1 period (20 years) by multiplying the initial probability distribution by the transition probability matrix. This formula is repeated by copying cells B5:D5 down to cells B16:D16.

The results show that, as time goes by, the probability that the stand is in the high-volume state H increases but the probabilities of the other states remain positive. A steady state is reached after 120 years, in the sense that the probability that the stand is in any one state remains constant over time.

The steady-state probabilities are independent of the initial state. You can check this by changing the initial probability distribution in cells B4:D4 of the spreadsheet (see Problem 16.1).

Forest Landscape Dynamics

The stand state probabilities just computed also have a landscape interpretation. The initial probability distribution can be viewed as the proportions of a large forest that are in each state. For example, the data in Figure 16.1 can be interpreted as meaning that 100% of the initial forest is occupied by stands in state L, with less than 400 m³/ha of timber. After 20 years, 40% of the forest would still be in that state while 60% would be occupied by stands in state M. After 40 years, 16% of the forest would consist of stands in state L, 42% of stands in state M, and 42% of stands in state H.

With time, the distribution of stands in this forest would converge to a steady state so that after about 120 years, 7% of the forest would be covered with stands in state L, 12% of stands in state M, and 82% of stands in state H. Thereafter, although the landscape would continue to change, the proportion of the forest area covered by stands in the different states would remain constant indefinitely.

One way of assessing the diversity of the forest landscape as it evolves over time is with an index like the Berger–Parker (BP) index introduced in Chapter 14. Here it has this expression:

$$\text{BP}_t = \frac{p_{Lt} + p_{Mt} + p_{Ht}}{\max(p_{Lt}, p_{Mt}, p_{Ht})} \qquad (16.2)$$

where p_{it} is the probability that a particular area of the forest is in state $i = L, M,$ or H at time t, which is also the proportion of the forest area in that state. The lowest value of BP is 1, which indicates that the entire forest is in the same state. The highest possible value is the number of states, 3 in this example, which indicates that an equal area of the forest is in each state.

The values of the Berger–Parker index in cells F4:F16 in Figure 16.1 show that, for a forest that starts entirely in state L and grows according to the transition probabilities in Table 16.2, the natural processes of biological forest growth and natural disturbances would lead to a rapid increase of landscape diversity, as portions of the forest move from state L to states M and H. After 40 years, however, diversity declines to a much lower steady state, reflecting the transition of most of the forest to state H. The steady-state value is independent of the initial state, but how fast the steady state is reached and the probability of each stand state in the interim period depend very much on the initial probability distribution of stand states (see Problem 16.1). Intuitively, it makes sense that the state of the forest 20 years from now will reflect its current state but that its state in a couple of centuries will have little to do with the current state.

MEAN RESIDENCE TIME OF STAND STATE

The mean, or expected, residence time of a stand state is defined as the average time that a stand will stay in that state *sequentially*, that is, without ever moving out of the state. Denoted as m_i, the expected residence time of each stand state i can be obtained from the probability that the stand will stay in the same state from one period to the next:

$$m_i = \frac{D}{(1 - p_{ii})} \qquad (16.3)$$

where D is the length of each period, in years, and p_{ii} is the probability that a stand in state i at the beginning of a period is still in that state at the end. For example, according to the data in Table 16.2, the expected residence time of a stand in state L (lowest volume) is:

$$m_L = \frac{20}{1 - 0.40} = 33.3 \text{ years}$$

The complete set of expected residence times for all the possible stand states in our example are in Table 16.3. It shows that a stand in the low-volume state would stay there for about 33 years, on average. A stand of medium volume would stay in that state for 29 years. The expected residence time for the high-volume state is much longer, 200 years, because once a stand is in that state, there is a high probability that it will stay there.

The results in Table 16.3 confirm the intuition that as the probability of staying in a particular state increases, the expected residence time should also increase. As the probability of staying in a particular state approaches 1, the expected residence time approaches infinity.

TABLE 16.3 Mean Residence Time Without Management

State i	20-year probability of staying p_{ii}	Expected residence time m_i (y)
L	.40	33.3
M	.30	28.6
H	.90	200.0

TABLE 16.4 Mean Recurrence Time Without Management

State i	Steady state probability π_i	Mean recurrence time m_{ii} (y)
L	.07	285.7
M	.12	166.7
H	.82	24.4

MEAN RECURRENCE TIME OF STAND STATE

In our example, the mean recurrence time, denoted by m_{ii}, is the average time it takes for a stand in a particular state, i, to *return* to that same state after exiting it. The mean recurrence time is inversely related to the steady-state probability of a particular state, π_i:

$$m_{ii} = \frac{D}{\pi_i} \qquad (16.4)$$

where D is the length of each time period (20 years in our example).

Table 16.4 shows the mean recurrence time for each stand state, given the steady-state probabilities of each state, obtained from cells B16:D16 of the spreadsheet in Figure 16.1. A stand in low-volume state L, under natural conditions, would take on average about 286 years to exit and return to that state. It would take less than one-tenth of that time for a stand to exit and return to the high-volume state H.

16.3 PREDICTING THE EFFECTS OF MANAGEMENT

A management policy is a rule that calls for a specific decision when the forest stand is in a particular state. As an example, consider a policy that lets the stand grow when it is in states L and M and harvests it when it is in state H, thereby returning it to state L.

TRANSITION PROBABILITIES

This policy changes the transition probabilities for natural growth shown in Table 16.2 into the transition probabilities for managed growth shown in Table 16.5. Only the last line of the tables is different. This last line of probabilities

TABLE 16.5 Transition Probabilities with Management

Begin state i	End state j		
	L	M	H
L	.40	.60	
M		.30	.70
H	.40	.60	

is the same as the first line, because the harvest policy changes a stand in state H to state L, and we know from Table 16.2 that the probability that a stand in state L would still be in state L after 20 years is .40 and that the probability that it would grow into state M is .60.

FOREST DYNAMICS WITH MANAGEMENT

Given the matrix of transition probabilities under management, G, the probability distribution of the stand state at in period t, p_t is obtained by postmultiplying the probability distribution of the stand state one period ago, p_{t-1}, by the transition probability matrix G. Thus, the prediction is made with the same recursive Equation (16.1) as for the unmanaged forest, but replacing matrix P by matrix G.

Figure 16.2 shows a spreadsheet that applies this new transition probability matrix to predict the future state of a stand that starts in state L (volume less than 400 m³/ha). The only difference between the spreadsheets in Figure 16.1 and in Figure 16.2 is in cells I6:K6.

The results in Figure 16.2 show that the effect of this particular harvest policy is to reduce the probability of stand state H and to increase the probabilities of the two other states. A steady state is reached after 140 years, in the sense that the probability that the stand is in any one state remains constant over time. The steady-state probabilities are again independent of the initial state.

As in the unmanaged case, the stand state probabilities in Figure 16.2 have a landscape-level interpretation, where the probability distribution in each period is the proportion of a large forest that would be in each state. Thus, beginning with a forest covered entirely by stands in state L and managing the forest according to the prescribed policy, the distribution of stands in this forest would converge over time to a steady state with 22% of the forest in state L, 46% in state M, and 32% in state H. According to the Berger–Parker index, the steady state of the managed forest would be more diverse than the steady state of the unmanaged forest, because the forest area would be more evenly distributed

	A	B	C	D	E	F	G	H	I	J	K
1	PROBABILITY OF STATES WITH MANAGEMENT										
2		State				BP		20-year probability			
3	Year	L	M	H		index		P	L	M	H
4	0	1.00	0.00	0.00		1.0		L	0.40	0.60	0.00
5	20	0.40	0.60	0.00		1.7		M	0.00	0.30	0.70
6	40	0.16	0.42	0.42		2.4		H	0.40	0.60	0.00
7	60	0.23	0.47	0.29		2.1					
8	80	0.21	0.46	0.33		2.2					
9	100	0.22	0.46	0.32		2.2					
10	120	0.21	0.46	0.32		2.2					
11	140	0.22	0.46	0.32		2.2					
12	160	0.22	0.46	0.32		2.2					
13	180	0.22	0.46	0.32		2.2					
14	200	0.22	0.46	0.32		2.2					
15	220	0.22	0.46	0.32		2.2					
16	240	0.22	0.46	0.32		2.2					
17											
18				Key cell formulas							
19	Cell		Formula						Copied to		
20	A5		=A4+20						A5:A16		
21	B5:D5*		{=MMULT(B4:D4,I$4:K$6)}						B5:D16		
22	F4		=SUM(B4:D4)/MAX(B4:D4)						F4:F16		
23	* See Fig. 16.1										

FIGURE 16.2 Spreadsheet to predict the probability of stand states and the landscape diversity index, with management.

by state. However, there would be less land occupied by high-volume stands (state H) in the managed vs the unmanaged forest.

Table 16.6 shows the mean residence time of each stand state under management. Comparing it with Table 16.3 shows that management has no effect on the mean time a stand stays sequentially in state L or M. This makes sense because neither state L nor state M triggers harvesting. However, managed stands stay in state H for only 20 years, on average, instead of 200 years for unmanaged stands. This shortening is also intuitively understandable, since the policy is to harvest a stand when it is in state H.

Comparing the mean recurrence times with and without management in Tables 16.7 and 16.4 shows that management shortens considerably the mean recurrence time of states L and M, while it lengthens the mean recurrence time

TABLE 16.6 Mean Residence Time with Management

State i	20-year probability of staying p_{ii}	Mean residence time m_i (y)
L	.40	33.3
M	.30	28.6
H	.00	20.0

TABLE 16.7 Mean Recurrence Time with Management

State i	Steady-state probability π_i	Mean recurrence time m_i (y)
L	.22	90.9
M	.46	43.5
H	.32	62.5

of state H. This makes sense because the policy is to harvest stands in state H, returning them to state L, thus shortening the recurrence time of state L and that of state M, which follows state L. Furthermore, since a stand in state H is always harvested, it cannot return to state H by the path H to M to H. It must go through the path H to L to M to H, which takes more time.

EXPECTED BIODIVERSITY PERFORMANCE

Forest management has both positive and negative effects. On the positive side, cutting trees and selling them for timber generates income, providing timber for various forest products, such as lumber and paper, and ultimately for housing and communication. On the negative side, harvests decrease the stock of standing trees, thus reducing some of the environmental benefits derived from the forest.

It is plausible that the contributions to biodiversity of the stand states L, M, and H increase with the standing volume of trees in each state. This would be especially true if H corresponded to an "old-growth" state, which is plausible given the mean residence time of 200 years for stand state H without management.

Assume that ecologists have scored the biodiversity value of stand states L, M, and H on a scale from 0 to 10. State L gets a score of 3, M a score of 7, and H a score of 10.

Then the expected, or average long-term biodiversity value, B, of a forest under a particular management policy is given by:

$$B = \pi_L B_L + \pi_M B_M + \pi_H B_H \qquad (16.5)$$

where π_i is the steady-state probability of state i under that policy and B_i is the biodiversity value of state i.

Thus without management the expected long-term biodiversity value, given the steady-state probabilities in Figure 16.1, is:

$$B = 0.07 \times 3 + 0.12 \times 7 + 0.82 \times 10 = 9.25$$

While with the management policy defined earlier, the long-term biodiversity value, based on the steady-state probabilities in Figure 16.2, is:

$$B = 0.22 \times 3 + 0.46 \times 7 + 0.32 \times 10 = 7.08$$

Note that since the steady-state probabilities are independent of the initial state, so is the long-term biodiversity value. Indeed, since the steady state is reached in a finite amount of time and the probabilities stay the same thereafter for an infinite length of time, the transient probabilities before the steady state have little weight. In the long run only the steady state matters.

EXPECTED RETURN PER UNIT OF TIME

Another effect of management is to generate timber income. For example, Table 16.8 shows that harvesting a stand in state H and thereby returning to state L generates a harvest volume of $817 - 259 = 558$ m³/ha and income of $7,254/ha.

Then the expected long-term periodic income that would be obtained with a policy that cuts a stand to state L if and only if it is in state H is given by an equation analogous to Equation (16.5):

$$R = \pi_L R_L + \pi_M R_M + \pi_H R_H \qquad (16.6)$$

where R_i is the immediate return obtained from a stand in state i under the stated policy. In this example, the expected long-term 20-year return is:

$$0.22 \times 0 + 0.46 \times 0 + 0.32 \times 7,254 = 2,321 \quad (\$/\text{ha})$$

TABLE 16.8 Immediate Return from Harvest

State	Volume (m³/ha)	Average volume (m³/ha)	Return with harvest to state L	
			m³/ha	$/ha
L	<400	259	0	0
M	400–700	603	344	4,472
H	>700	817	558	7,254

On a yearly basis this would mean an income of $116/ha. Because it is an average over many periods of equal weight, this result is independent of the initial stand state and depends only on the steady-state probabilities of stand states.

PRESENT VALUE OF EXPECTED RETURN

The expected yearly income is a useful, but incomplete, measure of the economic implications of a policy. A complete economic assessment must take into account that future returns have less value than immediate returns. How much less depends on how far in the future the return will occur and on the interest rate. If the yearly interest rate is r, the discount factor for the 20-year periods assumed in our example is:

$$d = \frac{1}{(1+r)^{20}}$$

which says that a $1 return in 20 years has a present value, d, that is less than $1 at any positive interest rate.

We seek the present value of the expected return from a stand, over an infinite horizon, given its initial state and a management policy. The computations use backward recursion. We assume the present value of the expected return with t periods to go before the planning horizon and then derive the present value of the expected return with $t + 1$ periods to go.

Let V_{it} be the present value of the expected return from a stand in state $i = L$, M, or H, managed with a specific harvesting policy and with t periods to go before the planning horizon. Then the present value of the expected return with $t + 1$ periods to go is:

$$V_{i,t+1} = R_i + d(p_{iL}V_{Lt} + p_{iM}V_{Mt} + p_{iH}V_{Ht}) \tag{16.7}$$

That is, for a stand in state i, the present value of the expected return with $t + 1$ periods to go is equal to the immediate return plus the present value of the expected return with t periods to go. Recall that p_{iL}, p_{iM}, and p_{iH} are the probabilities that a stand in state i moves to state L, M, and H in one period. So, the term in parentheses is the expected present value of the return for a stand in state i, managed according to the chosen policy, with t periods to go before the planning horizon.

The backward recursion with Equation (16.7) begins by setting the present value of expected return with $t = 0$ periods to go at some arbitrary level. For example:

$$V_{L0} = V_{M0} = V_{H0} = 0$$

	A	B	C	D	E	F	G	H	I	J	K	L
1	PRESENT VALUE WITH MANAGEMENT											
2	Years	Present value in state					20-year probability				Return	
3	to go	L	M	H			G	L	M	H	state	$/ha
4	0	0	0	0			L	0.40	0.60	0.00	L	0
5	20	0	0	7254			M	0.00	0.30	0.70	M	0
6	40	0	1914	7254			H	0.40	0.60	0.00	H	7254
7	60	433	2130	7687								
8	80	547	2269	7801				Discount factor			d	0.38
9	100	596	2315	7850				Interest rate (/y)			r	0.05
10	120	613	2333	7867								
11	140	620	2339	7874								
12	160	622	2342	7876								
13	180	623	2343	7877								
14	200	624	2343	7878								
15	220	624	2343	7878								
16												
17				Key cell formulas								
18	Cell		Formula								Copied to	
19	A5		=A4+20								A5:A15	
20	B5		=L$4+d*SUMPRODUCT(G$4:I$4,B4:D4)								B5:B15	
21	C5		=L$5+d*SUMPRODUCT(G$5:I$5,B4:D4)								C5:C15	
22	D5		=L$6+d*SUMPRODUCT(G$6:I$6,B4:D4)								D5:D15	
23	L8		=1/(1+L9)^20									

FIGURE 16.3 Spreadsheet to compute the present value of an initial stand state for a specific management policy.

We then calculate the present value of the expected return with $t = 1, 2, \ldots$ periods to go with Equation (16.7). Due to the discounting factor, d, as t increases, the present value of expected return V_{it} approaches a limit V_i^*, the present value of expected return for a stand that starts in state i and is managed according to the chosen policy for a very long (infinite) time.

Figure 16.3 shows a spreadsheet to obtain the present value of expected return for each initial stand state for a policy that harvests a stand if and only if it is in state H, thereby returning to state L. The transition probability matrix in cells G4:I6 reflects this policy, being the same as matrix **G** in Figure 16.2. The immediate returns for this policy are in cells L4:L6:

$$R_M = 0, \quad R_L = 0, \quad R_H = \$7,254$$

The interest rate in cell L9 is 5% per year, so the discount factor in cell L8 is $d = 1/1.05^{20} = 0.38$. Cells B4:D4 contain the present value of the stand in each state for $t = 0$ periods to go before the planning horizon, set at 0, an arbitrary value that does not affect the solution (see Problem 16.6.). Cells B5:D5 contain the formulas for Equation (16.7) when there is one 1 period (20 years) to go until the planning horizon. For example, the formula in cell D5 gives the present value of expected return for a stand in state H with one period (20 years) to go until the planning horizon:

$$V_{H,1} = R_H + d(p_{HL}V_{L0} + p_{HM}V_{M0} + p_{HH}V_{H0})$$
$$= 7254 + 0.38 \times (0.40 \times 0 + 0.60 \times 0 + 0.00 \times 0) = \$7{,}254/\text{ha}$$

The formulas in cells B5:D5 are copied down to cell B15:D15, to give the present value of the expected return for a stand in each initial state with up to 220 years to go. Note that the present value is the same for 200 or 220 years to go, so this is a close approximation to the present value with an infinite time horizon.

The results in cells B15:D15 of Figure 16.3 show that, for a stand managed under this policy over an infinite length of time, the present value of the expected return is $V_L^* = \$624/\text{ha}$ for a stand that starts in state L, $V_M^* = \$2{,}343/\text{ha}$ for a stand that starts in state M, and $V_H^* = \$7{,}878/\text{ha}$ for a stand that starts in state H.

In contrast with the expected return per period, the present value of the expected return depends on the initial stand state. The reason is that stands with higher initial volume per hectare produce more income early on. In fact, in this example the present value of the income from the stand in state H is not much larger than the value of the first harvest ($R_H = \$7{,}254$). On the other hand, although stands in states L and M produce no immediate income because they are never harvested, they do have a substantial present value, due to the return they will provide whenever they do reach state H. For each state, the present value of the expected return is the value of the trees and of the land on which they stand, for the chosen management policy.

16.4 SUMMARY AND CONCLUSION

Markov chain models are essentially tables of probabilities signifying the chances that a particular system changes from one state to another within a specified amount of time. They have been applied widely and effectively in forestry.

Even the simplest results of Markov chains can give insights on forest growth dynamics. In particular, they can help predict the effects of natural or human disturbances on forest landscapes. They can also be used to project the evolution of a forest stand over time through specific succession phases. The results of Markov chain theory, such as mean recurrence time and mean residence time, help clarify the dynamics of forest stands and their consequences for landscape diversity.

Markov chains are also useful to predict the effects of management policies under risk. Both economic and ecological criteria can be used. The long-term expected return per unit of time is independent of initial conditions, but the discounted return depends very much on initial conditions.

Despite their simplicity, Markov models are very general. They have sometimes been criticized for having no memory, the future depending exclusively on the present, regardless of how it came about. But this is not a valid criticism. There is no reason why the definition of a state could not include past change, if so desired. The state definition in a particular model may be too coarse for good prediction, but this does not question the Markovian principle. In fact, because all predictions, regardless of method, are based on current information, it can be argued that all scientific knowledge is Markovian.

In the next chapter, we shall study how to find best policies, according to specific criteria, for forest systems described by Markov chains.

PROBLEMS

16.1 (a) Set up your own spreadsheet model, like the one shown in Figure 16.1, to predict the probability of stand states and landscape diversity without management. Using the same data, verify that your results are the same.

(b) Change the data in the spreadsheet in Figure 16.1 so that the initial forest state is M. How does this change the steady-state probabilities as compared to those in Figure 16.1? How does this change the landscape diversity over time?

(c) Change the data in the spreadsheet in Figure 16.1 so that the initial forest state is H. How does this change the steady-state probabilities as compared to those in Figure 16.1? How does this change the landscape diversity over time?

(d) Change the data in the spreadsheet in Figure 16.1 so that the initial forest state is evenly distributed among the three states. How does this change the steady-state probabilities as compared to those in Figure 16.1? How does this change the landscape diversity over time?

(e) On the basis of parts (a)–(c), how does the initial state of the forest affect the steady-state probabilities? How does it influence the landscape diversity over time?

16.2 Consider a large forest area that grows without management according to the transition probabilities in Table 16.2 and that starts with 10% of its area in state L, 10% in state M, and 80% in state H.

(a) Use the model in Figure 16.1 to predict what fraction of the forest is in state L, in state M, and in state H over time.

(b) Compare the time it takes for this forest to reach a steady state with the time it takes for the forest in Figure 16.1. Explain the difference.

16.3 Consider a forest that grows without management according to this table of 20-year transition probabilities.

20-Year Transition Probabilities Without Management

Begin state i	End state j		
	L	M	H
L	.40	.60	.00
M	.00	.30	.70
H	.00	.05	.95

(a) Compare these probabilities with those in Table 16.2. Give plausible reasons for the differences.

(b) Modify the spreadsheet model in Figure 16.1 to reflect these new probabilities. Explain the changes in the steady-state probabilities and in the evolution of the landscape diversity over time. Is this consistent with your answer to part (a)?

(c) Calculate the mean residence and mean recurrence times without management for the transition probabilities given here. Compare these to the times shown in Tables 16.3 and 16.4. Explain the differences.

16.4 Consider a stand that grows without management according to the transition probabilities shown in Table 16.2. A harvest policy cuts a stand if and only if it is in state M, thereby changing it to state L. What is the transition probability matrix associated with this harvesting policy?

16.5 (a) Set up your own spreadsheet model, like the one shown in Figure 16.2, to predict the probability of stand states and landscape diversity with management. Using the same data, verify that your results are the same.

(b) Modify the spreadsheet model to reflect the same harvest policy as in Problem 16.4.

(c) How does this change the steady-state probabilities and the landscape diversity over time as compared to those in Figure 16.2? What is the reason for the differences?

(d) Calculate the mean residence and mean recurrence times with this management. Compare these to Tables 16.6 and 16.7. Explain the differences.

(e) Calculate the long-term expected biodiversity performance for this management policy. Use a biodiversity score of 3 for state L, 7 for state M, and 10 for state H. How does this compare with the expected biodiversity performance of the original management policy of cutting the stand to state L if and only if it is in state H?

16.6 (a) Set up your own spreadsheet model, like the one shown in Figure 16.3, to predict the present value of an initial stand for a specific management policy. Using the same data, verify that your results are the same.

(b) Change the present value by state at the horizon ($t = 0$ periods to go) from 0 to $1,000 for state L, $50 for state M, and $99 for state H. What is the effect on the present value of the expected return by state?

(c) Assuming the immediate returns in Table 16.8 and an interest rate of 5% per year, find the long-term expected periodic income of a stand managed according to the management policy specified in Problem 16.4. How does this differ from the expected periodic income from cutting a stand if and only if it is in state H?

ANNOTATED REFERENCES

Acevedo, M.F., D.L. Urban, and H.H. Shugart. 1996. Models of forest dynamics based on roles of tree species. *Ecological Modeling* 87:267–284. (Reduces a complex simulation model to a simple Markov chain).

Binkley, C.S. 1980. Is succession in hardwood forests a stationary Markov process? *Forest Science* 26(4):566–570. (Criticizes the Markov assumption, but see this chapter's conclusion for another viewpoint.)

Binkley, C.S. 1983. Private forestland use: Status, projections, and policy implications. Pages 51–70 in J.P. Royer and C.D. Risbrudt (Eds.), *Nonindustrial Private Forests, a Review of Economic and Policy Studies*. School of Forestry and Environmental Studies, Duke University, Durham. 398 pp. (Application of Markov chains to forest area changes.)

Bosh, C.A. 1971. Redwood: A population model. *Science* 172:345–349. (One of the earliest applications of Markov chain methods to model long-term forest growth).

Boychuk, D., and D.L. Martell. 1989. A Markov chain model for evaluating seasonal forest firefighter requirements. *Forest Science* 34(3):647–661. (Application of Markov chains to fire management.)

Brisbin, R.L., and M.E. Dale. 1987. Estimating tree-quality potential in a managed white oak stand by Markov chain analysis. *Canadian Journal of Forest Research* 17:9–16. (Markov chain model for evaluating interactions between thinning and wood quality.)

Frelich, L.E., and C.G. Lorimer. 1991. A simulation of landscape-level stand dynamics in the northern hardwood region. *Journal of Ecology* 79:223–233. (Application of Markov chains to predict forest changes at a landscape level.)

Hassler, C.C., R.L. Disney, and S.A. Sinclair. 1988. A discrete-state, continuous-parameter Markov process approach to timber-harvesting systems analysis. *Forest Science* 34(2):276–291. (Application of Markov chains to the interactions between two pieces of harvesting equipment.)

Kaya, I., and J. Buongiorno. 1987. Economic harvesting of uneven-aged northern hardwood stands under risk: A Markovian decision model. *Forest Science* 33(4):889–907. (Application of Markov chains with stochastic growth and prices.)

Lin, C. R., and J. Buongiorno. 1998. Tree diversity, landscape diversity, and economics of maple-birch forests: Implications of Markovian models. *Management Science* 44:1351–1366. (The first part predicts effects of catastrophic events on landscape diversity, models succession, and predicts economic and diversity consequences of current management.)

Parks, P.J., and R.J. Alig. 1988. Land base models for forest resource supply analysis: A critical review. *Canadian Journal of Forest Research* 18:965–973. (Reviews the potential of Markov chain models for predicting forest area changes.)

Roberts, M.R., and A.J. Hruska. 1985. Predicting diameter distributions: A test of the stationary Markov model. *Canadian Journal of Forest Research* 16:130–135. (Suggests that the transition probabilities may be invalid, but see the end of this chapter for another viewpoint.)

Usher, M.B. 1992. Statistical models of succession. Pages 215–247 in D.C. Glenn-Lewin, R.K. Peet, and T.T. Veben (Eds.), *Plant Succession: Theory and Prediction*. Chapman and Hall, London. (Good introduction to the use of Markov chains in modeling population dynamics and species succession).

Valentine, H.T., and G.M. Furnival. 1989. Projections with ingrowth by Markov chains. *Forest Science* 35(1):245–250. (This model also accounts for growth, death, and harvest of trees).

CHAPTER 17

Optimizing Forest Income and Biodiversity with Markov Decision Processes

17.1 INTRODUCTION

The last chapter defined Markov chain models and showed applications of the models to predict the evolution of stands and forests in the presence of uncertainty. In particular, we predicted how a forest would change over time without management. We also used the model to predict the consequences of a particular management policy. A policy is a rule that chooses a particular action when the stand is in a particular state.

In Chapter 16 the Markov model was used to predict the effect of particular management policies on long-term expected biodiversity performance and financial returns. These procedures are valuable to compare management alternatives. However, the possible number of management policies is infinite. It is then useful to have methods to determine, among all possible policies, those that are best according to specific criteria.

The purpose of this chapter is to present methods to optimize decisions for forest systems represented with Markov chain models. In particular we shall first study methods to maximize the present value of the income over an infinite horizon. Then we shall seek management policies that would maximize long-term expected biodiversity. Last, we shall learn how to develop compromise solutions, for example, by maximizing expected biodiversity while maintaining a prescribed level of timber income.

17.2 MARKOV DECISION PROCESS

As in Chapter 16, consider a forest composed of a set of stands in different states. The volume of timber per unit of land defines the stand state. It may be low, medium, or high (L, M, or H as in Table 16.1). Natural stand dynamics are described by the probability that a stand will move from one state to another over 20 years, as in Table 16.2.

As in the previous chapter, assume that decision makers consider stands for harvest every 20 years. But instead of either never harvesting or always harvesting in the same way, they may now choose between two options for stands in state M or H: Do nothing or harvest.

One effect of the decision to do nothing or harvest is to change the stand state and therefore the transition probabilities. If the decision is to do nothing, the stand will grow according to the transition probability matrix **N** in Table 17.1. If the decision is to harvest, the stand will grow according to transition probability matrix **C**. Matrix **C** means that the decision to harvest always brings the stand to state L, which has a 40% chance of still being in state L after 20 years and a 60% chance of being in state M.

Another effect of the decision is on the immediate returns. The data in Table 17.2 show that if the stand is in state H, the decision to harvest brings an

TABLE 17.1 20-Year Transition Probabilities According to Decision

N Begin state i	No cut End state j			C Begin state i	Cut End state j		
	L	M	H		L	M	H
L	.40	.60	.00	L	.40	.60	.00
M	.00	.30	.70	M	.40	.60	.00
H	.05	.05	.90	H	.40	.60	.00

TABLE 17.2 Immediate Return According to Decision

State i	Immediate return R_{ij} ($/ha)	
	No cut (n)	Cut (c)
L	0	0
M	0	4,472
H	0	7,254

immediate return $R_{Hc} = \$7{,}254/\text{ha}$, and that if the stand is in state M, the decision brings an immediate return $R_{Mc} = \$4{,}472/\text{ha}$. Harvesting when the stand is in state L brings no return, so $R_{Lc} = 0$.

A Markov chain model like this, with decisions that lead to different outcomes, is called a *Markov decision process model*.

17.3 MAXIMIZING DISCOUNTED EXPECTED RETURNS

As a first objective, we seek the policy that maximizes the present value of the expected income produced by a stand over an infinite horizon. A policy is a rule that specifies a decision for each stand state. The method of finding the best policy resembles the backward recursion used in Chapter 16 to find the present value of expected return from a particular policy. It differs by the presence of a decision in each stand state.

DYNAMIC PROGRAMMING FORMULATION

Let V_{it} be the highest present value of expected return from a stand in state $i = L$, M, or H, managed with a specific policy, with t periods to go before the planning horizon. Then the highest present value with $t + 1$ periods to go is related to the highest value with t periods to go by this equation:

$$V_{i,t+1} = \max[R_{in} + d(p_{iLn}V_{Lt} + p_{iMn}V_{Mt} + p_{iHn}V_{Ht}),$$
$$R_{ic} + d(p_{iLc}V_{Lt} + p_{iMc}V_{Mt} + p_{iHc}V_{Ht})] \qquad (17.1)$$

where p_{ijk} is the probability that a stand moves from state i to state j when the decision is k. For example, p_{iLn} is the probability that a stand moves from state i to state L if the decision is to do nothing. As in Chapter 16, d is the 20-year discount factor:

$$d = \frac{1}{(1+r)^{20}}$$

where r is the interest rate per year.

Equation (17.1) says that for a stand in state i, the highest present value of future returns with $t + 1$ periods to go is equal to the largest immediate return plus the discounted value of the highest expected future returns with t periods to go. This maximum is obtained by doing nothing (n) or harvesting (c). Thus, the best decision is n if:

$$(R_{in} + d(p_{iLn}V_{Lt} + p_{iMn}V_{Mt} + p_{iHn}V_{Ht}) \geq R_{ic} + d(p_{iLc}V_{Lt} + p_{iMc}V_{Mt} + p_{iHc}V_{Ht}) \quad (17.2)$$

otherwise, the best decision is c.

The backward recursive calculations of Equation (17.1) begin by setting the highest net present value of expected return with $t = 0$ periods to go at some arbitrary level. For example:

$$V_{L0} = V_{M0} = V_{H0} = 0 \quad (17.3)$$

We then calculate the best decision and the highest present value of expected return for $t = 1, 2, \ldots$ periods to go. Due to discounting, as t increases, V_{it} approaches a limit, V_i^*, the highest present value of the expected returns for a stand that starts in state i managed according to the best policy for a long (infinite) time.

Spreadsheet Solution

Figure 17.1 shows a spreadsheet to compute the management policy that maximizes forest value. Transition probability matrix **N** for the "no cut" decision is in cells B5:D7. Transition probability matrix **C** for the "cut" decision is in cells G5:I7. The immediate returns for each state and decision are in cells L5:M7. As in Chapter 16, the discount factor, d, in cell M9 assumes an interest rate of 5% per year.

Cells B10:D10 contain the present value by stand state when the horizon is reached ($t = 0$ periods to go). It is set at 0, an arbitrary value that does not affect the solution (see Problem 17.2). The formulas in cells B11:D11 compute the highest present value by stand state with Equation (17.1) when there is one period (20 years) to go. For example, the formula in cell D11 gives the highest present value of expected return for a stand in state H with 1 period (20 years) to go before the planning horizon:

$$V_{H1} = \max[R_{Hn} + d(p_{HLn}V_{L0} + p_{HMn}V_{M0} + p_{HHn}V_{H0}),$$
$$R_{Hc} + d(p_{HLc}V_{L0} + p_{HMc}V_{M0} + p_{HHc}V_{H0})]$$
$$= \max[0 + 0.38(0.05 \times 0 + 0.05 \times 0 + 0.90 \times 0),$$
$$7254 + 0.38(0.40 \times 0 + 0.60 \times 0 + 0.00 \times 0)] = \$7{,}254$$

The formulas in cells G11:I11 record the best decision according to Rule (17.2). For example, the formula in cell I11 gives the best decision for a stand

Optimizing Forest Income and Biodiversity with Markov Decision Processes   361

	A	B	C	D	E	F	G	H	I	J	K	L	M	N
1	PRESENT VALUE MAXIMIZING POLICY													
2		20-year probability										Immediate		
3		No cut (*n*)				Cut (*c*)						Return ($/ha)		
4	N	L	M	H		C	L	M	H		State	No cut	Cut	
5	L	0.40	0.60	0.00		L	0.40	0.60	0.00		L	0	0	
6	M	0.00	0.30	0.70		M	0.40	0.60	0.00		M	0	4472	
7	H	0.05	0.05	0.90		H	0.40	0.60	0.00		H	0	7254	
8	Years	Present value in state:				Best decision in:								
9	to go	L	M	H		L	M	H		Discount factor, d			0.38	
10	0	0	0	0						Interest rate, r (/y)			0.05	
11	20	0	4472	7254		*n*	*c*	*c*						
12	40	1011	5483	8265		*n*	*c*	*c*						
13	60	1392	5864	8646		*n*	*c*	*c*						
14	80	1536	6008	8790		*n*	*c*	*c*						
15	100	1590	6062	8844		*n*	*c*	*c*						
16	120	1611	6083	8865		*n*	*c*	*c*						
17	140	1618	6090	8872		*n*	*c*	*c*						
18	160	1621	6093	8875		*n*	*c*	*c*						
19	180	1622	6094	8876		*n*	*c*	*c*						
20	200	1623	6095	8877		*n*	*c*	*c*						
21	220	1623	6095	8877		*n*	*c*	*c*						
22														
23				Key cell formulas										
24	Cell	Formula											Copied to	
25	A11	=A10+20											A11:A21	
26	B11	=MAX(L$5+d*SUMPRODUCT(B$5:D$5,B10:D10),											B11:B21	
27		M$5+d*SUMPRODUCT(G$5:I$5,B10:D10))												
28	C11	=MAX(L$6+d*SUMPRODUCT(B$6:D$6,B10:D10),											C11:C21	
29		M$6+d*SUMPRODUCT(G$6:I$6,B10:D10))												
30	D11	=MAX(L$7+d*SUMPRODUCT(B$7:D$7,B10:D10),											D11:D21	
31		M$7+d*SUMPRODUCT(G$7:I$7,B10:D10))												
32	G11	=IF(L$5+d*SUMPRODUCT(B$5:D$5,B10:D10)>=											G11:G21	
33		M$5+d*SUMPRODUCT(G$5:I$5,B10:D10),"n","c")												
34	H11	=IF(L$6+d*SUMPRODUCT(B$6:D$6,B10:D10)>=											H11:H21	
35		M$6+d*SUMPRODUCT(G$6:I$6,B10:D10),"n","c")												
36	I11	=IF(L$7+d*SUMPRODUCT(B$7:D$7,B10:D10)>=											I11:I21	
37		M$7+d*SUMPRODUCT(G$7:I$7,B10:D10),"n","c")												
38	M9	1/(1+M10)^20												

FIGURE 17.1 Spreadsheet to compute the policy that maximizes present value.

in state H with 1 period to go. Because c gives the highest present value of expected returns, the best decision is to cut the stand.

The formulas in cells B11:I11 are copied down to cell B21:I21, to give the highest present value by stand state and the corresponding best decision for up to 220 years to go. The highest present value is the same for 200 or 220 years to go, so the present value at this point is approximately the highest present value with an infinite time horizon. The corresponding best policy is shown in cells G21:I21, and it is to harvest the stand if it is in state M or H. Note that in this example the algorithm already found the best policy at the first iteration with only 20 years to go. But the best policy is guaranteed only when the present value has converged to its maximum.

The results in cells B21:D21 of Figure 17.1 show that, for a stand that starts in state L and is managed with the best policy over an infinite length of time, the highest expected present value of returns is $V_L^* = \$1,623$/ha. For a stand that starts in state M it is $V_M^* = \$6,095$, and for a stand that starts in state H it is $V_H^* = \$8,877$/ha. These are the values of the land and trees, the forest values per unit area, under the best policy. They are higher than the forest values obtained in the last chapter with the policy that cut the stand if and only if it was in state H (see Figure 16.3).

17.4 MAXIMIZING LONG-TERM EXPECTED BIODIVERSITY

In Chapter 16 we studied how to compute the expected long-term biodiversity effect of a particular management policy (see Section 16.3). In particular we found that by harvesting a stand if and only if it was in state H, the expected long-term biodiversity value was B = 7.08.

We now consider alternative policies and seek the one that leads to the highest expected biodiversity in the long run. The possible stand states and corresponding decisions are still L, M, and H, as defined before. The possible decisions are to do nothing or to harvest, and harvests are possible only in states M and H, in which case they change the stand to state L.

Therefore, the transition probability matrices with and without harvest are the same as in Table 17.1. However, the rewards are different. Table 17.3 shows that the immediate effect of the cut is to reduce the stand biodiversity index from 10 to 3 when the stand is in state H, and from 7 to 3 when the stand is in state M.

In this context, there is no discounting: Biodiversity now is assumed to have the same value as biodiversity in the future. The objective is to maximize the expected, or average, biodiversity over an infinite horizon.

TABLE 17.3 Biodiversity Effect of Decision

State i	Diversity score	
	No cut (n)	Cut (c)
L	3	3
M	7	3
H	10	3

LINEAR PROGRAMMING FORMULATION

Let S_{ik} be the probability of being in state i and making decision k. We know from the results of Chapter 16 that over a very long time, this probability reaches a constant, or steady-state, value. Thus, the expression of long-term expected biodiversity that must be maximized is:

$$\max B = 3S_{Ln} + 7S_{Mn} + 10S_{Hn} + 3S_{Lc} + 3S_{Mc} + 3S_{Hc} \quad (17.4)$$

where S_{ik} is the probability that a stand is in state $i = L, M,$ or H and that the decision is $k = c$ (harvest) or n (do nothing). These probabilities are unknown. However, because they are probabilities, they must be nonnegative, and they must add up to 1:

$$S_{Ln}, S_{Mn}, \ldots, S_{Hc} \geq 0 \quad (17.5)$$

$$S_{Ln} + S_{Mn} + \cdots S_{Hc} = 1 \quad (17.6)$$

Furthermore, over a very long time period, the probability that a stand moves out of a state must be equal to the probability that the stand moves into that state. This implies:

$$S_{jn} + S_{jc} = (S_{Ln}P_{Ljn} + S_{Mn}P_{Mjn} + S_{Hn}P_{Hjn}) + (S_{Lc}P_{Ljc} + S_{Mc}P_{Mjc} + S_{Hc}P_{Hjc})$$
$$\text{for } j = L, M, \text{ or } H \quad (17.7)$$

where the expression on the left of the equality is the probability of exiting state j by either decision n or decision c. The first term on the right of the equality is the probability of returning to state j from other states through decision n. The second term on the right of the equality is the probability of returning to state j from other states through decision c.

The problem formed by Equations (17.4) to (17.7) is a linear program. The best solution is S_{ik}^*, the probability of being in state i and making decision k, under the best policy. The best decision is then derived from this solution as:

$$D_{in} = \frac{S_{in}^*}{S_{in}^* + S_{ic}^*}, \quad D_{ic} = \frac{S_{ic}^*}{S_{in}^* + S_{ic}^*} \quad \text{for } i = L, M, H \quad (17.8)$$

where D_{in} and D_{ic} are the probabilities of making decision n or c, respectively, when the stand is in state i and under the best policy to maximize the long-term average biodiversity.

In contrast with the case of the highest present value, the highest expected biodiversity is independent of the stand state. But as in the case of the highest present value, the best decision depends only on the stand state. And the best decision is also deterministic; that is, D_{in} and D_{ic} are always 0 or 1 and the decision is either n or c all the time. However, as will be seen in Section 17.5, the solution may not be deterministic with additional constraints.

SPREADSHEET SOLUTION

Figure 17.2 shows a spreadsheet to compute the management policy to maximize the long-term expected biodiversity. The bold entries are data; the rest are decision variables or depend on the decision variables. Cells B6:G8 contain the transition probability matrix with harvest and the transition probability matrix without harvest. These two matrices are the same as in Figure 17.1, but they are transposed, rows becoming columns, to simplify the formulas.

The decision variables of the linear program, the probabilities of being in a state and making a decision, are in cells B14:G14. Cells B16:G16 contain the biodiversity scores for each stand state and decision according to Table 17.3. Cell H16 contains the formula of the objective function, corresponding to Equation (17.4). Cells H17:H19 contain the left-hand side of the steady-state constraints of Equation (17.7) rewritten as:

$$(S_{Ln}P_{Ljn} + S_{Mn}P_{Mjn} + S_{Hn}P_{Hjn}) + (S_{Lc}P_{Ljc} + S_{Mc}P_{Mjc} + S_{Hc}P_{Hjc}) - S_{jn} - S_{jc} = 0$$

$$\text{for } j = L, M, H \quad (17.9)$$

Cell H20 contains the left-hand side of Constraint (17.6).

Cells B21:G21 contain the formulas to calculate the best decision according to Equation (17.8).

	A	B	C	D	E	F	G	H	I	J	K
1	MAXIMIZING EXPECTED BIODIVERSITY										
2	20-yr transition probability										
3		No cut (n)			Cut (c)						
4		Start state i									
5	End j	L	M	H	L	M	H				
6	L	0.40	0.00	0.05	0.40	0.40	0.40				
7	M	0.60	0.30	0.05	0.60	0.60	0.60				
8	H	0.00	0.70	0.90	0.00	0.00	0.00				
9		-1	0	0	-1	0	0				
10		0	-1	0	0	-1	0				
11		0	0	-1	0	0	-1				
12		Probability of state and decision:									
13		S_{Ln}	S_{Mn}	S_{Hn}	S_{Lc}	S_{Mc}	S_{Hc}				
14		0.07	0.12	0.82	0.00	0.00	0.00				
15		Biodiversity score:						Total			
16	State:	3	7	10	3	3	3	9.2	max		
17	L	-0.60	0.00	0.05	-0.60	0.40	0.40	0.0	=	0	
18	M	0.60	-0.70	0.05	0.60	-0.40	0.60	0.0	=	0	
19	H	0.00	0.70	-0.10	0.00	0.00	-1.00	0.0	=	0	
20		1	1	1	1	1	1	1.0	=	1	
21	Decision	1	1	1	0	0	0				
22											
23				Key cell formulas							
24	Cell	Formula							Copied to		
25	B17	=B6+B9							B17:G19		
26	H17	=SUMPRODUCT(B17:G17,B$14:G$14)							H17:H20		
27	B21	=B14/(B14+E14)							B21:D21		
28	E21	=E14/(B14+E14)							E21:G21		

FIGURE 17.2 Spreadsheet to maximize long-term expected biodiversity.

Figure 17.3 shows the Solver parameters to maximize the long-term expected biodiversity. The target cell, H16, contains the formula of the objective function. The Solver seeks the largest long-term biodiversity by changing cells B14:G14, the probabilities of being in a state and making a given decision. The optimization is done subject to:

The nonnegativity constraints: B14:G14 >= 0

The steady-state constraints corresponding to Equation (17.7): H17:H19 = K17:K19

The restriction that the decision variables must add up to unity: H20 = J20

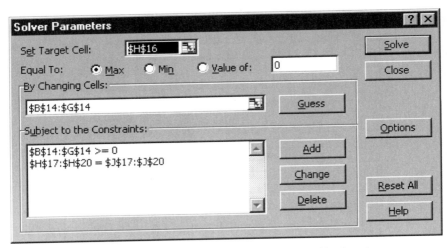

FIGURE 17.3 Solver parameters to maximize expected biodiversity.

The best solution, shown in Figure 17.2 in cells B14:G14, is:

$$S^*_{Ln} = .07, \quad S^*_{Mn} = .12, \quad S^*_{Hn} = .82, \quad S^*_{Lc} = S^*_{Mc} = S^*_{Hc} = 0$$

That is, under the best policy, the stand is in state L and not cut 7% of the time. It is in state M and not cut 12% of the time, and it is in state H and not cut 82% of the time.

The corresponding best policy, shown in Figure 17.2 in cells B21:G21, is:

$$D_{Ln} = D_{Mn} = D_{Hn} = 1 \quad \text{and} \quad D_{Lc} = D_{Mc} = D_{Hc} = 0$$

Thus, the probability of not harvesting is 1 regardless of the stand state, which means that the stand is never harvested. Letting nature take its course, we will observe the stand 7% of the time in state L, 12% of the time in state M, and 82% of the time in state H. The linear program has another best solution, but with the same long-term expected biodiversity (see Problem 17.4).

As in the case of present value maximization, the decision that maximizes long-term expected biodiversity depends only on the stand state. But in contrast to the highest present value, the highest long-term expected biodiversity is independent of the stand state. Cell H16 in Figure 17.2 shows a maximum long-term expected biodiversity score of 9.2, which is the same (apart from round-off errors) as that predicted in Chapter 16 for an unmanaged stand, as it should be.

IMPLICATIONS FOR THE FOREST LANDSCAPE

As noted in the previous chapter, the long-term probability of a stand state can also be interpreted as the fraction of a large forest that is in a particular state in the long run. Thus, an implication of the solution just obtained is that in a large forest that is left unmanaged, we should eventually observe on average 7% of it covered with stands in state L, 12% with stands in state M, and 82% with stands in state H. Of course, these are exactly the long-term probabilities of stand states that were obtained in the last chapter in predicting the stochastic evolution of a forest in a given initial state and with no harvesting (see Section 16.3).

17.5 MAXIMUM EXPECTED BIODIVERSITY WITH INCOME CONSTRAINT

The policy that maximizes the long-term expected biodiversity calls for never harvesting the stand. Contrasting this with the policy that maximized present value of timber harvest, we see that this policy has a high opportunity cost ($1,623/ha to $8,877/ha in present value of expected return, depending on the stand state). For some owners, this cost may not be acceptable, and some compromise solution would be called for.

A possible way of designing such a compromise is to seek a policy that maximizes long-term expected biodiversity while providing a specified income. To be consistent with the model framework, the income must be stated as a long-term expected periodic income. For example, assume that the owners of the forest we have studied so far want to maintain an expected yearly return of $50/ha in the long run. This is equivalent to a long-term expected periodic return of $1,000/ha for the 20-year periods used in our model.

For this to hold, the decision variables in the linear program that we have just solved must also satisfy this constraint:

$$0S_{Ln} + 0S_{Mn} + 0S_{Hn} + 0S_{Lc} + \underset{(\$/ha)}{4,472} S_{Mc} + \underset{(\$/ha)}{7,254} S_{Hc} \geq 1,000 \quad (\$/ha)$$

where the left-hand side of the constraint is the long-term expected income every 20 years. The coefficients are the returns from each stand state and decision shown in Table 16.8, while the variables are the probabilities of being in a state and making a decision, which we seek to optimize.

The remainder of the linear program is the same as in the previous section. Figure 17.4 shows a spreadsheet to compute the maximum expected long-term biodiversity with this additional constraint. This spreadsheet is the same as that

	A	B	C	D	E	F	G	H	I	J
1	MAXIMIZING EXPECTED BIODIVERSITY (INCOME >= 1000)									
2	20-yr transition probability									
3		No cut (n)			Cut (c)					
4		Start state i								
5	End j	L	M	H	L	M	H			
6	L	0.40	0.00	0.05	0.40	0.40	0.40			
7	M	0.60	0.30	0.05	0.60	0.60	0.60			
8	H	0.00	0.70	0.90	0.00	0.00	0.00			
9		-1	0	0	-1	0	0			
10		0	-1	0	0	-1	0			
11		0	0	-1	0	0	-1			
12		Probability of state and decision:								
13		S_{Ln}	S_{Mn}	S_{Hn}	S_{Lc}	S_{Mc}	S_{Hc}			
14		0.13	0.26	0.47	0.00	0.00	0.14			
15		Biodiversity score:						Total		
16	State:	3	7	10	3	3	3	7.3	max	
17	L	-0.60	0.00	0.05	-0.60	0.40	0.40	0.0	=	0
18	M	0.60	-0.70	0.05	0.60	-0.40	0.60	0.0	=	0
19	H	0.00	0.70	-0.10	0.00	0.00	-1.00	0.0	=	0
20		1	1	1	1	1	1	1.0	=	1
21	Income:	0	0	0	0	4472	7254	1000	>=	1000
22	Decision:	1	1	0.772	0	0	0.23		($/ha/20y)	
23										
24				Key cell formulas						
25	Cell	Formula						Copied to		
26	B17	=B6+B9						B17:G19		
27	H17	=SUMPRODUCT(B17:G17,B$14:G$14)						H17:H21		
28	B22	=B14/(B14+E14)						B22:D22		
29	E22	=E14/(B14+E14)						E22:G22		

FIGURE 17.4 Spreadsheet to maximize long-term expected biodiversity with income constraint.

in Figure 17.3, except for the addition of the income constraint in row 21. The corresponding Solver parameters are in Figure 17.5.

The best solution, shown in Figure 17.4 in cells B14:G14, shows that compared to the biodiversity maximizing policy, the probability of states L and M increase to .13 and .26, respectively, while the probability of state H decreases to .47 + .14 = .61. The maximum expected biodiversity decreases from 9.2 to 7.3, while the expected periodic income increases from 0 to $1,000/ha.

However, in contrast with the pure biodiversity maximizing policy, the best policy with the income constraint, shown in Figure 17.4 in cells B22:G22, is not deterministic. While states L and M still call for no harvesting, state H calls for no harvest 77% of the time and for a harvest 23% of the time. This more complicated policy can be implemented by drawing a random number every time a

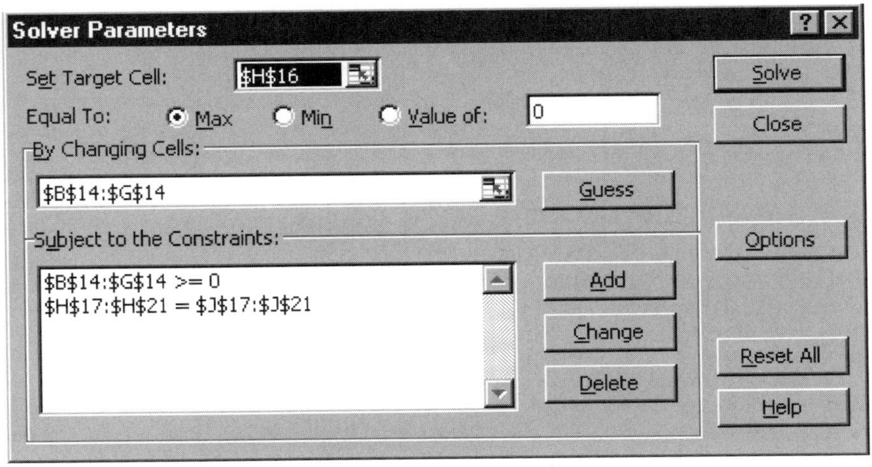

FIGURE 17.5 Solver parameters to maximize expected biodiversity with income constraint.

stand in state H is considered for harvest and cutting the stand to state L if the random number is greater than 0.77 and not cutting it otherwise.

17.6 SUMMARY AND CONCLUSION

Markov decision process models introduce decision making in forest systems that evolve according to a Markov chain. A policy is a rule that specifies a decision for each stand state.

A recursive method was used here to optimize the present value of expected returns from a stand state over an infinite horizon. The maximum present value depends significantly on the stand state. The best policy is stationary, independent of time and depending only on stand state. The best policy is also deterministic, calling always for a unique decision for each state.

Decisions that maximize a long-term expected value criterion, such as the long-term expected biodiversity of a stand or the long-term expected periodic income, were found by linear programming. The maximum expected return is independent of the stand state. The best policy is stationary and deterministic. Multiple-objective policies, such as maximizing long-term expected biodiversity with a floor on long-term expected income, can be found readily by adding appropriate constraints. But in that case, the best policies are not necessarily deterministic.

Markov decision process models are a very powerful approach to optimizing forest systems under risk. The approach consists first in transforming the initial system, possibly represented with a complex stochastic simulator, in a table of transition probabilities, as in a Markov chain, and then bringing the power of optimization to this simplified model.

PROBLEMS

17.1 Consider Recursive Equation (17.1) to maximize present value and Equation (17.2) to find the best policy.

(a) Write out these equations using the transition probabilities in Table 17.1, the immediate returns in Table 17.2, and an interest rate of 5% per year.

(b) Verify by hand the results obtained in the spreadsheet model to compute the best policy (Figure 17.1) for the first two iterations (20 years to go and 40 years to go).

17.2 (a) Set up your own spreadsheet model, like the one shown in Figure 17.1, to identify the policy that maximizes present value. Using the same data, verify that your results are the same.

(b) Change the maximum present value of expected returns with $t = 0$ periods to go from 0 to other values of your choice for states L, M, and H. How does this change the highest present value of expected returns over an infinite horizon?

17.3 Write out Constraint (17.9) with the transition probabilities in Table 17.1, and verify that they correspond to the formulas in rows 17–19 of the spreadsheet model shown in Figure 17.2.

17.4 (a) Set up your own spreadsheet model, like the one shown in Figure 17.2, to identify the policy that maximizes long-term expected biodiversity. Using the same data, verify that your results are the same. You may find two different solutions by running the Solver twice. Are the biodiversity implications of the two solutions different? (*Hint:* In Table 17.3, compare the effect on the diversity score of a policy that cuts in state L with one that does not).

(b) Based on the immediate returns in Table 17.2, use this spreadsheet model to find the policy that maximizes expected periodic income instead of the expected biodiversity. What is the effect of this policy on expected biodiversity?

(c) Add a constraint to this spreadsheet model to force the expected biodiversity score to be at least equal to 75% of the maximum value achievable without harvesting while maximizing expected periodic income.

17.5 (a) Set up your own spreadsheet model, like the one shown in Figure 17.4, to find the policy that maximizes long-term biodiversity with an income constraint. Using the same data, verify that you get the same results.

(b) Reduce the income constraint to $750/ha/20 y. What is the effect on the best policy and on the expected biodiversity?

ANNOTATED REFERENCES

Batabyal, A.A. 1998. On some aspects of the decision to conserve or harvest old-growth forests. *Journal of Environmental Management* 54:15–21. (Proposes Markov decision process models with expected discounted benefit and expected average net benefit criteria similar to those used in this chapter).

Buongiorno, J. 2001. Generalization of Faustmann's formula for stochastic forest growth and prices with Markov decision process models. *Forest Science* 47(4):466–474. (Uses solution methods similar to those in this chapter to maximize land expectation value in even-aged stands. Faustmann's formula is the limit of an MDP model when the transition probabilities approach 1 or 0.)

Kaya, I., and J. Buongiorno. 1987. Economic harvesting of uneven-aged northern hardwood stands under risk: A Markovian decision model. *Forest Science* 33(4):889–907. (Applies a Markov decision process model to uneven-aged forests, solved by successive approximation as in this chapter, and considers biological and economic risk.)

Kaya, J., and J. Buongiorno. 1989. A harvesting guide for uneven-aged northern hardwood stands. *Northern Journal of Applied Forestry* 6:9–14. (Field guide based on the results of Kaya and Buongiorno (1987).)

Lembersky, M.R. 1976. Maximum average annual volume for managed stands. *Forest Science* 22(1):69–81. (Expected return criterion optimized with a different technique than the linear programming used in this chapter).

Lembersky, M.R., and K.N. Johnson. 1975. Optimal policies for managed stands: An infinite-horizon Markov decision process approach. *Forest Science* 21(2):109–122. (Recursive methods different from those used in this chapter, applied to Douglas fir plantations, considering stochastic growth and prices.)

Lin, C.R., and J. Buongiorno. 1998. Tree diversity, landscape diversity, and economics of maple-birch forests: Implications of Markovian models. *Management Science* 44:1351–1366. (Investigates effects of catastrophic events on landscape diversity, models succession, and develops policies to optimize present value or tree diversity with successive approximation and linear programming methods presented in this chapter).

Norstrom, C.J. 1975. A stochastic model for the growth period decision in forestry. *Swedish Journal of Economics* 1975:330–337. (Uses MDP to find best rotation for even-aged stands that recognizes uncertainty in the future value of trees).

Szaro, R.C., and D.W. Johnston. 1996. *Biodiversity in Managed Landscapes: Theory and Practice.* Oxford University Press, Oxford, U.K. 778 pp. (Basic reference on the interaction between human management and biodiversity.)

Teeter, L.D., and J.P. Caulfield. 1990. Stand density management strategies under risk: Effects of stochastic prices. *Canadian Journal of Forest Research* 21:1373–1379. (Uses Markovian prices and dynamic programming to find best management.)

UNEP. 1995. *Global Biodiversity Assessment.* Cambridge University Press, Cambridge, U.K. 1140 pp. (Chapter 7 is an excellent reference on the measurement and monitoring of biodiversity.)

CHAPTER 18

Analysis of Forest Resource Investments

18.1 INTRODUCTION

Many of the things foresters do are investments for the future. In fact it is the very long time required to grow trees of commercial value that distinguishes forestry from most other economic activities.

Investment analysis is not new to us. In several of the previous chapters we studied decisions that involved future costs and returns. We routinely discounted these costs and returns to the present with interest rates, and compared alternatives based on their present value. In doing so, we were doing investment analysis without knowing it, just like Molière's Monsieur Jourdain was talking prose.

The purpose of this chapter is to study the rationale and methods of investment analysis in more detail. We shall start with a brief review of the theory of investment and interest rate. These concepts will then be used to justify the objective of present-value maximization, a criterion we used in past chapters. This will be followed by the study of other criteria that are commonly used to evaluate and compare investments. We shall then examine how inflation should be handled in investment analysis. Finally we shall make some suggestions regarding the choice of appropriate interest rates for forestry investments.

18.2 INVESTMENT AND INTEREST RATE

In past chapters, we have often suggested, without justifying it, that forest resource managers should compare the net present value of different courses of action, that is, the discounted value of future benefits minus future costs. To understand why this is appropriate, we need a broader view of the concepts of investment and interest rate.

To this end we shall use a model of a very simple economy, that of Robinson Crusoe. Wheat is his essential crop, and we shall ignore the rest. Every year, Robinson must decide how much wheat to consume to make bread and how much to set aside as seed. The amount of wheat he saves for planting is his investment. How much should he invest?

The factors that influence Robinson Crusoe's decision are summarized in Figure 18.1. The amount of wheat he consumes this year is C_0, and C_1 is the amount he will have next year.

PRODUCTIVITY OF INVESTMENTS

The amount of wheat that Robinson Crusoe invests depends in part on the yield from what he plants. In Figure 18.1 this is represented by curve AB, the

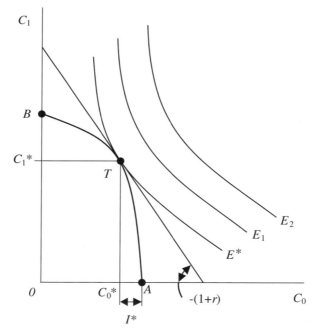

FIGURE 18.1 Optimum investment and interest rate.

Analysis of Forest Resource Investments   375

production possibility frontier of Robinson Crusoe. Point A refers to zero investment: Robinson consumes everything this year and has nothing the next. Point B is the case where he does not eat any bread this year and invests all his wheat in the future crop.

Curve AB is very steep close to point A. This shows that the first bushels of seed saved and planted have a high yield per bushel. However, as the quantity planted increases, the yield of each additional bushel of seed gradually decreases. Yields diminish because Robinson has limited resources. He can take great care of the first plot he plants, but the more he plants the more he loses to birds and drought. Ultimately, he reaches a point where he can barely recover the additional seed he plants. Of course, there is no reason for Robinson to invest that much. It is the high productivity of the first bushels planted, the fact that Robinson gets much more next year than he sacrifices now, that motivates him to set aside any of his wheat for planting. In sum, one of the reasons for making investments is their productivity, that they promise to return more in the future than what we sacrifice now to make them.

TIME PREFERENCE

Robinson Crusoe does not have all the wheat he wants, so it is a great sacrifice for him to put aside some of his crop. He is often tempted to give up part of the potential future harvest for a good loaf of bread now. Thus, a bushel of wheat next year is worth less to him than a bushel now.

This time preference for wheat is represented in Figure 18.1 by three *indifference curves*. The indifference curve E_1, for example, represents combinations of current and future wheat consumption among which Robinson is indifferent. The curve is steeper as current consumption, C_0, decreases. This means that the less wheat Robinson has now, the more valuable it is to him, so any additional sacrifice in current consumption must be compensated by a considerable increase in future consumption for Robinson to accept the trade.

Note that the farther away an indifference curve is from the origin, the higher the level of well-being associated with that curve. Consider any point on indifference curve E_1. There is at least one point on indifference curve E_2 that implies a higher level of both current and future consumption (check this). If more is better, Robinson must then prefer any combination of present and future consumption on E_2 than any on E_1.

OPTIMUM INVESTMENT

The best level of investment for Robinson Crusoe is the one that allows him to reach the highest possible indifference curve, given his production possibility frontier. Thus, the optimum investment is defined by the point where the production possibility frontier, AB, is tangent to an indifference curve. This is the point T in Figure 18.1, where AB is tangent to indifference curve E^*. The optimum investment is I^*, which corresponds to the optimum current consumption, C_0^*, and the optimum future consumption, C_1^*.

INTEREST RATE

The slope of the tangent of AB and E^* at T defines Robinson Crusoe's interest rate. To be precise, let r be the rate of interest expressed as a positive fraction per year. Then the following relation defines r:

$$\frac{dC_1}{dC_0} = -(1+r) \qquad (18.1)$$

where the derivative of C_1 with respect to C_0 is taken at T either along AB or along E^*. Equation (18.1) means that at the optimum investment level, a one-unit decrease in current consumption for purpose of investment is just compensated for by a $1 + r$ increase in future consumption. Thus, the interest rate is, by definition, the trade-off rate at which Robinson Crusoe is willing to exchange this year's consumption for next year's when he uses his resources in the best way to reach his highest possible level of well-being.

The investment decision in the simple Robinson Crusoe economy we have used so far can be generalized to decisions in very complex modern economies with many time periods and many goods and services. The principles remain the same. At least part of what every economic system produces can be either consumed or saved. What is saved can in turn be invested in a process that will produce something in the future. Investment is thus a sacrifice in current consumption. The sacrifice is induced by the net productivity of investments: They promise to return more than they cost. However, the attractiveness of investing is limited by people's time preference, the fact that one unit of anything tomorrow is worth less than one unit today. The rate at which current consumption is traded for future consumption at the optimum is the interest rate.

PRESENT-VALUE MAXIMIZATION

We are now ready to answer the question we posed at the beginning of the previous section. In making investment decisions, why should one maximize net present value?

Consider again the economy symbolized by Figure 18.1. Assume that we know the interest rate, r. Then the net present value of consumption over the

Analysis of Forest Resource Investments

two time periods is, by definition:

$$PV(C_0) = C_0 + \frac{C_1(C_0)}{1+r} \qquad (18.2)$$

The notations $PV(C_0)$ and $C_1(C_0)$ mean that PV and C_1 are functions of C_0. Different levels of current consumption, C_0, correspond to different levels of investment. These in turn lead to different levels of future consumption, C_1. In Figure 18.1 the relationship between future and current consumption is represented by the production possibility frontier AB.

To maximize PV, C_0 must be such that the first derivative of PV with respect to C_0 is zero:

$$\frac{dPV(C_0)}{dC_0} = 0$$

That is, by differentiating the right-hand side of equation (18.2):

$$1 + \frac{1}{1+r} \frac{dC_1(C_0)}{dC_0} = 0$$

or

$$\frac{dC_1(C_0)}{dC_0} = -(1+r) \qquad (18.3)$$

But this is just Equation (18.1), which must be true for the level of current consumption, C_0^*, which corresponds to the optimum investment, I^*. Therefore, maximizing present value does lead to the best level of investment, within the assumptions of our model.

Note, however, that the reasoning we have followed assumes that the "correct" interest rate, r, is known. In fact, it is difficult to choose an appropriate interest rate for investment analysis, but not much more difficult than for any other price. We shall return to this problem at the end of the chapter. Meanwhile, let us assume that we have a proper *guiding rate of interest* to evaluate forestry investments.

We now turn to the study of practical criteria for investment analysis. In doing this we shall apply the principle we have just established: To be any good, an investment criterion must be consistent with net present-value maximization.

18.3 INVESTMENT CRITERIA

NET PRESENT VALUE

The net present value (or present net worth) criterion extends Equation (18.2) to consider projects that generate costs and benefits over many years. Let B_t be the economic benefits in year t, and let C_t be the corresponding costs, both measured in dollars. In addition, let r be the guiding rate of interest, expressed as a fraction per year, and n be the duration of the project, in years. Then the present value of all costs is:

$$\text{PV(costs)} = \sum_{t=0}^{n} \frac{C_t}{(1+r)^t}$$

where the ratio $1/(1+r)^t$ is the discount factor, that is, the present value of one dollar occurring t years from the present. Similarly, the present value of all the benefits is:

$$\text{PV(benefits)} = \sum_{t=0}^{n} \frac{B_t}{(1+r)^t}$$

The net present value of the project is then the difference between the two:

$$\text{NPV} = \text{PV(benefits)} - \text{PV(costs)} = \sum_{t=0}^{n} \frac{B_t}{(1+r)^t} - \sum_{t=0}^{n} \frac{C_t}{(1+r)^t} \quad (18.4)$$

or, equivalently, it is the present value of the net returns in every year of the project:

$$\text{NPV} = \sum_{t=0}^{n} \frac{B_t - C_t}{(1+r)^t} \quad (18.5)$$

The present value criterion is extremely simple to apply: Any project with a positive net present value is worth doing.

Figure 18.2 shows an example of net-present-value calculation with a spreadsheet. The example deals with a reforestation project for a tract of land planted to ponderosa pine. The bold entries in the spreadsheet indicate the data. The other entries are the results of formulas. Costs are incurred to plant the land and subsequently for precommercial thinning, commercial thinning, and final harvesting. Benefits from thinnings occur at ages 30 and 50 years. The main benefit comes from a final harvest, expected at age 65.

Analysis of Forest Resource Investments

	A	B	C	D	E	F	G	H
1	PONDEROSA PINE REFORESTATION (NPV & B/C RATIO)							
2			Cost	Benefit	Discount	PV (cost)	PV (benefit)	NPV
3	Operation	Year	($/ha)	($/ha)	factor	($/ha)	($/ha)	($/ha)
4	Plant	0	1460	0	1.00	1460	0	-1460
5	Precommercial thin	10	610	0	0.74	454	0	-454
6	Thin	30	200	2270	0.41	82	935	853
7	Thin	50	200	2280	0.23	46	520	474
8	Final harvest	65	400	15500	0.15	59	2269	2211
9	Interest rate (/y)	0.03			Total	2100	3725	1624
10					Benefit-cost ratio		1.77	
11								
12				Key cell formulas				
13	Cell		Formula		copied to			
14	E4		=1/(1+$B$9)^B4		E4:E8			
15	F4		=C4*$E4		F4:G8			
16	F9		=SUM(F4:F8)		F9:H9			
17	G10		=G9/F9					
18	H4		=G4-F4		H4:H8			

FIGURE 18.2 Spreadsheet to compute the net present value and the benefit–cost ratio of a ponderosa pine reforestation project.

The formulas in cells E4:E8 calculate the value of the discount factor in various years. For example, with an interest rate of 3% per year, the present value of each dollar occurring 65 years from the present is only $0.15.

The total discounted value of costs, calculated in cell F9, is $2,100/ha, while that of benefits, in cell G9, is $3,725/ha. Therefore, the net present value of this project, in cell H9, is: $3,725 − 2100 = $1625/ha (due to rounding, the spreadsheet shows $1,624/ha). Since this net present value is positive, the project is worth doing, on purely financial grounds and based only on timber values.

BENEFIT–COST RATIO

The benefit–cost ratio (BC) of a project is the present value of the benefits expected throughout the life of the project, divided by the present value of the costs. With the definitions used earlier, this is:

$$BC = \frac{\sum_{t=0}^{n} B_t/(1+r)^t}{\sum_{t=0}^{n} C_t/(1+r)^t} \quad (18.6)$$

The benefit–cost ratio is dimensionless. For example, the benefit–cost ratio of the ponderosa pine reforestation project described in the spreadsheet in Figure 18.2 is given by the formula in cell G10 as 3,724/2,100 = 1.77.

A project that has a benefit–cost ratio of unity has a net present value equal to 0. To see this, set BC = 1 in Equation (18.6). This leads to:

$$\sum_{t=0}^{n} \frac{B_t}{(1+r)^t} = \sum_{t=0}^{n} \frac{C_t}{(1+r)^t}$$

That is, the present value of the benefits and of the costs are equal, and thus, from Equation (18.4), NPV = 0.

Similarly, a project that has a benefit–cost ratio greater than unity has a positive net present value. One with a benefit–cost ratio less than unity has a negative net present value.

We know already that only projects of positive NPV are worth doing. Thus, a project evaluated with the benefit–cost ratio criterion should be accepted only if its benefit–cost ratio is greater than 1. Used correctly, the benefit–cost ratio is equivalent to the net-present-value criterion in distinguishing good from bad projects.

INTERNAL RATE OF RETURN

The internal rate of return of a project is the interest rate such that the net present value of the project is equal to zero.

Given the definition of net present value of Equation (18.5), the internal rate of return is the interest rate that satisfies, for a given stream of benefits and costs, this equation:

$$\sum_{t=0}^{n} \frac{B_t - C_t}{(1+r)^t} = 0 \qquad (18.7)$$

Equation (18.7) also shows that the internal rate of return is the interest rate for which the present value of returns just balances the present value of costs.

Let r^* be the internal rate of return, that is, the particular value of r that solves Equation (18.7). Equation (18.7) is a polynomial equation of degree n in r; thus it may have up to n real roots. Therefore, the definition of the internal rate of return is ambiguous. But this is not crucial in most practical applications because the root we are seeking is an interest rate, that is, a number usually between 0 and 1. In fact, in that interval, the net present value of most projects decreases monotonically as the interest rate increases.

An example of this relation between net present value and interest rate is shown in Figure 18.3, which corresponds to another range of the spreadsheet in Figure 18.2. It uses the benefit and cost data in Figure 18.2, and computes the net present value for given interest rates ranging from 1% to 7% per year. The following steps obtain this table in Excel:

Analysis of Forest Resource Investments

	J	K
1	Interest	NPV
2	(/y)	($/ha)
3		1624
4	0.01	8697
5	0.02	4124
6	0.03	1624
7	0.04	239
8	0.05	-541
9	0.06	-985
10	0.07	-1242
11		
12	Key cell formulas	
13	Cell	Formula
14	K3	H9
15	K4:K10	see text

FIGURE 18.3 Spreadsheet to compute the net present value at various interest rates.

1. Set cell K3 equal to cell H9 in Figure 18.2, which contains the formula of the net present value.
2. Select the range of cells J3:K10.
3. On the data menu, click **Table**. In the **Column input cell** box, enter cell B9 in Figure 18.2, which contains the interest rate.

The results in Figure 18.3 show that the net present value decreases as the interest rate increases. This happens because, with the costs and benefits of this example, the formula of the net present value is:

$$NPV(r) = -1,460 - \frac{610}{(1+r)^{10}} + \frac{2,070}{(1+r)^{30}} + \frac{2,080}{(1+r)^{50}} + \frac{15,100}{(1+r)^{65}}$$

As the interest increases, the present values of future benefits and costs decreases but the initial cost remains the same.

Graphic Solution

A simple graphical method of determining the internal rate of return is to graph the function NPV(r) for a few values of r. The point where the graph intersects the horizontal axis is the internal rate of return. This was done in Figure 18.4 for the present example, which shows that the internal rate of return is about 4.3% per year, by definition, because at that rate the net present value is zero.

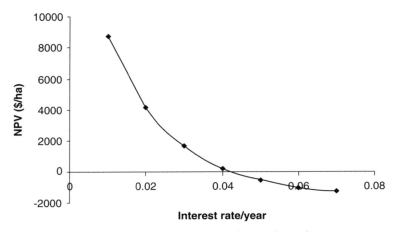

FIGURE 18.4 Graphic determination of internal rate of return.

Another use of Figure 18.4 is to derive the investment decision rule for the internal rate of return. The figure shows that the net present value of a project at a rate of interest smaller than the internal rate of return is positive. Conversely, the net present value at a rate of interest greater than the internal rate of return is negative.

Consequently, the rule is: Accept a project if its internal rate of return is greater than the guiding rate of interest. In that case, the net present value of the project is positive.

Spreadsheet Solution

A quicker and more precise way of finding the internal rate of return is with a spreadsheet. Figure 18.5 shows the setup to do this for our example. The spreadsheet is identical to that in Figure 18.2, but the interest rate is now a variable instead of an input parameter.

The internal rate of return is the interest rate that makes the net present value equal to zero. It can be found with the Excel Solver with the setting shown in Figure 18.6. The target cell is the net present value in cell H9, which must have the value of zero. Changing cell B9 contains the internal rate of return. The constraint states that the internal rate of return must be positive. In the options of the Solver, specify that the model is nonlinear. Indeed, the net present value is a highly nonlinear function of the interest rate.

The Solver results in Figure 18.5 show that the internal rate of return is 4.25% per year. At that interest rate, the net present value of the costs, $1,971/ha is just equal to the net present value of the benefits, the net present value of the reforestation project is 0, and the benefit–cost ratio is 1.

Analysis of Forest Resource Investments

	A	B	C	D	E	F	G	H
1	PONDEROSA PINE REFORESTATION, INTERNAL RATE OF RETURN							
2			Cost	Benefit	Discount	PV(cost)	PV(benefit)	NPV
3	Operation	Year	($/ha)	($/ha)	factor	($/ha)	($/ha)	($/ha)
4	Plant	0	1460	0	1.00	1460	0	-1460
5	Precommercial thin	10	610	0	0.66	402	0	-402
6	Thin	30	200	2270	0.29	57	651	594
7	Thin	50	200	2280	0.12	25	284	260
8	Final harvest	65	400	15500	0.07	27	1036	1009
9	Internal rate of return	0.0425			Total	1971	1971	0
10		(/y)			Benefit-cost ratio		1.00	
11								
12				Key cell formulas				
13	Cell		Formula		copied to			
14	E4		=1/(1+$B$9)^B4		E4:E8			
15	F4		=C4*$E4		F4:G8			
16	F9		=SUM(F4:F8)		F9:H9			
17	G10		=G9/F9					
18	H4		=G4-F4		H4:H8			

FIGURE 18.5 Spreadsheet to find the internal rate of return.

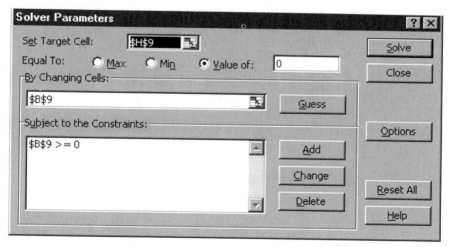

FIGURE 18.6 Solver parameters to find the internal rate of return.

SHORTCOMINGS OF BENEFIT–COST RATIO AND INTERNAL RATE OF RETURN

The conclusion of the previous section is that net present value, benefit–cost ratio, and internal rate of return all correctly discriminate projects that contribute to net present value from those that do not. For that purpose, it does not matter which criterion we choose. However, there may be reasons for which we cannot do all projects that have a positive NPV. Limited resources may force a choice among projects that could, if implemented, increase present value. In that case, ranking projects according to their benefit–cost ratio or internal rate of return may lead to the wrong decision.

Benefit–Cost Ratio

To illustrate how the benefit cost ratio criterion may fail to maximize present value, let us consider two projects, a and b. Only one of the two projects can be done. Let BC_i be the benefit–cost ratio, PVB_i be the present value of benefits, PVC_i be the present value of costs, and NPV_i be the net present value of project i ($i = a$ or b).

Assume the following data for the two projects:

Benefits: $PVB_a = \$150,000 \quad PVB_b = \$250,000$
Costs: $PVC_a = \$50,000 \quad PVC_b = \$100,000$

Thus, the benefit–cost ratios are:

$$BC_a = \frac{PVB_a}{PVC_a} = 3 \quad \text{and} \quad BC_b = \frac{PVB_b}{PVC_b} = 2.5$$

while the net present values are:

$$NPV_a = PVB_a - PVC_a = \$100,000 \quad \text{and} \quad NPV_b = PVB_b - PVC_b = \$150,000$$

Choosing the project with the highest benefit–cost ratio would lead to choosing project a. But this would be wrong, because project b has in fact a higher net present value. This shortcoming of the benefit–cost ratio stems from its being a ratio, which ignores the scale of a project.

Internal Rate of Return

Choosing projects of highest internal rate of return may also fail to maximize present value. Figure 18.7 illustrates how this may happen. The figure shows the graphs of the net present value of two projects, $NPV_a(r)$ and $NPV_b(r)$, as functions of the interest rate, r. The graphs cross at the interest rate r_m. By definition, the internal rate of return of project a is r_a^*, that of project b is r_b^*. Since r_b is greater than r_a, using the internal rate of return as the investment criterion leads to selecting project b.

Assume, however, that the proper guiding rate of interest is r_0. At that rate, the net present value of project a is greater than that of b, and project a should be selected.

Analysis of Forest Resource Investments

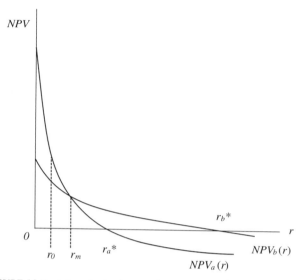

FIGURE 18.7 Internal rate of return inconsistent with net present value.

The graph shows that the internal rate of return criterion leads to the wrong decision when the guiding rate of interest is less than r_m, the rate at which the net-present-value graphs cross. If the guiding rate is greater than r_m or if the graphs do not cross, then the internal rate of return leads to a choice that is consistent with maximizing net present value.

In summary, projects with the highest benefit–cost ratio or the highest internal rate of return do not necessarily have the highest net present value. But this does not mean that benefit–cost ratios and internal rates of return are useless. First, as seen earlier, they separate correctly projects that are economical from those that are not. Second, they provide information that complements net present value. In particular, people are familiar with the concept of rates of return. For example, the statement that a eucalyptus plantation has a rate of return of 5% per year has more meaning to most people than the fact that its net present value is $750 per hectare.

18.4 CHOICE OF PROJECTS UNDER CONSTRAINTS

The previous section showed a sure way of choosing between two projects: Select the one with highest net present value. The problem is more difficult if there are several projects to choose from and several that can be done with the available resources. This is a situation that foresters face regularly. For example, in any given year, many tracts of land may be suitable for reforestation, but due to a limited stock of seedlings or to a limited budget or both, only a few of the tracts can be planted.

Projects Limited by One Resource

If a single resource limits the number of projects, then a simple method can be used to select those that maximize total net present value. Let R be the maximum amount of the constraining resource, say the number of seedlings in stock. Let R_i be the amount of resource needed for project i ($i = 1,\ldots, n$, where n is the number of projects). In our example R_i would be the number of seedlings necessary to plant the ith tract. Finally, let NPV_i be the net present value of project i if it is completely done. In our example, this would be the present value of the expected returns minus all the costs that will occur up to the final harvest on the ith tract of land.

Then define the *net present value–resource ratio* for project i, NPR_i as:

$$NPR_i = \frac{NPV_i}{R_i}$$

In our example, NPR_i would be the net present value per seedling. There is a simple way to determine the subset of projects that maximizes total net present value given the resource limit R. It consists in ranking all projects in decreasing order by net present value–resource ratios and then going down the list until the resource is exhausted. This procedure is a steepest-ascent algorithm (like the simplex method in Chapter 3). Each additional unit of a limiting resource is used in the project where it increases net present value the most.

This method works regardless of the resource involved, be it money, land, hours of labor, seedlings, etc. Suppose, to pursue the previous example, that the limiting resource is not the stock of seedlings but the budget for reforestation in the current year. Then NPR_i would be the ratio of net present value to the cost needed this year to reforest tract i. Note, however, that NPR_i is not the benefit–cost ratio, since the denominator is only this year's cost, not the present value of the costs incurred over the life of the project.

One limitation of this simple method is that it works perfectly only if projects are divisible, that is, if it is possible to do only part of a project. We shall return later to the case of indivisible projects.

Projects Limited by Many Resources

If the number of projects that can be done is limited by more than one resource, then a more general linear programming model must be used. The purpose of the model is to find the set of projects that maximizes total net present value without exceeding the available resources.

Let X_i be a decision variable that indicates to what extent project i should be done. There are n projects in total; thus $i = 1,\ldots, n$. Assume again that projects are divisible; thus X_i is a continuous variable of value between zero and 1. Project i is all done if $X_i = 1$, not done at all for $X_i = 0$, and partly done otherwise.

Let R_k be the amount of resource k available to do the projects. There are m such resources; thus $k = 1,\ldots, m$. Furthermore, let R_{ik} be the amount of resource k needed to complete project i.

Finally, let NPV_i be the net present value of project i if it is completed. We shall assume that the net present value of a project is directly proportional to its level of completion. Thus, the net present value of project i is $X_i \, \text{NPV}_i$.

Consequently, the expression of the objective function, the total net present value of all projects, is:

$$\max Z = \sum_{i=1}^{n} X_i \text{NPV}_i \qquad (18.8)$$

while the m resource constraints are:

$$\sum_{i=1}^{n} X_i R_{ik} \leq R_k \qquad k = 1,\ldots, m \qquad (18.9)$$

where the left-hand side of each inequality expresses the use of resource k by all projects. Thus each constraint states that the total amount of resource used cannot exceed what is available.

The best solution of this linear program $X_1^*, X_2^*, \ldots, X_n^*$ shows how much of each project should be done to maximize the total net present value.

INDIVISIBLE PROJECTS

The methods already discussed should suffice for most cases of project selection in forestry. However, there may be a few situations where projects are not divisible. That is, projects must be done completely or not at all. In that case the linear program of Equations (18.8) and (18.9) may have to be changed into an integer programming program. This can be done by changing the variables X_i to integer variables that take the value 1 if project i is selected or 0 otherwise.

If the number of projects is not too large, integer solutions can be found with spreadsheet optimizers such as the Excel Solver with a simple change of option (see Chapter 11). Otherwise, rounded solutions of the ordinary linear program may give results that are close to the best solution if many projects are being considered.

18.5 INFLATION AND INVESTMENT ANALYSIS

Up to now we have assumed that the net present value of a forest resource investment could be computed without ambiguity once appropriate prices and interest rates were known. However, the fact that prices change over time causes some difficulties. This is especially troublesome for forestry investments that take many years to mature.

NOMINAL AND REAL PRICE CHANGES

In considering prices over a specific time interval [0, t] one must distinguish *nominal* from *real* price changes. Nominal price changes are just changes of observed market prices. Real price changes refer instead to the change in the price of a good or service relative to the prices of all other goods and services.

To define real prices, we need an indicator of the general price level. In the United States, the producer price index can be used to that end for an intermediate good, such as timber. The consumer price index can be used to calculate the real price of a consumer good, such as forest recreation.

Let I_0 and I_t be the level of the producer price index in years 0 and t, respectively. Let P_0 and P_t be the nominal prices of the good of interest, say softwood lumber, in those years. Finally, let P_t' be the real price of softwood lumber at time t. Then, by definition:

$$\frac{P_t'}{P_0} = \frac{P_t/P_0}{I_t/I_0} \qquad (18.10)$$

That is, the real price change is equal to the nominal price change divided by the change in the general price level. Consequently, if there is no inflation or deflation, meaning $I_t = I_0$, then the real and the nominal price change are equal.

If there is inflation, then $I_t > I_0$ and the real price change is less than the nominal change. If there is deflation, the reverse is true. In the remainder of this section we will talk about inflation only, but it should be clear that deflation (a general decline of the price of all goods and services) has just the opposite effect of inflation; we just do not see it happen very often.

Example

Let us determine the real change of the price of timberland in Vilas County, in northern Wisconsin, between 1975 and 1994. Data from the Wisconsin Department of Revenue show that the average price of timberland in 1975 was about $532, while in 1994 it was $958. These are nominal prices, unadjusted for inflation. Meanwhile, data from the United States Bureau of Labor statistics show that

Analysis of Forest Resource Investments

the producer price index for all intermediate goods was 53 in 1975 and 126 in 1994. We use this as the index of inflation because timberland is an intermediate good: It is used to produce timber. If instead the land were (much more expensive) shoreline property, then the consumer price index would be a better standard of inflation.

$$\text{Nominal price change of timberland:} \quad \frac{P_{94}}{P_{75}} = \frac{958}{532} = 1.80$$

$$\text{Change of general price level:} \quad \frac{I_{94}}{I_{75}} = \frac{126}{53} = 2.38$$

Thus, over those years, the price of timberland has increased less than the price of all intermediate goods reflected by the producer price index. Therefore, the real price of timberland relative to the price of the producer goods had decreased from 1975 to 1994. To be precise, the real price of timberland has changed by:

$$\frac{P'_{94}}{P_{75}} = \frac{1.80}{2.38} = 0.76$$

Thus, the real price of timberland in 1994 in Vilas County was about 76% of what it was in 1975. While the nominal price of timberland increased by 80% during that interval, the real price decreased by 24%.

In making forest resource decisions it is real prices that matter, because they indicate the true changes in the value of resources relative to the value of other goods and services.

RATES OF PRICE CHANGE

Price changes are often expressed in rates per unit of time, e.g., percent per year, like interest rates. Let f be the rate of inflation, expressed as a fraction per year. Then, using the index I_t as the measure of inflation, the value of f over the interval $[0,t]$ satisfies the well-known compound interest formula:

$$I_t = I_0(1+f)^t$$

Similarly, let g be the nominal yearly rate of change of the price of the good or service of interest, timberland in our example. It is defined by this relationship:

$$P_t = P_0(1+g)^t$$

Last, let g' be the real rate of price change for timberland. Then, g' is also defined by:

$$P'_t = P_0(1+g')^t$$

Replacing I_t, P_t, and P'_t in Equation (18.10) by their new expressions leads to the following relationship between the rates of change:

$$(1+g')^t = \frac{(1+g)^t}{(1+f)^t}$$

or:

$$g' = \frac{1+g}{1+f} - 1 \tag{18.11}$$

Equation (18.11) may be simplified if the rates of real price change and the rate of inflation, g, and f are "small", say under 10%. This can be seen by multiplying both sides of Equation (18.11) by $1+f$, leading to:

$$g' + g'f = g - f \tag{18.12}$$

If g and f are small, say under 10% per year, the product $g'f$ can be neglected, giving the approximate formula:

$$g' \approx g - f \tag{18.13}$$

Thus, the rate of real price change is approximately equal to the nominal rate of price change minus the rate of inflation.

Example

Using the data of the previous example, the nominal rate of price change, g, for timberland in Vilas County from 1975 to 1994 was such that:

$$\frac{P_{94}}{P_{75}} = 1.80 = (1+g)^{19}$$

That is:

$$g = 1.80^{(1/19)} - 1 = 0.031 \text{ per year}$$

Similar calculations using the fact that $I_{94}/I_{75} = 2.38$ show that the rate of inflation, f, was:

$$f = 0.047 \text{ per year}$$

Analysis of Forest Resource Investments

So the real rate of price change of land was:

$$g' \approx g - f = -0.016 \text{ per year}$$

that is, a decrease in real price of 1.6% per year.

Equations (18.11) and (18.13) give the real rate of change of any price in terms of the nominal rate and of the rate of inflation. The same formulas also apply to obtain the real rate of interest, namely:

$$(1+r')^t = \frac{(1+r)^t}{(1+f)^t} \tag{18.14}$$

where r' and r are, respectively, the real and the nominal rate of interest during a period with a rate of inflation of f. If r and f are "small," the relationship is approximately:

$$r' \approx r - f \tag{18.15}$$

That is, the real rate of interest is approximately equal to the difference between the nominal rate of interest and the rate of inflation.

INVESTMENT CRITERIA IN NOMINAL OR REAL TERMS

We are now ready to show that forestry investments may be evaluated equivalently using nominal or real prices and interest rates. The results will be the same using either system, as long as we are consistent. That is, if nominal costs and benefits are used, then a nominal interest rate must also be used. A real interest rate is necessary if project evaluation is done with real costs and benefits. The points to remember are that the interest rate is essentially the price of money and that all prices in an investment analysis must be on the same basis, real or nominal.

To illustrate why this is so, consider a stand of black walnut trees currently of volume Q_0. This volume is expected to grow to Q_t in t years. The current price of the trees is P_0 per unit of volume. The nominal price is expected to be P_t by year t. Letting the trees grow instead of harvesting them immediately is an investment. The nominal net present value of this investment is, at a nominal interest rate r:

$$\text{NPV} = \frac{Q_t P_t}{(1+r)^t} - Q_0 P_0$$

where the first term on the right is the nominal present value of the future harvest and the second term is the opportunity cost of the trees we must leave standing to reap this harvest.

Note that the nominal rate of interest, r, is used in discounting. This is consistent with using the nominal price P_t. If, instead, we computed the real net present value, we would use the real interest rate r' and the real future price of black walnut P'_t in the analog formula:

$$\text{NPV}' = \frac{Q_t P'_t}{(1+r')^t} - Q_0 P_0$$

But real and nominal net present value are in fact equal, as can be seen by rewriting the nominal net present value as:

$$\text{NPV} = \frac{Q_t P_0 (1+g')^t (1+f)^t}{(1+r')^t (1+f)^t} - Q_0 P_0 = \frac{Q_t P'_t}{(1+r')^t} - Q_0 P_0 = \text{NPV}'$$

where, as before, g' is the real rate of price change between now and time t and f is the rate of inflation.

This result is general, extending to complex projects with many costs and benefits over many periods (see Problem 18.8). Net present value calculated in real or nominal terms is the same, as long as consistent prices and interest rates are used.

Similarly, benefit–cost ratios are the same in real or nominal terms, since they are simply the ratios of present values. Furthermore, either nominal or real internal rates of return can be used to make investment decisions. But the internal rate of return must be compared to a guiding rate of interest that is consistent with the prices used in calculating the internal rate of return.

Thus, if the internal rate of return of a project is computed at nominal prices, then the project should be accepted if the nominal guiding rate is less than the internal rate of return. Instead, if the internal rate of return is computed from real prices, then the guiding rate of interest must also be expressed in real terms, using Equation (18.14) or (18.15), before comparing it to the internal rate of return.

18.6 CHOOSING AN INTEREST RATE

Choosing an appropriate interest rate to evaluate forest resources investments is difficult, but probably not much more difficult than choosing any other price. The market can serve as a guide, especially for private firms, but considerable judgment and guess work are still needed.

Analysis of Forest Resource Investments

SHOULD THE INTEREST RATE BE ZERO?

Before suggesting practical ways of choosing interest rates, it is worth noting that—whether projects are public or private and the foresters' love of big trees notwithstanding—it is difficult to argue for a zero interest rate. To see this let us return to Figure 18.1 and to Equation (18.1), which defines the interest rate:

$$\frac{dC_1}{dC_0} = -(1+r)$$

A zero interest rate would imply $dC_1/dC_0 = -1$. Since dC_1/dC_0 is the first derivative along the production possibility frontier AB, this would imply that the net productivity of investments is zero. Every opportunity to invest and thereby reap more in the future than what is sacrificed now has been exhausted.

Furthermore, since dC_1/dC_0 is also the first derivative along the indifference curve E^*, this would imply that the time preference is zero, meaning that people are totally indifferent between current and future consumption. A situation where both a zero net productivity of investments and a zero time preference occur simultaneously is difficult to imagine, so a zero interest rate seems very unlikely.

PRIVATE INVESTMENTS

For a private firm, the interest rate is to a large extent determined by the market for capital. If the money needed for a project is borrowed, the guiding rate of interest should be at least equal to the rate charged by the lender. But it may be higher, depending on the other investment opportunities of the firm.

If, instead, all the money for an investment is to come from retained profits, then the guiding rate of interest is defined by the best investment opportunity open to the firm. A great deal of judgment must be used, however, since investments may differ widely in terms of timing and risk. For long-term forestry investments, like buying timberland and starting new plantations, it is the rate of return of alternative investments over long time periods that matter. For a few investments it is possible to calculate such rates from published statistics.

For example, Figure 18.8 shows a spreadsheet with the yields (annual rates of return, in percent per year) of Aaa corporate bonds in the United States from 1970 to 1994. Column B shows the nominal yield, r, while column C shows the yearly rates of change of the consumer price index, f, a standard measure of inflation. The data come from the *Economic Report of the President, 2001*.

	A	B	C	D
1	Aaa BOND YIELDS (%/y)			
2		Nominal yield	CPI change	Real yield
3	Year	r	f	r'
4	1970	8.0	5.7	2.2
5	1971	7.4	4.4	2.9
6	1972	7.2	3.2	3.9
7	1973	7.4	6.2	1.2
8	1974	8.6	11.0	-2.2
9	1975	8.8	9.1	-0.2
10	1976	8.4	5.8	2.5
11	1977	8.0	6.5	1.4
12	1978	8.7	7.6	1.1
13	1979	9.6	11.3	-1.5
14	1980	11.9	13.5	-1.4
15	1981	14.2	10.3	3.5
16	1982	13.8	6.2	7.1
17	1983	12.0	3.2	8.6
18	1984	12.7	4.3	8.1
19	1985	11.4	3.6	7.5
20	1986	9.0	1.9	7.0
21	1987	9.4	3.6	5.6
22	1988	9.7	4.1	5.4
23	1989	9.3	4.8	4.3
24	1990	9.3	5.4	3.7
25	1991	8.8	4.2	4.4
26	1992	8.1	3.0	5.0
27	1993	7.2	3.0	4.1
28	1994	8.0	2.6	5.2
29	1995	7.6	2.8	4.7
30	1996	7.4	3.0	4.2
31	1997	7.3	2.3	4.8
32	1998	6.5	1.6	4.9
33	1999	7.0	2.2	4.7
34	Average	9.1	5.2	3.7
35				
36		Key cell formulas		
37	Cell	Formula		Copy to
38	D4	=((1+B4/100)/(1+C4/100)-1)*100		D4:D33
39	B34	=AVERAGE(B4:B33)		C34:D34

FIGURE 18.8 Spreadsheet to compute the real yield of Aaa corporate bonds.

The real rate of return for every year, r', in column D, was computed with Equation (18.14) instead of the simpler Equation (18.15) because some of the rates were large. The results show that, while the nominal yield in 1970 was 8%, the real yield was only 2.2% due to the inflation of 5.7% that took place during that year. In the years 1974, 1975, 1979, and 1980, the nominal yield was less than inflation and investors in Aaa bonds actually incurred a loss.

The average nominal yield over the period was 9.1% per year; but with an average rate of inflation of 5.2%, this left a real rate of return of only 3.8%.

Analysis of Forest Resource Investments

This is probably less than what most people would expect, which shows the importance of adjusting for inflation. Many forest resource projects are capable of yielding comparable real rates of return.

Investments are often financed from funds from a variety of sources. The interest rates used in evaluating investments should reflect this. For example, if a corporation plans to buy forest land by borrowing part of the money and financing the rest from profits, then the rate of interest should be a weighted average of the rates appropriate for the two sources of fund; that is:

$$r = r_b + (1 - B)r_p$$

where:

r = effective rate
r_b = interest rate on borrowed funds
r_p = rate of return on the best possible use of profits
B = fraction of funds that are borrowed.

Public Projects

In the public sector, such as on lands administered by the United States Forest Service and the Bureau of Land Management, one possibility is to use the rate of return on government bonds, r_g, as the guiding rate of interest. Proponents of this argue that if projects do not return at least that much, then the country would be better off if the government repaid the national debt, since this would be equivalent to a public investment returning the rate r_g.

The rates of return of U.S. treasury securities with maturities of 3–10 years do not differ much from the returns on Aaa corporate bonds listed in Figure 18.8, that is, around 4% in real terms. This is due to the fact that the risks of these investments are similarly low. To raise money, the government must pay a return on its bonds that is comparable to the rate earned in the private sector.

Nevertheless, the rate of return on government bonds should probably be viewed as a lower bound on the interest rate applicable to public investments, because many people cannot afford to buy such bonds. This implies that for a large group of people, current needs are such that they are not willing to trade current consumption for future consumption at the rate r_g (see Section 18.1). The guiding rate of interest appropriate for this group of people is not easy to determine. However, the fact that there are many of them implies that the rate of discount used to evaluate forest resource investments on public lands should be no less than 4% in real terms.

18.7 SUMMARY AND CONCLUSION

The basic criterion to evaluate forestry investments is net present value: the difference between the discounted value of returns and costs. Benefit–cost ratio and internal rate of return lead to decisions that are consistent with net present value only when all projects under consideration can be done.

If only one resource is limiting, projects should be ranked according to their net present value–resource ratio. If more than one resource is limiting, the correct choice of projects requires the solution of a programming problem.

Investments can be evaluated in real or nominal terms. However, the guiding rate of interest must be consistent with the price system used: real if other prices are real, nominal otherwise.

The guiding rate of interest is critical in assessing the value of investment. For private firms, this depends on their sources of financing and their investment opportunities. For public agencies, the rate of return on government bonds can be viewed as a lower bound on the appropriate guiding rate of interest.

PROBLEMS

18.1 (a) Set up your own spreadsheet model, like the one in Figure 18.2, to calculate the net present value and benefit–cost ratio for the ponderosa pine reforestation project.

(b) Using the same data, verify that your results are the same as in Figure 18.2.

(c) Change the interest rate, the costs, and the benefits, and study the effect on the net present value and on the benefit–cost ratio.

18.2 (a) Add a spreadsheet like the one in Figure 18.3 to the spreadsheet developed in Problem 18.1 to determine the sensitivity of the NPV to the rate of interest. Using the same data, verify that your results are the same as in Figure 18.3.

(b) Add a column to this spreadsheet to show the sensitivity of the benefit–cost ratio to the interest rate.

(c) Make a chart, similar to Figure 18.4, of the NPV and of the benefit–cost ratio against the interest rate.

(d) At what guiding rate of interest does the benefit–cost ratio equal 1? What is the value of the NPV at this interest rate?

18.3 (a) Set up your own spreadsheet models, like those shown in Figures 18.2 and 18.5, to calculate a forestry project's net present value, benefit–cost ratio, and internal rate of return. Verify the results obtained in the text. Use these models to determine how the three economic criteria change if:

Analysis of Forest Resource Investments

(b) Prices increase, doubling the benefits from the thinnings and the final harvest.

(c) Prices decrease, halving the benefits from the thinnings and the final harvest.

18.4 A Peace Corps forester in Central America must evaluate three short-rotation fuelwood projects. The costs and benefits associated with each project and the years in which they will occur are shown in the table. Assume a guiding rate of interest of 3% per year. All data are in real terms.

	Project A			Project B			Project C	
Year	Benefit	Cost	Year	Benefit	Cost	Year	Benefit	Cost
0	$0	$4,500	0	$0	$10,000	0	$0	$4,100
5	0	2,000	7	1,500	3,000	4	500	1,500
9	2,500	1,000	10	3,500	1,500	8	2,000	5,00
15	15,000	2,500	15	24,000	4,500	15	11,000	1,000

(a) Use spreadsheets like the ones in Figures 18.2 and 18.5 to calculate each project's net present value, benefit–cost ratio, and internal rate of return.

(b) Which projects should the forester undertake if he had enough resources to do all of them?

(c) Assume that the forester has enough money to do only one project. Which project looks best according to each criterion you calculated in part(a)? Explain any differences you observe.

(d) Which criterion should the forester use to select the best project? Why?

18.5 A reforestation manager in the Pacific Northwest has five areas that need to be replanted, but he does not have enough seedlings to plant all of them up to the silvicultural standard of 750 trees/ha. The size of each area of each clear-cut is given in the table, along with an estimate of the net present value of the plantations.

Land tract	Area (ha)	Net present value
A	150	$12,750
B	290	21,170
C	220	21,780
D	340	21,346
E	120	12,240

(a) Given only 200,000 seedlings, which areas should the forester replant?

(b) Assume that the reforestation policy is to reforest a tract of land or not reforest it at all. Set up an integer programming problem in a spreadsheet to find the tracts that maximize present value. With this policy, would all the seedlings be used?

18.6 Consider Problem 18.5 again, and assume that in addition to having only 200,000 seedlings, the forester has a $24,000 budget for labor and that such labor costs $80 per person-day. Labor requirements to replant each tract are given in the table. (These requirements vary depending on a tract size, slope, soil type, and degree of site preparation.) Each tract may be partially reforested.

Land tract	Person-days for planting
A	50
B	70
C	80
D	110
E	50

(a) Set up a linear programming model on a spreadsheet to decide which tracts should be replanted to maximize net present value, subject to the limited budget and seedlings. For the best solution, how much of each tract is reforested? How much of the budget and seedlings is used?

(b) Redo this problem, assuming that the reforestation policy is to reforest each tract completely or not at all. How does this change the total net present value, the tracts reforested, and the use of seedlings and budget?

18.7 The nominal producer price index for softwood lumber in the United States was 206.5 in 1997 but fell to 178.6 by 2000 as the economy slowed. Over the same period, the producer price index for intermediate goods rose from 125.6 to 129.

(a) What was the annual nominal percentage price change for softwood lumber from 1997 to 2000?

(b) What was the annual percentage price change of all intermediate goods?

(c) What was the annual real percentage price change of softwood lumber?

18.8 Assume that all the data for the analysis of the ponderosa pine project in Figure 18.2 are in real terms. Create a new spreadsheet, linked to that in Figure 18.2, to show the data and the results in nominal terms for any given rate of inflation. The new spreadsheet should contain (1) formulas to calculate nominal costs and benefits based on the real costs and benefits in Figure 18.2 and the rate of inflation, and (2) a formula to calculate the nominal interest rate based on the real interest rate and the rate of inflation. Verify that the net present value and benefit–cost ratio computed in nominal terms are equal to those computed in real terms.

ANNOTATED REFERENCES

Ashton, M.S., R. Mendelsohn, B.M.P. Singhakumara, G.V.S. Gunatilleke, I.A.U.N. Gunatilleke, and Alexander Evans. 2001. A financial analysis of rain forest silviculture in southwestern Sri Lanka. *Forest Ecology and Management* 154:431–441. (Investment analysis of alternative silvicultural regimes.)

Browder, J.O., E.A.T. Matricardi, and W.S. Abdala. 1996. *Ecological Economics* 16:147–159. (Investment analysis of mahogany plantations on farms, coffee plantations, or in pure stands.)

Brown, B.G., and A.H. Murphy. 1988. On the economic value of weather forecasts in wildlife suppression mobilization decisions. *Canadian Journal of Forest Research* 18:1641–1649. (Investment analysis of management information.)

Bullard, S.H., and J.E. Gunter. 2000. Adjusting discount rates for income taxes and inflation: A three-step process. *Southern Journal of Applied Forestry* 24(4):193–195. (Discusses importance of consistency in treatment of real and nominal prices and interest rates.)

Busby, R.L., J.H. Miller, and M.B. Edwards, Jr. 1998. Economics of site preparation and release treatments using herbicides in Central Georgia. *Southern Journal of Applied Forestry* 22(3):156–162. (Application of investment analysis to computation of LEV for stands subjected to alternative site preparation and release treatments.)

Clark, C.W. 1988. Clear-cut economies (should we harvest everything now?). *The Sciences* :16–20. (Argues that economics leads to total harvest of a resource that grows at a rate inferior to the real interest rate.)

Cubbage, F.W., J.M. Pye, T.P. Holmes, and J.E. Wagner. 2000. An economic evaluation of fusiform rust protection research. *Southern Journal of Applied Forestry* 24(4):77–85. (Investment analysis of forest pathology research.)

Davis, L.S., K.N. Johnson, P.S. Bettinger, and T.E. Howard. 2001. *Forest Management: To Sustain Ecological, Economic, and Social Values*. McGraw-Hill, New York. 804 pp. (Chapter 7 discusses present values and internal rates of return, and Chapter 9 discusses benefit–cost ratios.)

Dubois, M.R., G.R. Glover, T.J. Straka, and M.O. Sutton. 2001. Historic and projected economic returns to alternative site preparation treatments: The Fayette study. *Southern Journal of Applied Forestry* 25(2):53–59. (Investment analysis of site preparation treatments.)

Fabrycky, W.J., G.J. Thuesen, and D. Verma. 1998. *Economic Decision Analysis*. Prentice Hall, Upper Saddle River, NJ. 384 pp. (Basic reference on investment analysis.)

Fight, R.D., N.A. Bolon, and J.M. Cahill. 1993. Financial analysis of pruning Douglas fir and ponderosa pine in the Pacific Northwest. *Western Journal of Applied Forestry* 10(1):58–61.

Garrett, H.E., W.J. Rietveld, and R.F. Fisher. 2000. *North American Agroforestry: An Integrated Science and Practice*. American Society of Agronomy, Madison, WI. 402 pp. (Chapter 9 discusses investment analysis for agroforestry enterprises and references a variety of agroforestry case studies.)

Gregory, G.R. 1987. *Resource Economics for Foresters*. Wiley, New York. 477 pp. (Chapter 11 discusses investment analysis in forestry.)

Howard, A.F., and H. Temesgen. 1997. Potential financial returns from alternative silvicultural prescriptions in second-growth stands of coastal British Columbia. *Canadian Journal of Forest Research* 27:1483–1495. (Investment analysis of alternative silvicultural regimes.)

Hueting, R. 1991. *Ecological Economics* 3:43–57. (Argument against using discounting in analysis of investments in rain forest areas.)

Klemperer, W.D. 1996. *Forest Resource Economics and Finance*. McGraw-Hill, New York. 551 pp. (Chapters 4–6 discuss investment analysis in forestry.)

Loomis, J.B. 1993. *Integrated Public Lands Management: Principles and Applications to National Forests, Parks, Wildlife Refuges, and BLM Lands*. Columbia University Press, New York. 474 pp. (Chapter 6 discusses investment analysis in the context of land management agency decision making.)

Macmillan, D.C., D. Harley, and R. Morrison. 1998. Cost-effectiveness analysis of woodland ecosystem restoration. *Ecological Economics* 27:313–324. (Investment analysis of a restoration project.)

Portney, P.R., and J.P. Weyant, eds. 1999. *Discounting and Intergenerational Equity*. Resources for the Future, Washington, DC. 186 pp. (Collection of papers discussing what interest rate, if any, should be used in environmental analysis.)

Straka, T.J., R.L. Ridgway, R.H. Tichenor, Jr., R.L. Hedden, and J.A. King. 1997. Cost analysis of a specialized gypsy moth management program for suburban parks. *Northern Journal of Applied Forestry* 14(1):32–39. (Application of investment analysis to a recreation management situation where benefits are difficult to estimate.)

CHAPTER 19

Econometric Analysis and Forecasting of Forest Product Markets

19.1 INTRODUCTION

All the models we have studied so far make bold assumptions regarding the context in which a forest operates. For example, we often made assumptions regarding the future price of the commodities or services produced by a forest. These assumptions are critical; changing them may alter considerably the way the forest should be managed.

Unfortunately, economic forecasting is still as much an art as a science, but some progress is being made. Many econometric models have been developed of the markets for forest products, including timber, water, recreation, and other goods and services. Since the demand and supply conditions define much of the context in which forest resource decisions are being made, forest managers should know the general principles upon which these models are built. It is important that managers be aware of the potential value of econometric models as well as of their limitations.

The purpose of this chapter is to provide a brief introduction to econometric analysis and forecasting. We shall do this mostly with an example: a simple model of the pulpwood market in the United States, built to predict the demand, supply, and price of pulpwood.

19.2 ECONOMETRICS

An econometric model consists of equations that give a quantitative explanation of the changes in economic variables. Econometrics combines economic theory and statistical methods. The theory suggests the form of the equations relating the variables of interest, while the statistical methods provide the means to estimate the parameters of the equations based on observations.

Econometric models may be weak due to the limitations of economic theory (we may not know how the economy really works), to the limitations of the statistical methods used to estimate their parameters, or to the limitations of the data needed to use these methods. Economic data are often poorly defined, inaccurate, or altogether nonexistent.

Nevertheless, econometric modeling of forest product markets has progressed enormously in recent decades. Rigorous empiricism is undoubtedly one of the best ways to analyze and forecast forest product markets. This is in part because a formal model is clear: It shows explicitly all the assumptions that have been made; such clarity facilitates understanding, communication, and progress.

19.3 A MODEL OF THE UNITED STATES PULPWOOD MARKET

To avoid too much abstraction, throughout this chapter we shall use the case of the market for pulpwood in the United States. The model is a simplified version of a model proposed originally by Leuschner (1973) for the state of Wisconsin.

The model has two objectives: (1) to understand the forces that determine the demand, supply, and price of pulpwood; (2) to forecast the demand for pulpwood and its price in the next five years. Forecasts of this kind would be valuable to forest owners and managers. Demand forecasts would help them determine whether they should produce more or less pulpwood. Price forecasts are essential to help them predict the profitability of forest operations, such as the thinning of forests to produce pulpwood.

STRUCTURAL EQUATIONS

The theoretical model describing the pulpwood market in the United States consists of the following equations:

Demand:	$D_t = a + bK_t + u_t$	(19.1)
Supply:	$S_t = c + dP_t + v_t$	(19.2)
Equilibrium:	$D_t = S_t$	(19.3)

where:

> t = a particular year
>
> S = quantity of pulpwood supplied in the United States
>
> D = quantity of pulpwood demanded in the United States
>
> K = capacity of pulp mills
>
> P = price of pulpwood
>
> a, b, c, d = unknown parameters
>
> u, v = random disturbances

These *structural* equations describe our view of how various forces interact to determine the supply, demand, and price of pulpwood. The form of the structural equations of the model may be shaped in part by economic theory, previous studies, or personal experience.

The first equation is our hypothesis concerning the demand for pulpwood. It says that in a given year the quantity of pulpwood demanded is a linear function of pulping capacity. One would expect demand to increase as capacity of production increases, and therefore the coefficient b should be positive. Note that the equation suggests that demand for pulpwood is not influenced by price, presumably because the cost of pulpwood is small compared to other costs of producing pulp. The fixed costs of pulp manufacturing are particularly high, so pulp mills are usually run at or near capacity, even when pulp prices are low, in order to generate revenues to defray the fixed costs. Thus, all the pulpwood necessary to run a mill at capacity is bought, regardless of the cost. We shall not try to test whether price does in fact influence demand, but we shall test whether the data are consistent with the proposed hypothesis.

The second structural equation is a classic supply equation: Quantity supplied in a given year is a linear function of the price of pulpwood. As the price of pulpwood increases, one expects the supply to increase, and thus the coefficient d should be positive. This hypothesis is consistent with the organization of pulpwood production in United States: There are many small independent pulpwood producers, as assumed in the classical model of supply in a perfectly competitive market.

The first two equations are stochastic. They hold only approximately. This is reflected by the disturbances u_t and v_t, which are assumed to be small and random.

The third structural equation instead is an identity, supposed to hold exactly. It states that the quantity of pulpwood demanded in a given year is equal to the quantity supplied. This identity allows a simplification of the model. Defining the quantity of pulpwood demanded and supplied as $Q_t = S_t = D_t$ leads to:

Demand: $\quad Q_t = a + bK_t + u_t \quad$ (19.4)

Supply: $\quad Q_t = c + dP_t + v_t \quad$ (19.5)

Types of Variables

The capacity of pulp mills, K_t, is assumed to be determined by forces not explained within the model. Presumably, capacity is determined by the demand for paper and other factors, but this relationship is not clarified in this model. Capacity is an *exogenous* variable.

The other variables, Q_t and P_t, are determined jointly by the system of equations when the parameters, the disturbances, and the exogenous variable K_t are given. They are called *endogenous* variables. That Q_t and P_t are determined jointly is illustrated in Figure 19.1, in which the pulpwood supply is represented by the upward-sloping straight line, indicating that pulpwood supply responds positively to price. The vertical line represents pulpwood demand, meaning that demand is independent of price. At price P_t the quantity demanded is equal to the quantity supplied; thus P_t is the *equilibrium* price. An increase in capacity, K_t, other things being equal, causes the vertical demand

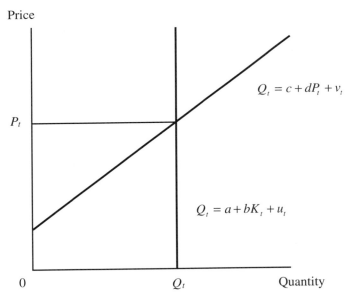

FIGURE 19.1 Demand and supply of pulpwood, and equilibrium price.

Reduced Form

There is an algebraic analog to this graphical explanation of the equilibrium quantity and price. The system of Equations (19.4) and (19.5) can be solved for the endogenous variables, Q_t and P_t, as a function of the exogenous variable, K_t, and the parameters and disturbances. The two equations lead to:

$$c + dP_t + v_t = a + bK_t + u_t$$

and

$$P_t = \frac{a-c}{d} + \frac{b}{d}K_t + \frac{u_t - v_t}{d}$$

Therefore, the structural equations imply the following quantity and price equations:

$$\text{Quantity:} \quad Q_t = a + bK_t + u_t \qquad (19.6)$$

$$\text{Price:} \quad P_t = e + gK_t + z_t \qquad (19.7)$$

where:

$$e = \frac{a-c}{d}, \quad g = \frac{b}{d}, \quad \text{and} \quad z_t = \frac{u_t - v_t}{d}$$

Equations (19.6) and (19.7) are the *reduced form* of the model. They show that both the quantity of pulpwood (demanded and supplied) and the price of pulpwood change simultaneously as pulping capacity changes. Reduced-form equations can be used to predict future values of the endogenous variables Q_t and P_t. It is clear, however, that these forecasts will be inexact. There are three sources of error: (1) Predictions of the exogenous variable, K_t, must be made; (2) the parameters of the reduced-form equations must be estimated; (3) forecasts must be made of the disturbances u_t and v_t. It is generally assumed that the future value of the disturbances is equal to their expected value, zero, so we actually make forecasts only of the expected values of Q_t and P_t.

The reduced-form equations can also be used to recover the coefficients of the structural equations. Once a, b, e, and g have been estimated, the supply equation is fully defined by:

$$d = \frac{b}{g} \quad \text{and} \quad c = a - de$$

We shall see later why it is preferable to estimate the supply equation in this indirect way.

19.4 DATA

To estimate the value of the coefficients in Reduced-Form Equations (19.6) and (19.7), we need data on the variables Q_t, P_t, and K_t. Rarely do we have data that describe exactly the variables of interest. Often, they are only rough approximations, many sources must be reconciled, and bold assumptions must be made to arrive at some estimate of the desired variables. For the present example, we shall use the following data sources:

> *For the price of pulpwood*, we shall use the data on the price of Southern pine pulpwood, collected for many years by the Louisiana Department of Agriculture, Office of Marketing (Howard, 2001). A large part of the U.S. pulp industry is in the South, so we assume that this price is a good *proxy* variable for the price of pulpwood in the country.
> *For the quantity of pulpwood supplied and demanded*, we shall use data on pulpwood receipts at U.S. pulpmills. The data come from the Forest Resources Association for recent years and from Forest Service estimates for 1996 and earlier years (Ince, 2001).
> *For pulping capacity*, we shall use the data on annual capacity of production. These data came from the American Forest and Paper Association annual capacity surveys (Ince, 2001).

The spreadsheet in Figure 19.2 shows all the data used in this chapter. In all, there are 30 observations, covering the period 1970–1999.

19.5 MODEL ESTIMATION

To get accurate forecasts of pulpwood price and quantity (demanded and supplied) with Reduced-Form Equations (19.6) and (19.7), we need accurate estimates of the parameters a, b, e, and g. A common way of estimating equations such as Equations (19.6) and (19.7) is the ordinary least squares (OLS) method.

THE ORDINARY LEAST SQUARES METHOD

Consider the reduced-form equation of Quantity Equation (19.6). The OLS method consists in finding the coefficients a and b such that the sum of the squared residuals, SSR, is a minimum. That is, given a series of n observations

	A	B	C	D
1	Year	K_t	Q_t	P_t
2	1970	40.5	156.2	7.46
3	1971	41.4	154.9	7.31
4	1972	42.4	158.9	6.99
5	1973	44.0	167.9	6.77
6	1974	44.4	178.8	6.64
7	1975	44.9	151.3	6.43
8	1976	45.9	168.4	6.43
9	1977	47.0	173.2	6.41
10	1978	47.8	177.2	6.55
11	1979	48.7	186.7	6.92
12	1980	50.1	194.4	6.73
13	1981	51.6	190.5	7.57
14	1982	52.3	179.5	8.38
15	1983	52.2	196.5	8.60
16	1984	53.5	199.4	9.97
17	1985	54.1	197.5	8.63
18	1986	55.1	211.0	7.05
19	1987	56.3	215.1	7.90
20	1988	57.9	217.0	8.75
21	1989	59.0	215.0	9.58
22	1990	59.9	213.3	9.01
23	1991	61.2	211.6	10.47
24	1992	62.3	215.7	11.75
25	1993	61.9	211.0	12.35
26	1994	62.2	215.6	11.45
27	1995	63.0	225.7	11.45
28	1996	63.1	210.8	10.94
29	1997	64.5	219.0	11.01
30	1998	64.1	215.2	13.25
31	1999	63.2	207.2	10.93
32	K_t=pulping capacity (million t/y)			
33	Q_t=pulpwood deliveries (million m³/y)			
34	P_t=pulpwood price ($/m³)			

FIGURE 19.2 Spreadsheet with data to estimate the demand-and-supply model.

on the variables Q_t and K_t, we compute:

$$\text{SSR} = \sum_{t=1}^{n} u_t^2 = \sum_{t=1}^{n} (Q_t - a - bK_t)^2$$

and find the values of a and b that make SSR as small as possible. These calculations are performed routinely in spreadsheets with functions that do *regression analysis*.

Subject to some assumptions, listed shortly, the estimates of a and b obtained by the OLS method are best linear estimates of the parameters being sought. This means that:

> The expected value of the estimates is equal to the value of the parameters. That is, on average, the estimates are equal to the parameters.
>
> Among all possible linear unbiased estimates of a and b, the OLS estimates have minimum variance. In that sense, then, OLS estimates are the most accurate that we can get.

But these properties of OLS estimates hold only under the following conditions:

> The expected value of the residuals is zero, and their variance is constant.
> The residuals are uncorrelated over time; for example, u_t is independent of u_{t-1} for all values of t.
> The independent variable K_t is predetermined; in particular, it is independent of u_t and v_t.

In addition, the residuals must be normally distributed to make statistical tests. In that case, the ratio of each OLS coefficient to its standard error has Student's t distribution.

Part of the work of building an econometric model is to make sure that these assumptions hold. If they do not, the proposed model is incorrect and must be changed, or some other method of estimation is needed.

REGRESSION ANALYSIS WITH EXCEL

Regression analysis can be done with the regression procedure in Excel. To access the regression tool, choose in the Excel menu: Tools, then Data Analysis, and then Regression.

To estimate the pulpwood Quantity Equation (19.6), complete the regression dialog box as in Figure 19.3. The Input Y range contains the data for the *dependent*, or explained, variable. These are the data for pulpwood receipts, in cells C1:C31 of the worksheet in Figure 19.2. The Input X range contains the data for the *independent*, or explanatory, variable. These are the data for pulp mill capacity, in cells B1:B31 in the same worksheet. The Labels box has been checked to indicate that the first row of the data range contains column labels, Q_t for pulpwood deliveries and K_t for pulping capacity. "New Worksheet Ply"

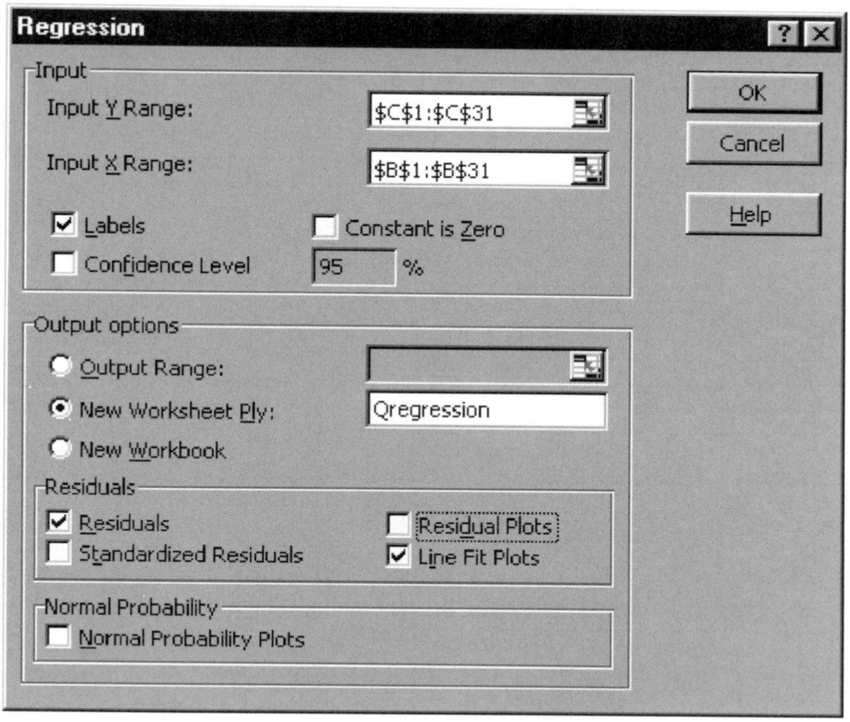

FIGURE 19.3 Regression dialog box for pulpwood quantity equation.

has been selected to direct the output to a new worksheet named "Qregression." "Residuals" has been checked, to obtain the residuals, u_t. Last, the "Line Fit Plots" box has been checked, to obtain a plot of the regression line.

Figure 19.4 shows the results of the estimation of the pulpwood Quantity Equation (19.6) with the Excel regression procedure. The most important results are the estimated coefficients a (intercept) and b (on K_t) in cells B17:B18. They indicate that the equation that best fits the data, in the sense of minimizing the sum of the squared residuals, is:

$$Q_t = \underset{(10.50)}{48.2} + \underset{(0.19)}{2.72} K_t \qquad (19.8)$$

The numbers in parentheses below this equation are the standard errors of the coefficients, in cells C17:C18 in Figure 19.4. The smaller the standard errors, the more precise are the estimates of the coefficients. Another useful statistic is

	A	B	C	D	E	F	G
1	SUMMARY OUTPUT						
2							
3	*Regression Statistics*						
4	Multiple R	0.94					
5	R Square	0.88					
6	Adjusted R Square	0.87					
7	Standard Error	8.08					
8	Observations	30					
9							
10	ANOVA						
11		*df*	*SS*	*MS*	*F*	*Significance F*	
12	Regression	1	12930.78	12930.78	197.85	0.00	
13	Residual	28	1829.98	65.36			
14	Total	29	14760.76				
15							
16		*Coefficients*	*Standard Error*	*t Stat*	*P-value*	*Lower 95%*	*Upper 95%*
17	Intercept	48.20	10.50	4.59	0.00	26.68	69.71
18	Kt	2.72	0.19	14.07	0.00	2.32	3.11
19							
20							
21							
22	RESIDUAL OUTPUT						
23							
24	*Observation*	*Predicted Qt*	*Residuals*				
25	1	158.292322	-2.092321522				
26	2	160.738885	-5.838884664				

FIGURE 19.4 Regression results for pulpwood quantity equation.

the adjusted $R^2 = 0.87$, in cell B6 in Figure 19.4. It indicates how much of the variance of the dependent variable, Q_t, is explained by the independent variable, K_t. Here, variations in capacity accounted for 87% of the variance of the quantity of pulpwood receipts from 1970 to 1999.

Figure 19.5 shows a plot of the regression line of pulpwood deliveries against capacity. It suggests that the linear relation between pulpwood delivery and capacity is a plausible model.

The same regression procedures applied to Price Equation (19.7), with the price and capacity data in Figure 19.2, lead to the results in Figure 19.6. The OLS regression for the pulpwood price is:

$$P_t = -3.82 + 0.23 K_t \quad (19.9)$$
$${\scriptstyle(1.33)(0.02)}$$

The corresponding adjusted R^2 is 0.76, indicating a less good fit for the price equation than for the quantity equation. The plot of the best-fitting line in Figure 19.7 confirms this.

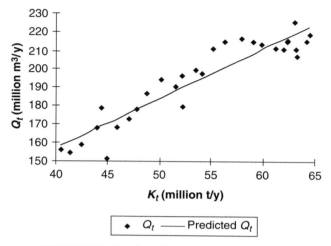

FIGURE 19.5 Best-fitting line for pulpwood quantity.

	A	B	C	D	E	F	G
1	SUMMARY OUTPUT						
2							
3	*Regression Statistics*						
4	Multiple R	0.87					
5	R Square	0.77					
6	Adjusted R Square	0.76					
7	Standard Error	1.03					
8	Observations	30					
9							
10	ANOVA						
11		*df*	*SS*	*MS*	*F*	*Significance F*	
12	Regression	1	96.01	96.01	91.28	0.00	
13	Residual	28	29.45	1.05			
14	Total	29	125.46				
15							
16		*Coefficients*	*Standard Error*	*t Stat*	*P-value*	*Lower 95%*	*Upper 95%*
17	Intercept	-3.82	1.33	-2.87	0.01	-6.55	-1.09
18	Kt	0.23	0.02	9.55	0.00	0.18	0.28
19							
20							
21							
22	RESIDUAL OUTPUT						
23							
24	*Observation*	*Predicted Pt*	*Residuals*				
25	1	5.67	1.79				
26	2	5.88	1.43				

FIGURE 19.6 Regression results for price equation.

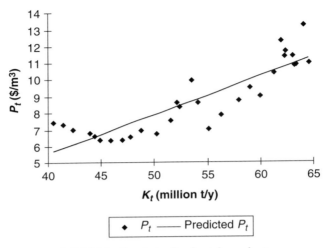

FIGURE 19.7 Best-fitting line for pulpwood price.

Before trying to infer anything further from the quantity and price equations, we must check whether the residuals satisfy the OLS assumptions.

DIAGNOSTIC CHECKS

Are Errors Correlated?

One of the most pervasive problems of econometric analysis with time-series data is *autocorrelation*, which means that the residual in any year is not independent of the residual in the previous year.

Plotting the residuals in chronological order and watching for systematic patterns can reveal autocorrelation. For example, the estimated residuals for the pulpwood quantity equation are:

$$u_t = Q_t - (48.2 + 2.72 K_t) \qquad (19.10)$$

The residuals for the years 1970 and 1971 are in cells C25:C26 in Figure 19.4. The complete series of residuals for 1970–1999 are plotted in Figure 19.8. The corresponding residuals for the price equation are plotted in Figure 19.9. There seems to be some positive autocorrelation, because several sequential residuals have the same sign. Variants of OLS exist to correct for serial correlation. However, the most useful approach is to find the theoretical reason for the autocorrelation, often a missing explanatory variable, and to change the model accordingly. But here, the serial correlation does not seem serious enough to warrant complicating the model or estimation method.

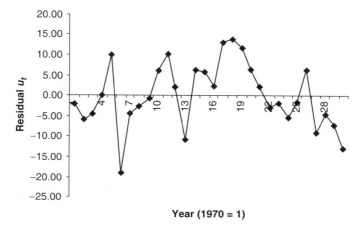

FIGURE 19.8 Time plot of the residuals of the quantity equation.

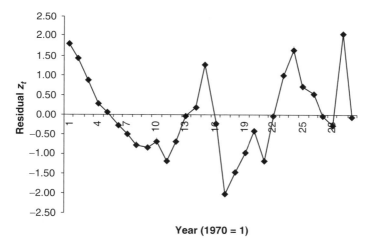

FIGURE 19.9 Time plot of the residuals of the price equation.

Is the Variance of Errors Constant?

One of the assumptions of OLS is that the variance of the residuals is constant. This is sometimes violated by a tendency of the residuals to become larger, in absolute value, as the value of the variables increases. In our example, however, there is no such systematic pattern. The spread of the residuals around the regression line seems to be independent of the level of capacity (see Figures 19.5 and 19.7).

19.6 INFERENCES, FORECASTING, AND STRUCTURAL ANALYSIS

Hypothesis Testing and Confidence Intervals

The estimated pulpwood Quantity Equation (19.8) suggests that as pulping capacity increases by 1 million tons per year, the quantity of pulpwood demanded and supplied increases by about 2.72 million m^3. This coefficient is just an estimate that would be different if it were estimated with a different data set, but the standard error of this coefficient is small, about 0.19. The ratio of the coefficient to its standard error is 2.72/0.19 = 14.1. This ratio, shown in cell D18 of the Excel regression output in Figure 19.4, has a Student's t distribution. There is approximately a 5% chance that the coefficient is zero if its t ratio is 2. Since the actual t ratio is much larger than that, the probability that the coefficient is zero is very small. Indeed, this probability, shown in cell E18 in Figure 19.4, is less than .005.

Viewed another way, a 95% confidence interval for the elasticity of demand with respect to capacity is:

$$(2.72 - 2 \times 0.19, 2.72 + 2 \times 0.19) = (2.34, 3.10)$$

There is a 95% chance that the true coefficient lies in this interval. This 95% confidence interval, computed more accurately by the Excel regression procedure, is shown in cells F18:G18 in Figure 19.4.

Applying the same principles to the estimated pulpwood Price Equation (19.9), we observe that the coefficient with respect to capacity is 0.23. Thus, an increase in pulping capacity of 1 million tons per year leads to an increase in pulpwood price of $0.23/m^3. Cells C18:G18 of the Excel regression output in Figure 19.6 show that the standard error of this coefficient is 0.02, the t ratio is 9.55, the probability that the coefficient is zero is less than .005, and the 95% confidence interval of the coefficient is (0.18, 0.28).

Forecasting

The American Paper Institute publishes annual forecasts of capacity of the pulp industry in different regions of the United States. This is essentially a compendium of the expansion plans of most pulp mills for the next four years.

Given the estimates of Reduced-Form Equations (19.8) and (19.9), it is a simple matter to forecast the implications of the expected growth of capacity on pulpwood demand and price. For example, let us assume that the American

Paper Institute predicts a pulping capacity in the United States of 60 million tons/year in 2005. This would imply an expected 5% decrease in capacity relative to 1999. From the reduced-form pulpwood Quantity Equation (19.8), the expected demand and supply of pulpwood in 2005 would be:

$$Q_{2005} = 48.2 + 2.72 \times 60 \approx \$211 \text{ million}/y$$

Similarly, the forecast of the price of pulpwood in 2005, would be obtained from the reduced-form Price Equation (19.9) as:

$$P_{2005} = -3.82 + 0.23 \times 60 \approx \$10/m^3$$

It is clear that the quality of the forecast depends not only on the quality of the model, but also on the accuracy of the forecasts of capacity. Because capacity plans may not be realized, one would usually compute several conditional forecasts of price and demand for different values of future capacity.

STRUCTURAL ANALYSIS

The reduced-form equations do not show how U.S. pulpwood suppliers respond to changes in the price of pulpwood. To learn this, we must go back to Supply Equation (19.5) in the revised structural model. In that equation, coefficient d shows by how much the pulpwood supply changes if the price changes by one unit. The value of this coefficient can be recovered from the reduced-form equations. We saw earlier that $d = b/g$. And from the estimates of the reduced-form equations we know that $b = 2.72$ and $g = 0.23$. Therefore:

$$d = \frac{2.72}{0.23} = 11.83$$

which shows that the total U.S. supply of pulpwood increases by about 12 million m^3 per year as the real price (net of inflation) increases by $1 per m^3.

We can also recover the constant of the supply equation. We have already seen that $c = a - de$; therefore:

$$c = 48.2 - 11.83(-3.82) = 93.39$$

So the estimate of the supply equation is:

$$Q_t = 93.39 + 11.83 P_t$$

A keen reader may wonder why the supply equation was not estimated directly by OLS using Equation (19.5). The reason is that in that equation, price P_t is not independent of the residual, v_t. This is made clear by Reduced-Form Price Equation (19.7). A change in v_t changes z_t, which changes P_t. Thus, the OLS assumption that the explanatory variable is independent of the residual is violated.

Applying OLS to the reduced-form equations avoids this problem because independent variable K_t is exogenous, thus, by definition, independent of the residuals in our model.

19.7 CONCLUSION

To make forecasts of the demand, supply, and price of pulpwood in the United States we developed a quantitative model of that market. We started with structural equations of demand and supply based on economic theory. From them we derived reduced-form equations that expressed quantity demanded and price as functions of a single exogenous variable: pulping capacity. We then estimated these equations by ordinary least squares, using annual data on pulpwood receipts, pulpwood prices, and pulping capacity. The reduced-form equations were used to make conditional forecasts of price and demand, given published forecasts of the exogenous variables.

Other, more mechanical methods of forecasting exist that rely mostly on observing the patterns in a variable over time. However, it is advantageous to use models based on theory. More confidence can be placed on a relationship between variables if there is a plausible explanation of why the variables are related.

However, the sophistication of the theory is necessarily limited by available data. There is little point in adding variables to a model if they cannot be measured. The quantity and quality of data available on the forestry sector limit the usefulness of some complex estimation methods. The OLS method used throughout this chapter remains the workhorse for practical applications.

In addition to theory, data, and methods, the quality of forecasts in the forest sector depends on the accuracy of the forecasts of the exogenous variables that drive the sector, like the pulping capacity in the example of this chapter. Forest resource managers have generally little means of projecting those exogenous variables on their own, and they must instead rely on other experts.

PROBLEMS

19.1 (a) Copy the data in Figure 19.2 into a spreadsheet. Use the Excel regression procedure to obtain OLS estimates of the pulpwood Quantity Equation (19.6) and of the pulpwood Price Equation (19.7). Verify that your results are the same as in Figures 19.4 and 19.6.

(b) Chart the best-fitting regression equations. Verify that your results are the same as in Figures 19.5 and 19.7.

(c) Chart the time plots of the residuals for the price and the quantity equation. Verify that your results are the same as in Figures 19.8 and 19.9.

(d) Repeat parts (a) through (c), but use only the data from 1980 through 1999. What differences do you observe in the results? What might be the reason for the differences?

19.2 Consider the set of structural Equations (19.1) to (19.3) in the model of U.S. pulpwood supply and demand. This model might be improved by modifying the Demand Equation (19.1) to allow price to affect the quantity demanded. The new set of structural equations would then be:

Demand: $D_t = a + bK_t + eP_t + u_t$

Supply: $S_t = c + dP_t + v_t$

Equilibrium: $D_t = S_t$

where:

t = a particular year
D = quantity of pulpwood demanded in the U.S.
P = price of pulpwood in the U.S.
S = quantity of pulpwood supplied in the U.S.
K = capacity of pulp mills in the U.S.
a, b, c, d, e = unknown parameters
u, v = random disturbances

(a) Write the reduced-form equations expressing the endogenous variables P_t and Q_t as functions of the exogenous variable K_t and the unknown parameters $a, b, c, d,$ and e.

(b) How do these reduced-form equations differ from Equations (19.6) and (19.7)?

19.3 In the text, the OLS estimates of the parameters of the reduced-form pulpwood Price Equation (19.6) and Quantity Equation (19.7) could be used to derive estimates of the parameters of the structural Demand and Supply Equations (19.1) and (19.2). When this is possible, the structural equations are said to be "identified." It is not always possible to identify the parameters of the structural equations. Given OLS estimates of the coefficients of the reduced-form equations derived in Problem 19.2, could you derive estimates of all, some, or none of the structural equation parameters $a, b, c, d,$ and e? (Hint: Think of each estimate of a coefficient in a reduced-form equation as providing an equation of the form

$f(a, b, c, d, e)$ = estimated coefficient

Are there enough equations to calculate all the parameters?)

19.4 A forest products company needs to decide between investing in a new lumber mill or a new plywood mill. Since either mill would represent a major capital investment with a useful life of several decades, the company's officers have asked their forest economist to evaluate how lumber and plywood consumption will grow in the future. To do this, she will try to forecast U.S. future consumption with the following models:

Lumber consumption: $L_t = a + bt + u_t$

Plywood consumption: $P_t = c + dt + v_t$

where:

t = a particular year
L = consumption of lumber
P = consumption of plywood
a, b, c, d = unknown parameters
u, v = random disturbances

(a) Use the data in the table and the Excel OLS regression procedure to estimate the unknown parameters in each equation.

Year	Lumber (10^6 m³/y)	Plywood (10^4 m³/y)	Housing starts (10^3/y)
1977	94.5	2,042	2,279
1978	97.8	2,128	2,312
1979	92.8	1,953	2,037
1980	76.6	1,611	1,534
1981	71.5	1,605	1,341
1982	73.6	1,551	1,311
1983	93.6	1,890	2,008
1984	101.1	1,945	2,051
1985	104.4	2,007	2,029
1986	112.1	2,098	2,051
1987	119.3	2,280	1,856
1988	114.4	2,200	1,706
1989	113.2	2,123	1,574
1990	105.4	2,036	1,381
1991	98.8	1,824	1,185
1992	107.0	1,901	1,411
1993	107.1	1,917	1,542
1994	112.4	1,965	1,761
1995	111.6	1,955	1,694
1996	116.8	1,911	1,840
1997	120.0	1,811	1,828
1998	123.2	1,889	1,990
1999	128.3	1,956	2,023

Sources: Plywood data are from Howard, J.L. 1999. *U.S. Timber Production, Trade, Consumption, and Price Statistics.* FPL-GTR-116, USDA Forest Products Laboratory, Madison, WI, updated by personal communication from J.L. Howard. Lumber and housing start data are from the *Western Wood Products Association Statistical Yearbook.*

(b) How much of the variation in per capita consumption of lumber and plywood is explained by these simple models?
(c) What do the signs on the estimates of parameters b and d indicate?
(d) What are the 95% confidence intervals for each of these parameters?
(e) Does either model seem to have a problem with autocorrelation? with the variance of the errors?
(f) Use the estimated models to forecast the per capita consumption of lumber and plywood in the year 2010.

19.5 The forecasting models estimated in Problem 19.4 only project trends in consumption and do not suggest a reason for the change in consumption. In the United States, lumber and plywood are used primarily to build houses. Therefore, models that make lumber and plywood consumption functions of housing starts seem plausible. In particular, consider the following models:

$$\text{Lumber consumption:} \quad L_t = a + eH_t + u_t$$
$$\text{Plywood consumption:} \quad P_t = c + fH_t + v_t$$

where:

t = a particular year
L = consumption of lumber in year t
P = consumption of plywood
H = number of housing starts
a, b, e, f = unknown parameters
u, v = random disturbances

(a) Use the data in the table in Problem 19.4 and the Excel OLS regression procedure to estimate the unknown parameters in each equation.
(b) How much of the variation in per capita consumption of lumber and plywood is explained by these simple models? Is this better or worse than the models estimated in Problem 19.4?
(c) What do the signs on the estimates of parameters e and f indicate?
(d) What are the 95% confidence intervals for each of these parameters?
(e) Does either model seem to have a problem with autocorrelation? with the variance of the errors?
(f) Try to correct for the autocorrelation in the lumber equation by reestimating it with both housing starts and time as explanatory variables. What do you observe?
(g) Use the estimated models to forecast the per capita consumption of lumber and plywood in the year 2010, assuming that there will be 2,135 housing starts in that year. Compare these forecasts to those made in part (f) of Problem 19.4.

ANNOTATED REFERENCES

Adams, D.M., and R.W. Haynes. 1989. A model of national forest timber supply and stumpage markets in the western United States. *Forest Science* 35(2):401–424. (Econometric model of the market for federal timber.)

Ahn, S., A.J. Plantinga, and R.J. Alig. 2000. Predicting future forestland area: A comparison of econometric approaches. *Forest Science* 46(3):363–376. (Compares different econometric specifications for equations to predict changes in land use.)

Buongiorno, J., and T. Young. 1984. Statistical appraisal of timber with an application to the Chequamegon National Forest. *Northern Journal of Applied Forestry* 1(4):72–76. (Describes a single-equation econometric model to predict the value of auctioned timber.)

Daniels, S.E., and R. Johnson. 1991. Estimating and comparing demand functions for personal use Christmas tree cutting at seven Utah sites. *Western Journal of Applied Forestry* 6(2):42–46. (Econometric model of demand for recreation.)

Englin, J., D. Lambert, and W.D. Shaw. 1997. A structural equations approach to modeling consumptive recreation demand. *Journal of Environmental Economics and Management* 33:33–43. (Describes a two-equation econometric model of recreational fishing demand.)

González-Cabán, A., and C.W. McKetta. 1986. Analyzing fuel treatment costs. *Western Journal of Applied Forestry* 1(4):116–121. (Application of econometrics to estimating costs of hazard reductions based on site characteristics.)

Hill, R.C., Griffiths, W., and G. Judge. 2001. *Undergraduate Econometrics*. Wiley, New York. 402 pp. (Basic reference on econometrics.)

Howard, J.L. 2001. *U.S. Timber Production, Trade, Consumption, and Price Statistics 1965 to 1999*. Research Paper FPL-RP-595. USDA Forest Service, Forest Products Laboratory, Madison, WI. (Source of the price data used in this chapter).

Ince, P.J. 2001. *Pulp and Paper Statistics*. Unpublished electronic database. USDA Forest Service, Forest Products Laboratory, Madison, WI. (Source of data on pulpwood deliveries and pulping capacity used in this chapter.)

Klemperer, W.D. 1996. *Forest Resource Economics and Finance*. McGraw-Hill, New York. 551 pp. (Chapter 2 reviews concepts of supply, demand, and market equilibrium.)

Latta, G.S., and D.M. Adams. 2000. An econometric analysis of output supply and input demand in the Canadian softwood lumber industry. *Canadian Journal of Forest Research* 30:1419–1428. (Application of econometrics to estimating regional demand and supply.)

Leuschner, W.A. 1973. An econometric analysis of the Wisconsin aspen pulpwood market. *Forest Science* 19(1):41–46. (Regional econometric model with structure similar to that discussed in this chapter.)

Marshall, P., T. Szikszai, V. LeMay, and A. Kozak. 1995. Testing the distributional assumptions of least squares regression. *The Forestry Chronicle* 71(2):213–218. (Gives more information on the assumptions of regression analysis.)

Montgomery, C.A. 2001. Modeling the United States housing sector. *Forest Science* 47(3):371–389. (Econometric model of the most important source of demand for solid wood products.)

Olson, K.W., R.L. Moomaw, and R.P. Thompson. 1988. Redwood National Park expansion: Impact on old-growth redwood stumpage prices. *Land Economics* 64(3):269–275. (Econometric analysis of timber market impacts of preserving old-growth.)

Ragsdale, C.T. 1998. *Spreadsheet Modeling and Decision Analysis: A Practical Introduction to Management Science*. South-Western College Publishing, Cincinnati, OH. 742 pp. (Chapter 9 shows how to use spreadsheets to estimate regression models, and Chapter 11 shows how to use spreadsheets to estimate time series models.)

APPENDIX A

Compounding and Discounting

The purpose of this appendix is to review the compounding and discounting formulas, with finite and infinite time horizons, used in the book.

FUTURE VALUE AT COMPOUND INTEREST

With an interest rate, r, expressed as a fraction per year, \$1 invested for one year earns \$$r$ in interest, so the capital at the end of the year is:

$$1 + r$$

This amount invested for another year yields the interest:

$$r(1 + r)$$

So the capital after 2 years is:

$$(1 + r) + r(1 + r) = (1 + r)(1 + r) = (1 + r)^2$$

Similarly, the capital after 3 years is:

$$(1 + r)^2 + r(1 + r)^2 = (1 + r)(1 + r)^2 = (1 + r)^3$$

Continuing in this fashion shows that \$1 invested at the yearly interest rate r gives at the end of n years the capital:

$$(1 + r)^n \tag{A.1}$$

PRESENT VALUE AT COMPOUND INTEREST

The present value of \$1 paid in n years, given a yearly interest rate r, is:

$$\frac{1}{(1+r)^n} \tag{A.2}$$

Indeed, Compounding Equation (A.1) shows that the amount $1/(1 + r)^n$ invested for n years gives $1:

$$\frac{1}{(1+r)^n}(1+r)^n = 1$$

INFINITE PERIODIC DISCOUNTING

The present value of a constant periodic income, say $1, paid every D years for an infinite number of periods is:

$$PV = \frac{1}{(1+r)^D - 1} \tag{A.3}$$

To show this, note that by definition and by application of the basic discounting formula of Equation (A.2), the expression of PV is:

$$PV = \frac{1}{(1+r)^D} + \frac{1}{(1+r)^{2D}} + \frac{1}{(1+r)^{3D}} + \cdots \tag{A.4}$$

where the fist term on the right is the present value of $1 paid D years from now, the second term is the present value of $1 paid $2D$ years from now, and so on to infinity. The sum of this infinite series is finite because for any positive interest rate, the general term

$$\frac{1}{(1+r)^{mD}}$$

approaches 0 as the integer m increases to infinity. So each additional $1 contributes less and less to the sum of Equation (A.4) because it is paid later and later, and eventually it contributes nothing.

To simplify the calculation of the sum of the infinite series of Equation (A.4), let

$$d = \frac{1}{(1+r)^D} \tag{A.5}$$

With this notation, Equation (A.4) can be rewritten as:

$$PV = d + d^2 + d^3 + \cdots \tag{A.6}$$

Compounding and Discounting

Multiplying this equation left and right by d gives:

$$d\text{PV} = d^2 + d^3 + \cdots \tag{A.7}$$

Then, subtracting Equation (A.7) from Equation (A.6), term by term, gives:

$$\text{PV}(1 - d) = d$$

That is:

$$\text{PV} = \frac{d}{1-d} = \frac{1}{\frac{1}{d}-1}$$

Substituting d with its expression in Equation (A.5) gives the final formula for the present value of $1 paid every D years:

$$\text{PV} = \frac{1}{(1+r)^D - 1}$$

In the special case where the $1 is paid every year, that is, $D = 1$, the formula simplifies to:

$$\text{PV} = \frac{1}{r}$$

APPENDIX B

Elements of Matrix Algebra

The purpose of this appendix is to explain the matrix notations and operations used in some chapters. For example, in Chapter 8 we described the growth of an uneven-aged stand with the following equations:

$$\begin{aligned} y_{1,t+1} &= 0.92 y_{1t} - 0.29 y_{2t} - 0.96 y_{3t} + 109 \\ y_{2,t+1} &= 0.04 y_{1t} + 0.90 y_{2t} \\ y_{3,t+1} &= \phantom{0.04 y_{1t} + {}} 0.02 y_{2t} + 0.90 y_{3t} \end{aligned} \qquad (B.1)$$

To simplify further manipulations, we shall write this system of equations with matrices and vectors. A matrix is a table of numbers, variables, or algebraic expressions. A vector is a matrix that has only one row or column.

For example, the coefficients of y_{1t}, y_{2t}, y_{3t} in Equations (B.1) constitute the following matrix of three rows and three columns:

$$\mathbf{G} = \begin{bmatrix} 0.92 & -0.29 & -0.96 \\ 0.04 & 0.90 & 0 \\ 0 & 0.02 & 0.90 \end{bmatrix}$$

And the variables and the constant term constitute the three following column vectors:

$$\mathbf{y}_{t+1} = \begin{bmatrix} y_{1,t+1} \\ y_{2,t+1} \\ y_{3,t+1} \end{bmatrix} \qquad \mathbf{y}_t = \begin{bmatrix} y_{1t} \\ y_{2t} \\ y_{3t} \end{bmatrix} \qquad \mathbf{c} = \begin{bmatrix} 109 \\ 0 \\ 0 \end{bmatrix}$$

With these notations, the system of Equations (B.1) can be written as:

$$\mathbf{y}_{t+1} = \mathbf{G} \mathbf{y}_t + \mathbf{c} \qquad (B.2)$$

where $\mathbf{Gy}_t$ is the product of matrix $\mathbf{G}$ by vector $\mathbf{y}_t$; that is:

$$\mathbf{Gy}_t = \begin{bmatrix} 0.92 & -0.29 & -0.96 \\ 0.04 & 0.90 & 0 \\ 0 & 0.02 & 0.90 \end{bmatrix} \begin{bmatrix} y_{1t} \\ y_{2t} \\ y_{3t} \end{bmatrix}$$

The general law of multiplication of matrices is as follows: Given a matrix $\mathbf{A}$ of m rows and p columns and a matrix $\mathbf{B}$ of p rows and n columns, the product $\mathbf{C} = \mathbf{AB}$ is a matrix of m rows and n columns of which each element c_{ij} is obtained by multiplying each element of row i in $\mathbf{A}$ by the ith element in column j of $\mathbf{B}$ and adding up the products.

Therefore, the product $\mathbf{Gy}_t$ is the following column vector:

$$\mathbf{Gy}_t = \begin{bmatrix} 0.92 y_{1t} - 0.29 y_{2t} - 0.96 y_{3t} \\ 0.04 y_{1t} + 0.90 y_{2t} \\ 0.02 y_{2t} + 0.90 y_{3t} \end{bmatrix}$$

Furthermore, matrices with the same number of rows and columns are added by adding their corresponding elements. Thus, the right-hand side of Equation (B.2) is obtained by adding vector $\mathbf{c}$ to vector $\mathbf{Gc}$, element by element, leading to the following vector:

$$\mathbf{Gy}_t + \mathbf{c} = \begin{bmatrix} 0.92 y_{1t} - 0.29 y_{2t} - 0.96 y_{3t} + 109 \\ 0.04 y_{1t} + 0.90 y_{2t} \\ 0.02 y_{2t} + 0.90 y_{3t} \end{bmatrix}$$

Equation (B.2) means that each element of this vector is equal to the corresponding element of vector $\mathbf{y}_{t+1}$. Therefore, Equation (B.2) is indeed the same as the system of Equations (B.1). It is just written in a more compact way with matrices and vectors.

Similarly, the algebraic equation of the harvest volume per hectare:

$$\underset{(m^3/ha)}{Z_Q} = 0.20 \quad \underset{(m^3/tree)}{h_{1t}} \underset{(trees/ha)}{+ 1.00 h_{2t} + 3.00 h_{3t}}$$

(where h_{it} is the number of trees harvested in size class i at time t) has an equivalent matrix notation:

$$Z_Q = \mathbf{v}\mathbf{h}_t$$

Elements of Matrix Algebra

where **v** is a row vector containing the data on the volume per hectare in each age class:

$$\mathbf{v} = [0.20 \quad 1.00 \quad 3.00]$$

and $\mathbf{h}_t$ is a column vector containing the variables that stand for the number of trees in each age class at time t:

$$\mathbf{h}_t = \begin{bmatrix} h_{1t} \\ h_{2t} \\ h_{3t} \end{bmatrix}$$

INDEX

0-1 variables (see binary variables)

A
abstraction 2
activities
  critical 239, 247–248
  definition 236–237
  duration 236, 250
  network representation 235, 237–238
  precedence relationships 236–237
  slack 238, 246–248
additive return function 277
additivity assumption 23
adjacency constraints 220
adjusted R^2 410
age-classes 90
algorithms
  definition 4–5
  graphical 29–36, 205–206
  minimum spanning tree 207–209
  ordinary least squares 406–414
  recursive 262, 269–270, 278, 359–360
  simplex 36–42
  steepest ascent 39–42
alternative land uses 112–113
applications
  approaching a desirable stand structure 192–195
  assigning foresters to jobs 210–214
  best thinning 260–269
  consultant's problem 204–207
  converting southern hardwoods to pine 55–65

  converting to a desirable forest landscape 189–192
  designing an efficient road network 214–225
  deterministic simulation of even-aged management 314–324
  deterministic simulation of uneven-aged management 291–300
  keeping the river clean 14–16, 21–22, 33–36, 46–47
  loblolly pine forest 71–83
  logging okume 207–209
  maximizing discounted expected returns 359–362
  maximizing long-term expected biodiversity 362–369
  minimizing the risk of losing an endangered species 277–282
  natural forest growth 338–344
  poet and his woods 10–14, 20–21, 29–33, 36–46
  predicting the effects of management 344–351
  pulpwood market in the U.S. 402–416
  river pollution control revisited 180–189
  short-leaf pine forest 90–104
  simulation of catastrophic events 324–332
  slash-burn project 236–252
  stochastic simulation of uneven-aged management 300–307
  trimming paper sheets 270–277

approaching a desirable stand structure
   (application)
  model formulation 193–194
  problem definition 192–193
  spreadsheet formulation 194–195
arcs 235, 237–238
area-control
  constraint 74, 313–316
  definition 70–71
assigning foresters to jobs (application)
  model formulation 211–212
  problem definition 210
  spreadsheet formulation 212–214
assignment model 210–214
assumptions of linear programming
  additivity 23
  determinism 23–24
  divisibility 23
  proportionality 22–23
autocorrelation 408, 412–413

B

basal area 135
basic feasible solution 37–42
basic variables 38–42
benefit-cost ratio
  definition 379–380
  real vs. nominal terms 391–392
  shortcomings 383–385
Berger, W.H. 294
Berger-Parker index 294, 342
best solution 5, 32–33, 35–42, 47–49
best thinning (application)
  dynamic programming solution
     262–267
  model formulation 260–262
  myopic solution 262
  problem definition 260
  spreadsheet formulation 267–269
big trees (see large trees)
binary variables 206–225
biodiversity
  Berger-Parker index 294, 342
  expected 347–348
  landscape 97, 101–103, 122, 342,
     347–351
  MaxiMin criterion 97, 165, 294
  minimizing the risk of losing an endangered
     species (application) 277–282
  structural 164–165

biological
  risk 300–301
  variables 3
boolean variables (see binary variables)
branches (see arcs)

C

carbon storage 318–321
cardinal weights 187
catastrophic events
  fires 338
  storms 325–327, 338
clear-cuts 54, 78
column activity 19
compartment (see stand)
compounding & discounting 421–423
computers 4–5
confidence intervals 305–307, 414
connecting locations (see logging okume)
constant variance 408, 413–414
constraints
  definition 10
  equality & inequality 18
  independent vs. redundant 38
  nonnegativity 13, 16, 19
  resource availability 9–10, 19, 385–387
consultant's problem (application)
  feasible region 205–206
  graphical solution 205–206
  model formulation 204–205
  problem definition 204
  rounded solution 206–207
consumer price index 388
continuous
  cover forestry (see uneven-aged forest
     management)
  variables 203–204
control (see area-control or volume-control)
converting southern hardwoods to pine
   (application)
  general formulation 64–65
  model formulation 56–58, 60–61
  problem definition 55
  solutions 58–60
  spreadsheet formulation 61–63
converting to a desirable forest landscape
   (application)
  model formulation 190
  problem definition 189–190
  spreadsheet formulation 190–192

Index   431

costs
 fixed 163–164, 295
 opportunity 110, 112–113, 122–123, 155–156, 166–167
 start-up 222–224
CPM (see PERT/CPM)
criteria
 benefit-cost ratio 379–380, 383–385, 391–392
 internal rate of return 380–385, 391–392
 land expectation value 110–111, 156–157
 MaxiMin 97, 165, 294
 net present value 374, 376–379, 391–392
critical
 activities 239, 247–248
 path 246–248
 path method (see PERT/CPM)
Crusoe (see Robinson Crusoe economy)
Curtis, F.H. 55
cutting cycle 160–164, 293–294, 297–298

D
Dantzig, G. 10
Davis, J.B. 236
decision
 variables 11–12, 15, 56, 72, 91, 140–143, 205, 211, 216–217, 221, 260–262, 269, 278
 yes-or-no 203–204
deflation 388–389
demand
 downward-sloping 321–324
 elasticity of 322
dependent variable 408
designing an efficient road network (application)
 adjacency constraints 220
 divisible projects 221–222
 meeting one of several goals 224–225
 model formulation 216–219
 problem definition 214–216
 spreadsheet formulation 219–220
 startup costs 222–224
detached coefficient matrix (see tableau)
determinism assumption 23–24
deterministic simulation of even-aged management (application)
 downward-sloping demand 321–324
 model formulation 316–318
 performance indicators 317–318

 problem definition 314–315
 sensitivity analysis 319–321
deterministic simulation of uneven-aged management (application)
 performance indicators 294–295
 problem definition 291
 sensitivity analysis 297–300
 spreadsheet formulation 291–293
 varying the cutting cycle 293–294
development of models 5
diagnostic checks 412–414
diameter distribution (see inverse-J distribution)
discounting (see compounding & discounting)
discrete random events 290, 325–327
distribution
 exponential 325
 inverse-J 133
 standard normal 251
 Student's t 408, 414
 uniform 301–302, 326, 328
disturbances 403
divisibility assumption 23
divisible projects 221–222
downward-sloping demand 321–324
duality
 dual of the poet's problem 43–46
 dual of the river pollution problem 46–47
 dual problem 42–43, 81
 primal problem 42–43
 shadow prices 44–47, 81
 theorem 43
duration (see activity duration)
dynamic even-aged growth model 90–93
dynamic programming
 additive vs. multiplicative return functions 277
 best thinning (application) 260–269
 general formulation 269–270
 maximizing discounted expected returns (application) 359–362
 minimizing the risk of losing an endangered species (application) 277–282
 networks 260–262, 271–272
 recursive solution 262, 269–270, 278, 359–360
 stages, states, and decisions 259–262, 269, 271, 278
 trimming paper sheets (application) 270–277

432

E

earliest times
  definition 238
  model formulation 239–241
  spreadsheet formulation 241–243
econometrics
  definition 402
  exogenous vs. endogenous variables 404
  forecasting 401–402, 414–415
  hypothesis testing 414
  ordinary least squares 406–416
  recovering structural parameters 405, 415
  reduced form equations 405
  structural analysis 415–416
  structural equations 402–404
economic
  cutting cycle 160–164
  forecasting 401–402, 414–416
  risk 302, 327–328
  rotation 111–112
  theory 4, 402
  variables 3
elasticity of demand 322
endangered species (see minimizing the risk of losing an endangered species (application) & biodiversity)
endogenous variable 404
equality constraints 18
equilibrium price 404
even-aged forest management
  best thinning (application) 260–269
  clear-cuts 54, 78
  converting southern hardwoods to pine (application) 55–65
  converting to a desirable forest landscape (application) 189–192
  definition 54–55
  deterministic simulation of even-aged management (application) 314–324
  dynamic even-aged growth model 90–93
  loblolly pine forest (application) 71–83
  maximizing discounted expected returns (application) 359–362
  maximizing long-term expected biodiversity (application) 362–369
  natural forest growth (application) 338–344
  nondyanmic general formulation 123–125
  regulation 55, 69–71, 94–95
  rotation 70, 77–78, 111–112, 319–321
  short-leaf pine forest (application) 90–104
  simulation of catastrophic events (application) 324–332
even-flow policy 119–121
examples (see applications)
Excel Solver
  integer programming 219–220
  linear programming 20–21, 47–50
  optimizers 5
  regression analysis 408–409
  shadow prices 49–50
exogenous variable 404
expected value
  biodiversity 347–348
  returns 348–351
experiments 3, 289–290
exponential distribution 325
extreme points 37–42

F

Faustmann, M. 110
feasible region 30–31, 34–35, 205–206
finish time 238
fires 338
fixed cost 163–164
flow constraints (see timber flow constraints)
forecasting (see economic forecasting)
forest products markets
  downward-sloping demand 321–324
  pulpwood market in the U.S. (application) 402–416
forest value
  definition 109–110
  estimation 113–116
  optimizing 116–119
formal vs. informal models 2
fundamental theorem of linear programming 38
FV (see forest value)

G

Gantt chart 248–249
goal programming
  approaching a desirable stand structure (application) 192–195
  cardinal weights 187
  constraints 196
  converting to a desirable forest landscape (application) 189–192
  general formulation 195–198

Index 433

goal variables 181, 196–197
goal vs. linear programming 198–199
goal weights 182–184
multiple goals 17
objective function 182–184, 196–198
ordinal weights 187–189, 197–198
relative deviations 183–184, 197
river pollution control revisited
   (application) 180–189
satisficing vs. optimizing 180
spreadsheet formulation 184–187
variables 181, 196–197
good models 5–6
graphical solution 29–36, 205–206
growth models
  dynamic even-aged 90–93
  markov chain 338–340
  uneven-aged 133–143, 147–149
guiding rate of interest 377

H

harvest flow (see timber flow)
homogeneity of units 12
hypothesis testing 414

I

independent constraints 38, 60–61
independent variable 408
indifference curves 375
indivisible projects 216–217, 387
inequality constraints 18
inflation 388–391
ingrowth (see recruitment)
Institute for Operations Research and Management Science 4
integer programming
  assigning foresters to jobs (application) 210–214
  consultant's problem (application) 204–207
  designing an efficient road network (application) 214–225
  feasible region 205–206
  rounded solutions 204, 206–207, 387
interest rates
  Aaa bonds 393–395
  choosing 392–395
  different sources of funds 395
  government bonds 395

guiding 377
public vs. private investments 393–395
Robinson Crusoe economy 376
zero 393
internal rate of return
  definition 380–383
  real vs. nominal terms 391–392
  shortcomings 383–385
International Union of Forestry Research Organizations 4
inverse-J distribution 133
investment analysis
  benefit-cost ratio 379–380, 383–385
  choosing an interest rate 392–395
  compounding & discounting 421–423
  consumer & producer price indexes 388
  different sources of funds 395
  divisible vs. indivisible 387
  Faustmann formula (see land expectation value)
  forest value 109–110, 113–119
  guiding rate of interest 377
  inflation & deflation 388–391
  interest rates 377, 392–395
  internal rate of return 380–385
  investment vs. current consumption 376
  land expectation value 110–111, 156
  many limiting resources 386–387
  net present value 374, 376–379
  net present value-resource ratio 386
  one limiting resource 386
  optimum 375–376
  production possibilities frontier 374–375
  public vs. private 393–395
  real vs. nominal terms 388–392
  Robinson Crusoe economy 374–377
  shortcomings of benefit-cost ratio and internal rate of return 383–385
IRR (see internal rate of return)

K

keeping the river clean (application)
  dual problem 46–47
  graphic solution 33–36
  model formulation 15–16
  problem definition 14–15
  spreadsheet formulation 21–22
Kirby, M. 214

L

la futaie jardinée (see uneven-aged forest management)
land expectation value 110–111, 156–157
landscape diversity 97, 101–103, 122, 342, 347–351
large trees 145–147, 149, 298–300
latest times
  definition 238
  model formulation 243–245
  spreadsheet formulation 245–246
LEV (see land expectation value)
limited resources (see resources)
linear programming
  additivity assumption 23
  basic feasible solution 37–42
  basic vs. non basic variables 38–42
  best solutions 32–33, 35–42, 47–49
  column activity 19
  converting southern hardwoods to pine (application) 55–65
  decision variables 11–12, 15, 56, 72, 91, 140–143
  determinism assumption 23–24
  divisibility assumption 23
  duality 42–47, 81
  equality and inequality constraints 18
  Excel Solver 20–21, 47–50
  extreme points 37–42
  feasible region 30–31, 34–35
  fundamental theorem 38
  goal vs. linear programming 198–199
  graphical solution 29–36
  independent vs. redundant constraints 38, 60–61
  keeping the river clean (application) 14–16, 21–22, 33–36, 46–47
  loblolly pine forest (application) 71–83
  maximization vs. minimization 18
  negative variables 18
  non negativity constraints 13, 16, 19
  objective function 12, 15, 57, 72, 95–97, 115–116, 124–125, 143, 158
  poet and his woods (application) 10–14, 20–21, 29–33, 36–46
  proportionality assumption 22–23
  row activity 19–20
  sensitivity analysis 23, 33
  shadow prices 44–47, 49–50, 81, 146
  short-leaf pine forest (application) 90–104
  simplex algorithm 36–42
  slack variables 36–42
  spreadsheet formulation 5, 20–22
  standard formulation 17–19
  tableau 58
loblolly pine forest (application)
  clear-cut area 78
  different rotations 77–78
  dual solution 81
  general formulation 82–83
  model formulation 72–75
  problem definition 71–72
  solutions 76–77
  spreadsheet formulation 78–81
logging okume (application)
  manual solution 207–209
  problem definition 207
Loucks, D.P. 69

M

Magurran, A.E. 294
management science 4, 10
management unit (see stand)
markets (see forest products markets)
Markov chains
  dynamics 339–340
  matrix algebra 425–427
  mean recurrence time 344
  mean residence time 343
  natural forest growth (application) 338–344
  predicting the effects of management (application) 344–351
  spreadsheet formulation 340–342
  states 338
  transition probabilities 339
Markov decision processes
  definition 358–359
  maximizing discounted expected returns (application) 359–362
  maximizing long-term expected biodiversity (application) 362–369
mathematical models 2–5
mathematical programming
  dynamic programming 259–282
  goal 179–198
  integer 203–225
  linear (see linear programming)
  mixed integer 204, 221–225
  simulation 289–307, 313–332

Index

matrix algebra 425–427
MaxiMin criterion 97, 165, 294
maximization vs. minimization 18
maximizing discounted expected returns (application)
   model formulation 359–360
   problem definition 359
   spreadsheet formulation 360–362
maximizing long-term expected biodiversity (application)
   income constraint 367–369
   landscape dynamics 367
   model formulation 363–364
   problem definition 362
   spreadsheet formulation 364–366
mean recurrence time 344
mean residence time 343
minimization vs. maximization 18
minimizing the risk of losing an endangered species (application)
   dynamic programming solution 278–281
   model formulation 278
   problem definition 277–278
   spreadsheet formulation 281–282
minimum-spanning tree algorithm 207–209
mixed-integer programming
   definition 204
   designing an efficient road network (application) 221–225
models
   abstraction 2
   development of 5
   formal vs. informal 2
   model I vs. model II 67, 87, 107
   spreadsheets 5
   nature of 2–3
   mathematical 2–5
   systems 3–4
   good 5–6
multiple goals or objectives 17, 179–198
multiplicative return function 277
myopic solution 262

N

natural forest growth (application)
   landscape dynamics 342–343
   mean recurrence time 344
   mean residence time 343
   model formulation 339–340

   problem definition 338–339
   spreadsheet formulation 340–342
nature of models 2–3
Nautiyal, J.C. 90
negative variables 18
net present value
   definition 374, 376–379
   real vs. nominal terms 391–392
   resource ratio 386
networks
   arcs & nodes 235, 237–238
   best thinning (application) 260–269
   designing an efficient road network (application) 214–225
   logging okume (application) 207–209
   project activities (PERT/CPM) 235, 237–238
   slash-burn project (application) 236–252
   trimming paper sheets (application) 270–277
nominal (see real vs. nominal)
nonbasic variable 38–42
nondeclining even-flow (see even-flow)
nonnegativity constraints 13, 16, 19
normal distribution (see standard normal distribution)
NPV (see net present value)

O

objective function 12, 15, 57, 72, 95–97, 115–116, 124–125, 143, 158, 182–184, 187–189, 196–198, 205, 212, 217, 223
operations research 4, 10
opportunity cost 110, 112–113, 122–123, 155–156, 166–167
optimal solution (see best solution)
optimizer (see Excel Solver)
ordinal weights 187–189, 197–198
ordinary least squares
   adjusted R^2 410
   autocorrelation 408, 412–413
   confidence intervals 414
   constant variance 408, 413–414
   dependent vs. independent variables 408
   diagnostic checks 412–414
   disturbances 403
   Excel 408–409

ordinary least squares (*continued*)
  pulpwood market in the U.S. (application) 402–416
    residuals 408–409, 412–414
    standard errors 409
    Student's t statistic 408, 414
    sum of squared residuals 406–408
organizations
  Institute for Operations Research and Management Science 4
  International Union of Forestry Research Organizations 4
  Society of American Foresters Operations Research Working Group 4

P
Parker, F.L. 294
Pearse, P.H. 90
PERT (see PERT/CPM)
PERT/CPM
  activities 236–237
  activity duration 236, 250
  activity slack 238, 246–248
  CPM 236
  critical activity 239, 247–248
  critical path 247–248
  earliest times 238–243
  finish times 238
  Gantt chart 248–249
  latest times 238, 243–246
  network representation 235, 237–238
  PERT 236, 249–252
  precedence relationship 236–237
  probability of completion 251–252
  project duration 251–252
  slash-burn project (application) 236–252
  start times 238
  uncertainty 249–252
poet and his woods (application)
  dual problem 43–46
  graphic solution 29–33
  model formulation 11–14
  problem definition 10–11
  sensitivity analysis 33
  simplex solution 36–42
  spreadsheet formulation 20–21
precedence relationship 236–237

predicting the effects of management (application)
  expected performance 347–351
  model formulation 345
  problem definition 344–345
  spreadsheet formulation 345–347
present value (see net present value)
price
  equilibrium 404
  indexes 388
  inflation 388–391
  real vs. nominal 388–391
  risk 302, 327–328
  shadow 44–47, 49–50, 81, 146
  trend 302
primal (see duality)
private vs. public investments 393–395
probability of completion 251–252
problems (see applications)
producer price index 388
production possibilities frontier 374–375
programming (see mathematical programming)
project evaluation and review technique (see PERT/CPM)
projects (see investments)
  divisible 221–222
  duration 251–252
  indivisible 216–217, 387
  probability of completion 251–252
proportionality assumption 22–23
pulpwood market in the U.S. (application)
  confidence intervals 414
  diagnostic checks 412–414
  equilibrium price 404
  forecasting 414–416
  hypothesis testing 414
  model estimation 406–412
  problem definition 402
  recovering structural parameters 405, 415
  reduced form equations 405
  structural analysis 415–416
  structural equations 402–404
pure integer programming (see integer programming)

Q
q-ratio management 291–293

Index

R

random variable 301–302, 326, 328
ratios
   benefit-cost 379–380, 383–385, 391–392
   net present value-resource 386
   q 291–293
real vs. nominal
   prices 388–389
   rates of change 389–391
   terms for investment criteria 391–392
recovering structural parameters 405, 415
recruitment 134–135, 300–301
recurrence time (see mean recurrence time)
recursive algorithm 262, 269–270, 278, 359–360
reduced form equations 405
redundant constraints 38, 60–61
regeneration (see recruitment)
regression analysis (see ordinary least squares)
regulation 55, 69–71, 94–95
relative deviations 183–184, 197
replication 304–307, 331
residence time (see mean residence time)
residuals 408–409, 412–414
resources
   limiting 9–10, 19, 385–387
   net present value-resource ratio 386
risk
   biological 300–301
   economic 302, 327–328
   fires 338
   storms 325–327, 338
river pollution control revisited (application)
   model formulation 181–184
   ordinal weights 187–189
   problem definition 180–181
   relative deviations 183–184
   spreadsheet formulation 184–187
R^2 (see adjusted R^2)
roads (see transportation)
Robinson Crusoe economy 374–377
rotation 70, 77–78, 111–112, 319–321
rounded solution 204, 206–207, 387
row activity 19–20

S

satisficing vs. optimizing 180
selection management (see uneven-aged management)
sensitivity analysis 23, 33, 297–300, 319–321
shadow prices 44–47, 49–50, 81, 146
shelterwood 54
short-leaf pine forest (application)
   general formulation 103–104
   model formulation 91–97
   problem definition 90
   spreadsheet formulation 97–103
simplex algorithm 36–42
simulation
   confidence intervals 305–307
   deterministic 290–300, 314–324
   deterministic simulation of even-aged management (application) 314–324
   deterministic simulation of uneven-aged management (application) 290–300
   discrete random events 290, 325–327
   experiment 289–290
   replication 304–307, 331
   simulation of catastrophic events (application) 324–332
   stochastic 290, 300–307, 324–332
   stochastic simulation of uneven-aged management (application) 300–307
simulation of catastrophic events (application)
   price risk 327–328
   problem definition 324–325
   replication 331
   spreadsheet formulation 328–332
   storms 325–327
size-classes 133
slack
   activity 238, 246–248
   variable 36–42
slash-burn project (application)
   critical path 246–248
   earliest times 238–243
   Gantt chart 248–249
   latest times 238, 243–246
   network representation 237–238
   problem definition 236–237
   uncertainty 249–252
social variables 3
Society of American Foresters Operations Research Working Group 4
solutions
   basic feasible 37–42
   rounded 204, 206–207, 387
   best 5, 32–33, 35–42, 47–49

Solver (see Excel Solver)
spreadsheets 5
SSR (see sum of squared residuals)
stages 259–262, 269, 271, 278
stand
  age-classes 90
  basal area 135
  dynamics 136–137, 140–143
  even-aged 54
  inverse-J distribution 133
  size-classes 133
  state 90, 134, 338
  uneven-aged 131–132
standard error 409
standard normal distribution 251
start times 238
start-up costs 222–224
states
  in dynamic programming models 259–262, 269, 271, 278
  steady 94–95, 137–138, 142–145, 148, 156–164, 168–171
  variables 90, 134, 338
steady state 94–95, 137–138, 142–145, 148, 168–171
steepest-ascent algorithm 39–42
stochastic simulation of uneven-aged management (application)
  biological risk 300–301
  confidence intervals 305–307
  economic risk 302
  problem definition 300
  replication 304–307
  spreadsheet formulation 302–304
storms 325–327, 338
structural
  analysis 415–416
  equations 402–404
Student's t statistic 408, 414
sum of squared residuals 406–408
summation notation 19
sustainability 93–95, 143–144
systems models 3–4

T

tableau 58
theorem of linear programming 38
thinning 54 (see also best thinning (application))
timber flow
  even-flow 119–121
  increasing 73
time preference 375
transition probabilities 339
transportation
  designing an efficient road network (application) 214–225
  logging okome (application) 207–209
trimming paper sheets (application)
  dynamic programming solution 271–275
  model formulation 271
  problem definition 270
  spreadsheet formulation 275–277

U

uncertainty (see risk)
uneven-aged forest management
  approaching a desirable stand structure (application) 192–195
  conversion to desired steady state 168–171
  cutting cycle 160–164, 293–294, 297–298
  definition 131–132
  deterministic simulation of uneven-aged management (application) 290–300
  economic steady state 156–164
  fixed costs 163–164, 295
  growth model 133–143, 147–149
  inverse-J distribution 133
  large trees 145–147, 298–300
  matrix formulation 147–149, 171–173
  maximizing discounted expected returns (application) 359–362
  maximizing long-term expected biodiversity (application) 362–369
  maximizing production 143–144
  maximizing revenue 159–164
  natural forest growth (application) 338–344
  poet and his woods (application) 10–14, 20–21, 29–33, 36–46
  q-ratio management 291–293
  recruitment 134–135, 300–301
  stand dynamics 136–137, 140–143
  steady state 137–138, 142–145
  stochastic simulation of uneven-aged management (application) 300–308
uniform distribution 301–302, 326, 328
units (see homogeneity of units)

Index

V

value (see forest value)
variable
  basic vs. nonbasic 38–42
  binary 206–225
  biological 3
  continuous 203–204
  decision 11–12, 15, 56, 72, 91, 140–143, 205, 211, 216–217, 221, 260–262, 269, 278
  dependent vs. independent 408
  economic 3
  endogenous vs. exogenous 404
  goal 181, 196–197
  integer 203–225
  negative 18
  random 301–302, 326, 328
  slack 36–42
  social 3
  state 90, 134, 338

volume-control
  constraint 74
  definition 70–71

W

weights
  cardinal 187
  goal 182–184
  ordinal 187–189, 197–198
  relative deviations 183–184, 197

Y

yes-or-no decisions 203–204

Z

zero-or-one variables (see binary variables)